W0079279

Progress in Molecular and Subcellular Biology

Series Editors: Ph. Jeanteur, I. Kostovic, Y. Kuchino, W.E.G. Müller (Managing Editor), A. Macieira-Coelho, R.E. Rhoads

23

Springer-Verlag Berlin Heidelberg GmbH

Heinz C. Schröder   Werner E.G. Müller (Eds.)

Inorganic Polyphosphates

Biochemistry, Biology, Biotechnology

With 78 Figures

 Springer

Professor Dr. Dr. Heinz C. Schröder
Professor Dr. Werner E.G. Müller

Institut für Physiologische Chemie
Abteilung für Angewandte Molekularbiologie
Johannes Gutenberg-Universität
Duesbergweg 6
D-55099 Mainz
Germany

ISBN 978-3-540-65303-5

Library of Congress Cataloging-in-Publication Data.
Inorganic polyphosphates / Heinz C. Schröder, Werner E.G. Müller
    p. cm.—(Progress in molecular and subcellular biology; 23)
    Includes bibliographical references and index.
    ISBN 978-3-540-65303-5   ISBN 978-3-642-58444-2 (eBook)
    DOI 10.1007/978-3-642-58444-2
    1. Polyphosphates–Physiological   effect.   2. Polyphosphates-Metabolism.   3. Polyphosphates–
Biotechnology.   I. Schröder, Heinz C. (Heinz Christoph)   II. Müller, Werner E.G.   III. Series.
QH506.P76   no. 23
[QP535.P1]
572-8 s–dc21
[572'.553]

This work is subject to copyright. All rights are reserved, whether the whole part of the material is concerned, specifically the rights of translation, reprinting reuse of illustrations, recitation, broadcasting, reproduction on microfilm or in any other way, and storage in data Banks. Duplication of this publication or parts thereof is permitted only under the provisions of the German Copyright Law of September 9, 1965, in its current version, and permissions for use must always be obtained from Springer-Verlag. Violations are liable for prosecution under the German Copyright Law.

© Springer-Verlag Berlin Heidelberg 1999
Originally published by Springer-Verlag Berlin Heidelberg New York in 1999

The use of general descriptive names, registered names, trademarks, etc. in this publication does not imply, even in the absence of a specific statement, that such names are exempt from the relevant protective laws and regulations and therefore free for general use.

Production: Pro Edit GmbH, D-69126 Heidelberg, Germany
Cover design: Meta Design, Berlin, Germany
Typesetting: Mitterweger Werksatz GmbH, D-68723 Plankstadt, Germany
Computer to film: Saladruck, Berlin, Germany
SPIN: 10636861   39/3137   5 4 3 2 1 0 – Printed in acid-free paper

Preface

Amongst the biopolymers found in living organisms, inorganic polyphosphates have been ignored for a long time, although these energy-rich molecules most likely existed on earth a long time before nucleic acids, proteins and polysaccharides appeared during prebiotic evolution. The research on inorganic polyphosphates started in the early 1950s, closely connected with the research on nucleic acids. Several researchers, including J.P. Ebel, F.M. Harold, A. Kornberg, P. Langen, K. Lohmann, M. and O. Szymona, E. Thilo, J.M. Wiame and A. Yoshida, worked on this topic at that time. Moreover, the main enzymes involved in polyphosphate metabolism, the polyphosphate synthetase (Kornberg et al. 1956), the polyphosphate:AMP phosphotransferase (Dirheimer and Ebel 1965), the polyphosphate glucokinase (Szymona and Ostrowski 1964), and polyphosphatases (Mattenheimer 1956; Muhammed et al. 1959) were discovered during that period. Most of these early experiments were done in bacteria and yeast, but also in animals (Grossmann and Lang 1962). The reason why inorganic polyphosphates became a forgotten area of research 10 to 15 years later remains unclear, perhaps due to the lack of proper methods to study these polymers. Only a few researchers, including I.S. Kulaev and H.G. Wood, continued to work on this subject. However, about 10 years ago, an increasing number of scientists became interested in inorganic polyphosphates again. The development of modern molecular biological methods and special techniques for analysing polyphosphates may have contributed to this new and rapidly increasing interest. Moreover, it has now been definitively shown that inorganic polyphosphates are also present in cells of higher eukaryotes. Furthermore, inorganic polyphosphates have attracted particular attention in biotechnology as a tool for phosphorus removal from wastewater as well as in medicine due their presence in human-pathogenic bacteria and in bone.

The state of knowledge about inorganic polyphosphates has been summarized in one comprehensive monograph (Kulaev 1979) and in a few review articles (more recently: Kulaev and Vagabov 1983; Wood and Clark 1988; Kornberg 1994, 1995). Here, we try to combine, in one book, the most important aspects of current research on inorganic polyphosphates. The 15 chapters of this book deal among others with the metabolism and function of inorganic polyphosphates in bacteria, yeast and higher eukaryotes, the development of methods for investigation of these polymers and the possible applications of inorganic polyphosphates in biotechnology and medicine.

We would like to express our graditude to all authors who contributed to this book for their cooperation, help and patience.

Mainz, Germany
May 1999

H.C. Schröder
W.E.G. Müller

References

Dirheimer G, Ebel JP (1965) Caractérisation d'une polyphosphate-AMP-phosphotransférase dans *Corynebacterium cerosis*. CR Acad Sci Paris 260:3787–3790

Grossmann D, Lang K (1962) Anorganische Poly- und Metaphosphatasen sowie Polyphosphate im tierischen Zellkern. Biochem Z 336:351–370

Kornberg A (1994) Inorganic polyphosphate: a molecular fossil come to life. In: Torriani-Gorini AM, Yagil E, Silver S (eds) Phosphate in microorganisms: cellular and molecular biology. American Society for Microbiology, Washington, DC, pp 204–209

Kornberg A (1995) Inorganic polyphosphate: toward making a forgotten polymer unforgettable. J Bacteriol 177:491–496

Kornberg A, Kornberg SR, Simmes ES (1956) Metaphosphate synthesis by an enzyme from *Escherichia coli*. Biochim Biophys Acta 20:215–227

Kulaev IS (1979) The biochemistry of inorganic polyphosphates. Wiley, New York

Kulaev IS, Vagabov VM (1983) Polyphosphate metabolism in microorganisms. Adv Microb Physiol 24:83–171

Mattenheimer H (1956) Die Substratspezifität „anorganischer" Poly- und Metaphosphatasen. I. Optimale Wirkungsbedingungen für den enzymatischen Abbau von Poly- und Metaphosphaten. Hoppe-Seylers's Z Physiol Chem 303:107–114

Muhammed A, Rodgers A, Hughes DE (1959) Purification and properties of a polymethaphosphatase from *Corynebacterium xerosis*. J Gen Microbiol 20:482–495

Szymona M, Ostrowski W (1964) Inorganic polyphosphate glucokinase of *Mycobacterium phlei*. Biochim Biophys Acta 85:283–295

Wood HG, Clark JE (1988) Biological aspects of inorganic polyphosphates. Annu Rev Biochem 57:235–260

Contents

Inorganic Polyphosphate: A Molecule of Many Functions
A. Kornberg

Research in Inorganic Polyphosphates: The Beginning
P. Langen

Metabolism and Function of Polyphosphates in Bacteria and Yeast
I.S. Kulaev, T.V. Kulakovskaya, N.A. Andreeva, and L.P. Lichko

Inorganic Polyphosphate in Eukaryotes:
Enzymes, Metabolism and Function
H.C. Schröder, B. Lorenz, L. Kurz, and W.E.G. Müller

Cyclic Condensed Metaphosphates in Plants and the Possible Correlations Between Inorganic Polyphosphates and Other Compounds

R. Niemeyer

Polyphosphate Glucokinase

N.F.B. Phillips, P.C. Hsieh, and T.H. Kowalczyk

Cytoplasmic Inorganic Pyrophosphatase
A.A. Baykov, B.S. Cooperman, A. Goldman, and R. Lahti

**Polyphosphate/Poly-(R)-3-hydroxybutyrate Ion Channels
in Cell Membranes**
R.N. Reusch

Inorganic Polyphosphate Regulates Responses of *Escherichia coli*
to Nutritional Stringencies, Environmental Stresses and Survival
in the Stationary Phase
N.N. Rao and A. Kornberg

From Polyphosphates to Bisphosphonates
and Their Role in Bone and Calcium Metabolism
H. Fleisch

**Methods for Investigation of Inorganic Polyphosphates
and Polyphosphate-Metabolizing Enzymes**
B. Lorenz and H.C. Schröder

Definitive Enzymatic Assays in Polyphosphate Analysis
D. Ault-Riché and A. Kornberg

Study of Polyphosphate Metabolism in Intact Cells
by 31-P Nuclear Magnetic Resonance Spectroscopy
K.Y. Chen

Polyphosphate-Accumulating Bacteria and Enhanced Biological Phosphorus Removal
G.J.J. Kortstee and H.W. van Veen

Genetic Improvement of Bacteria for Enhanced Biological Removal of Phosphate from Wastewater
H. Ohtake, A. Kuroda, J. Kato, and T. Ikeda

Inorganic Polyphosphate:
A Molecule of Many Functions

A. Kornberg[1]

1
Introduction

Inorganic polyphosphate (polyP) is a linear polymer of many tens or hundreds of orthophosphate (P_i) residues linked by high-energy, phosphoanhydride bonds (Fig. 1). PolyP is formed from P_i by dehydration at an elevated temperature and was likely involved in prebiotic evoluton. Of greatest interest is that polyP is found in every living thing – bacteria, fungi, protozoa, plants and animals (Kulaev 1979; Wood and Clark 1988; Kornberg 1994). Yet, for lack of any known function, it was for a long time dismissed as a "molecular fossil" and is still ignored in all textbooks of biology, biochemistry and chemistry.

We now have evidence to show that polyP has numerous and varied biological functions depending on where it is – species, cell, subcellular compartment – and when it is needed. Among these funcitons are: substitution for ATP in kinase reactions, reservoir of P_i, chelation of metals (e.g., Mn^{2+}, Mg^{2+}, Ca^{2+}), buffer against alkali, capsule of bacteria, competence for bacterial transformation, disposal of pollutant phosphate, and, of special interest, the physiologic adjustments to growth, development, stress and deprivation.

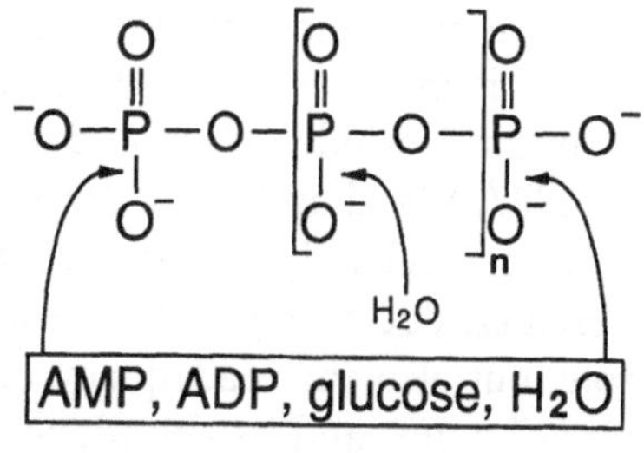

Fig. 1. Inorganic polyP as a substrate. Chains are attacked at their termini by AMP, ADP, glucose or H_2O, catalyzed respectively by polyP-AMP phosphotransferase, polyP kinase, polyP glucokinase and exopolyphosphatases, and internally by endopolyphosphatases

AMP, ADP, glucose, H_2O

---

[1] Department of Biochemistry, Stanford University School of Medicine, Stanford, California 94305-5307, USA.

Progress in Molecular and Subcellular Biology, Vol. 23
H. C. Schröder, W. E. G. Müller (Eds.)
© Springer-Verlag Berlin Heidelberg 1999

2
Metachromatic Granules Are Inorganic PolyP

PolyP was first isolated from yeast (Liebermann 1890). Later it was seen as metachromatic granules in microorganisms (particles stained pink by basic blue dyes); it was called "volutin" (Meyer 1904). For some time they were mistaken for nucleic acids. With the advent of electron microscopy, these particles, viewed under the electron beam, were highly refractive and quickly disappeared. It was then that they were distinguished from chromatin and identified as a novel compound, polyP (Wiame 1947; Ebel 1949). Like other polyanions (e.g., heparin), polyP shifts the absorption of a bound basic dye, such as toluidine blue, to a shorter wavelength (630 to 530 nm).

Historically, the polyP particle was recognized as a diagnostic feature of medically important bacteria, such as *Corynebacterium diphtheriae*. Decades later, polyP became of interest in biochemistry in connection with the major biochemial riddle of the 1940s – how P_i is fixed by an anhydride bond to ADP in aerobic (oxidative) phosphorylation. Such studies led first to the source of inorganic pyrophosphate (Kornberg 1993) and then to a curiosity about how the many more phosphoanhydride-linked residues in polyP were assembled. Although *Escherichia coli*, a major source of biochemical insights, lacks any visible particles of polyP, it still proved to be a rich source of an enzyme which makes polyP (later named polyP kinase) and also catalyzes the conversion of polyP to ATP (Kornberg et al. 1956; S.R. Kornberg 1957):

$$nATP \leftrightarrow polyP_n + nADP \qquad (1)$$

3
Occurence and Enzymology of PolyP

PolyP is prominent in many organisms, especially in the vacuole of yeast where it may represent 10 to 20 % of the cellular dry weight. Studies by Schmidt, Harold, Kulaev and others (Schmidt 1951; Harold 1967; Kulaev et al. 1987; Wood and Clark 1988) disclosed the ubiquity of polyP and identified a few related enzyme activities. Yet, polyP remained a largely forgotten polymer, due to not only the lack of evidence for any essential metabolic role, but also the inadequacy of methods to establish the authenticity and size of polyP, and its abundance at very low concentrations.

In addition to the staining and appearance of the granular accumulations observed in the light and electron microscopes, NMR analysis has been used to identify polyP in intact cells. However, as described in Chapter 12, NMR detection requires high concentrations and fails to measure the polyP in aggregates and in metal complexes. Otherwise, identification of polyP rests on crude and cumbersome separations in cell extracts from the known phosphate-containing polymers, followed by determination of the acid-lability characteristic of phosphoanhydride bonds (i.e., conversion to P_i in 7 min at 100 °C in 1 M HCl). These assay methods are not sufficiently quantitative to be conclusive, especially for low concentrations of polyP.

Enzymology offers an attractive route to analysis as well as to physiologic functions. Enzyme isolation has on many occasions revealed a novel mechanism, sometimes an insight into a metabolic or biosynthetic pathway. Now, the purified enzyme has also opened the route of reverse genetics: its peptide sequence leads to its gene and thereby the means to knock out the gene or overexpress it. By manipulating expression of the gene and the cellular levels of its product, phenotypes are created which may provide clues to metabolic functions. More immediate and decisive, as in the studies of polyP, the enzyme can be a unique and invaluable reagent for analytic and preparative work, as described in Chapter 12. Toward this end, several enzymes have been purified and used in studies of polyP metabolism (Ahn and Kornberg 1990; Akiyama et al. 1993; Crooke et al. 1994; Wurst and Kornberg 1994).

One enzyme, the polyP kinase (PPK), purified to homogeneity from *E. coli*, catalyzes the readily reversible conversion of the terminal (γ) phosphate of ATP to polyP (Ahn and Kornberg 1990) (Eq. 1). The enzyme, a tetramer of 80-kDa subunits bound to cell membranes, and possessing phosphohistidyl active sites (Kumble et al. 1996), is responsible for the processive synthesis of long polyP chains (ca. 750 residues) in vitro; labeling with ^{32}P provides such chains for use as substrates and standards. With ADP in excess, PPK quantitatively converts polyP to ATP. PPK also converts GDP, UDP and CDP to their respective triphosphates and is thus a nucleoside diphosphate kinase (Kuroda and Kornberg 1997) that may function auxiliary to the principal cellular kinase. PPK has still another capacity to transfer inorganic pyrophosphate to GDP to form the linear guanosine tetraphosphate (ppppG) (Kuroda and Kornberg 1997; Tzeng and Kornberg, unpubl. observ.).

These multiple functions of PPK, responsible for synthesis and utilization of polyP, show different responses to radiation inactivation, denaturants, pH, temperature and Mg^{2+} levels. Site-directed mutagenesis and truncation of genetic regions have also revealed discrete polypeptide domains for some of these functions, indicating a high degree of complexity in the structure of this multisubunit enzyme (Tzeng and Kornberg, unpubl. observ.).

A second *E. coli* enzyme, exopolyphosphatase (PPX) encoded by a gene in the *ppk* operon (see below), hydrolyzes the terminal residues of polyP to P_i processively with a strong preference for a long-chain substrate (Akiyama et al. 1993). Another exopolyPase (sc PPX1), isolated from *Saccharomyces cerevisiae*, is the most powerful of these analytic reagents, releasing 30 000 P_i residues per min per enzyme molecule at 37 °C (Wurst and Kornberg 1994). It acts with about 40 times the specific activity of the *E. coli* PPX and exhibits a far broader size range among substrate polyP chains (i.e., 3 to 1000 residues). Cloning the gene for this polyPase enabled the enzyme to be overproduced in *E. coli* (Wurst and Kornberg, unpubl. observ.). Yeast mutants lacking PPXI show no growth phenotype, but several other polyPases are known in yeast (Wurst and Kornberg 1994, Andreeva et al. 1998) which might replace its function.

Application of the potent PPXI to remove the polyP that contaminates DNA preparations from yeast and other polyP-rich organisms (Rodriguez 1993) may solve a problem that has bedeviled the action of restriction nucleases and the use of shuttle vectors for expression of fungal genes in *E. coli*.

Based on new and sensitive assays employing PPK and PPX, polyP occurrence and levels in cells and tissues have been ascertained with more definitive assurance (Table 1). In this rather random sampling, the presence of polyP in all life forms has been reaffirmed, as well as the extremes of levels in eukaryotes (i.e., from 26 µM in liver to 120 mM in yeast). The value cited for *E. coli*, which applies to stationary phase cells, may be amplified 1000-fold in response to certain nutritional deficiencies and environmental stresses (see Chap. 9). It is notable that this very low level of 100 µm or less is essential (without any amplification) for the various adaptations to stresses in the stationary phase and for the very survival of the organism.

With respect to the sources of polyP, it is imperative as with analysis of other cellular constituents, especially in eukaryotes, to distinguish subcellular compartments: nucleus, mitochondria, lysosomes (vacuoles), other vesicular entities and the cytosol. A dramatic example is the yeast vacuole which may contain 99 % or more of the cellular polyP and mask the significant remainder in the mitochondria and the nucleus.

Other enzymes available as reagents for analysis of polyP are the glucokinases which attack the terminal residues of the polyP chain with glucose (Hsieh et al. (1993); (Fig. 1) and a phosphotransferase which attacks the termini with AMP (Bonting et al. 1991). Another potentially available reagent is an endopolyPase purified from yeast and also identified in mammalian cells (Kumble and Kornberg 1996). For lack of cloning of the yeast endopolyPase gene, this reagent is not yet in hand, nor have the physiologic consequences of its removal been ascertained.

Table 1. Occurence of polyP in various cells and tissues

Eukaryotes		Prokaryotes	
Protozoa		Bacteria	
Fungi		*E. coli*:	100 µM
S. cerevisiae:	120 mM	*A johnsonii*:	200 mM
Plants		Archaea	
Animals			
Rat liver	26 µM		
Cytosol	12 µM		
Nucleus	89 µM		

4
Biosynthesis of PolyP

The only pathway for the synthesis of polyP that has been established is the polymerization of the terminal phosphate of ATP through the action of PPK in *E. coli* (Ahn and Kornberg 1990). The gene encoding the kinase is part of an operon in which the gene for PPX is located immediately downstream (Akiyama et al. 1993). Interruption of the operon produces mutants, which, for lack of long-chain polyP, are defective in survival in the stationary phase (Chap. 9). How the operon is regulated to balance two counteracting enzymes, roughly equal in activity, needs to be explained. For that matter, the complex metabolic networks and the enzymes responsible for the dynamics of accumulation and depletion of polyP in *E. coli* are still largely unknown.

Determinations of the sequences of many bacterial genomes have revealed the most remarkable conservation of the sequence of PPK (Figs. 2–4). Included among these genomes are the sequences of some major pathogens (see Sect. 8.3).

Although a PPK activity has been purified from other bacteria (Robinson et al. 1987; Tinsley et al. 1993) and reported in yeast (Kulaev et al. 1987), such an enzyme action has yet to be proven in animal systems. While PPK as the device to produce long polyP chains has been validated in *E. coli* by genetic studies, mutants lacking this enzyme may still make tiny amounts of a short polyP chain, about 60 residues long, by an undefined pathway (Castuma and Kornberg, unpubl. results).

Several other plausible routes for the biosynthesis of polyP need to be considered: from ADP by reversal of an AMP phosphotransferase, from acetyl P, from 1,3-diphosphoglycerate, from dolichyl pyrophosphate (Shabalin and Kulaev 1989), and, of special interest, by proton motive forces, known to fix P_i in inorganic pyrophosphate (Nyren et al. 1991) as well as in ATP.

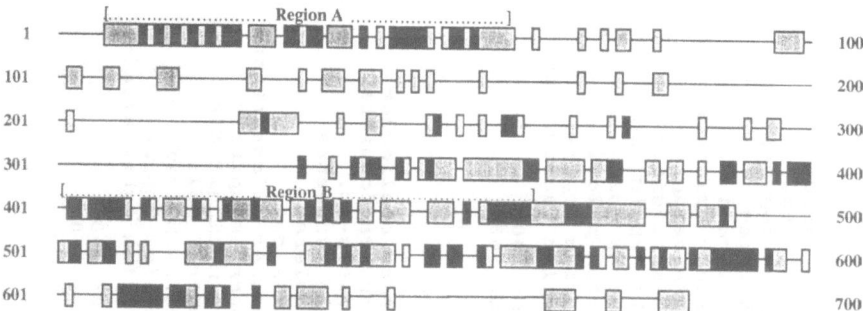

Fig. 2. Identity of polyphosphate kinase (PPK) among 12 microorganisms. Based on the length of *E. coli* (687 a.a.), 100% identity is represented by *black*, over 60% identity by *grey*. Partial *S. typhimurium* and *S. coelicolor* PPK sequences are not included in this figure

Region A

Deinococcus radiodurans	10-FLNRELSWLAFNERVLAEARDERNPLLERLKYVAICGSNLDEFFMVRVAGVHRQ-63	[2]
Synechocystis-sp.	19-YFNRELSWLAFNQRVLHEGLDDRTPLLERLKFLAIFCSNLDEFFMVRVAGLKQQ-72	[3]
Klebsiella aerogenes	7-YIEKELSWLAFNERVLQEAADKSNPLIERMRFLGIYSNNLDEFYKVRFAELKRR-60	[4]
Vibrio cholerae	7-YIDKELSWLSFNERVLQEAADKTVPLIERIRFLGIFSNNLDEFYKVRFADVKRQ-60	[2]
Escherichia coli	7-YIEKELSWLSFNERVLQEAADKSNPLIERMRFLGIYSNNLDEFYKVRFAELKRR-60	[5]
Acinetobacter calcoaceticus	18-YINRELSILDFHLRVLEQAVDPLHPLLERMNFLLIFSRNLDEFFEIRVAGVLEQ-71	[6]
Pseudomonas aeruginosa	53-YIHRELSQLQFNIRVLEQALDESYPLLERLKFLLIFSSNLDEFFEIRIAGLKKQ-10	[7]
Neisseria meningitidis	7-ILCRELSLLAFNRRVLAQAEDKNVPLLERLRFLGCIVSSNLDEFFEVRMAWLKRE-60	[8]
Salmonella typhimurium	7-YIEKELSWLAFNERVLQEAADKSNPLIERMRFLGIYSNNLDEFYKVRFAELKRR-60	[9]
Campylobacter coli	8-FLNRELSWLRFNSRVLDQC-SRPLPLLERLKFVAIYCTNLDEFYMIRVAGLKQL-61	[10]
Helicobacter pylori	4-FFNRELSWLAFNTRVLNEAKDESLPLLERLKFLAIYDTNLDEFYMIRVAGLKQL-57	[11]
Mycobacterium leprae	49-YLNRESSWLDFNARVLALAADNSLPLLERAKFLAIFASNLDEFYMVRVAGLKRR-102	[12]
Mycobacterium tuberculosis	7-YLNRELSWLAFNARVLALAADKSMPLLERAKFLAIFASNLDEFYMVRVAGLKRR-60	[13]

CONSENSUS: 100%   * * * *   * ***   * ** **   *   ***** *
CONSENSUS: >60%   ++++ + +++++   +++++ +++ +++   + + + +++++

Region B

Streptomyces coelicolor	?-IELKARFDESANIKWARKLEESGCHVVYGLVGLKTHCKLSLVVRQEGETLRRYSHVGTGNY-?	[1]
Deinococcus radiodurans	392-VELKARFDEQRNISWARKLERAGAHVVYGITGLKTHAKVTLVVRREEGGLRRYVHVGTGNY-452	
Synechocystis-sp.	433-VELKARFDEENNINWARKLEQYGVHVVYGLVGLKTHTKTVLVVRQEGPDIRRYVHIGTGNY-493	
Klebsiella aerogenes	397-VELQARFDEEANIHWARRLTEAGVHEVIFSAPGLKIHAKLFLISRKEGDDVVRYAHIGTGNF-457	
Vibrio cholerae	398-VELQARFDEEANIEWSRILTDAGVHVVIFGVPGMKIHAKLLLITRKEGDEFVRYAHIGTGNF-458	
Escherichia coli	400-VELQARFDEEANIHWAKRLTEAGVHEVIFSAPGLKIHAKLFLISRKENGEVVRYAHIGTGNF-460	
Acinetobacter calcoaceticus	407-IELRARFDEESNIAVANVLQEAGAVVVYGVGVKTHAKMIMVVRRENNKLVRYVHLGTGNY-467	
Pseudomonas aeruginosa	446-VELRARFDEESNLQLASRLQQAGAVVIYGVGFKTHAKMMLILRREDGELRRYAHLGTGNY-506	
Neisseria meningitidis	400-VELMARFDEANNVNWAKQLEEAGAHVVYGYKVHAKMALVIRREDGVLKRYAHLGTGNY-460	
Campylobacter coli	392-VELKARFDEENNLHWAKALENAGAHVIYGITGFKVHAKLLITKKTDNQLRHFTHLSTGNY-452	
Helicobacter pylori	397-VELKARFDEESNLHWAKALERAGALVVYGVFKLKVHAKMLLITKKTDNQLRHFTHLSTGNY-457	
Mycobacterium leprae	453-VEIKARFDEQANIRWARALEHAGVHVVYGIVGLKTHCKTCLVVRREGPTIRRYCHIGTGNY-513	
Mycobacterium tuberculosis	400-VEIKARFDEQANIAWARALEQAGVHEVAYGIVGLKTHCKTALVVRREGPTIRRYCHVGTGNY-460	

CONSENSUS: 100%   * *****   *   * * *   *   * *   *   ***   *   ***
CONSENSUS: >60%   + ++   + + +++   + + ++ +++   +++ +   ++   ++++ ++++   +   +

Fig. 4. Evolutionary tree of PPK sequences. Percentage of amino acid identiy was calculated by GeneWork software using individual alignment for a pair of species, for species within a group and between groups

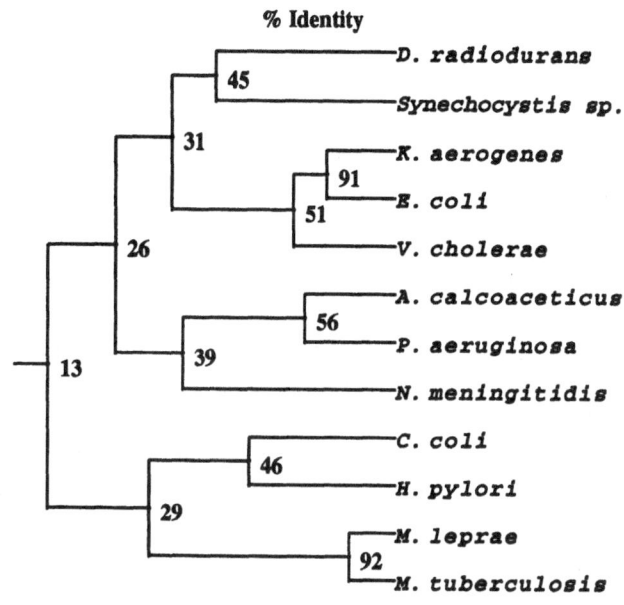

% Identity

◄ Fig. 3. Homology of PPK sequences. Asterisks in the consensus represent 100 % identity; crosses represent over 60 % identity. Ref. *1* Villar C (1997, unpubl. data); *2* TIGR database (1998); *3* Kaneto T, Sato S, Kotani H, Tanaka A, Asamizu E, Nakamura Y, Miyajima N, Hirosawa M, Sugiura M, Sasamoto S, Kumura T, Hosouchi T, Matsuno A, Muraki A, Nakazaki N, Naruo K, Okumura S, Shimpo S, Takeuchi C, Wada T, Watanabe A, Yamada M, Tabata S (1996) Sequence analysis of the genome of the unicellular cynanobacterium *Synechocystis* sp. Strain PPC6803. II. Sequence determination of the entire genome and assignment of potential protein-coding regions. DNA Research 3: 109–36; *4* Kato J, Yamamota T, Yamada K, Ohtake (H (1993) Cloning, sequence and characterization of the polyphosphate kinase-encoding gene (*ppk*) of *Klebsiella aerogenes*. Gene 137: 237–42; *5* Akiyama M, Crooke E, Kornberg A (1992) The polyphosphate kinase gene of *Escherichia coli*. Isolation and sequence of the *ppk* gene and membrane location of the protein. J Biol Chem 267: 22556–61; *6* Geissdorfer W, Ratajczak A, Hillen W (1998) Transcription of *ppk* from *Acinetobacter* sp. Strain ADP1, encoding a putative polyphosphate kinase, is induced by phosphate starvation. Appl Environ Microbiol 64: 896–901; *7* Ishige K, Kameda A, Noguchi T, Shiba T (1998) The polyphosphate kinase gene of *Pseudomonas aeruginosa*. DNA Research 30: 157–62; *8* Tinsley CR, Manjula BN, Gotschlich EC (1993) Purification and characterization of polyphosphate kinase from *Neisseria meningitidis*. Infect Immunol 61: 3701–10; *9* Kim KS, Fraley CD, Kornberg A (1998) Genbank database; *10* Park SF (1997) EMBL database; *11* Tomb JF, White O, Kerlavage AR, Clayton RA, Sutton GG, Fleischmann RD, Ketchum KA, Klenk HP, Gill S, Doughery BA, Nelson K, Quackenbush J, Zhou L, Kirkness EF, Peterson S, Loftus B, Richardson D, Dodson R, Khalak HG, Glodek A, McKenney K, Fitzgerald LM, Lee N, Adams MD, Venter JC, Krogh A, McLean J, Moule S, Murphy L, Oliver K, Osborne J, Quail MA, Rajandream MA, Rogers J, Rutter S, Seeger K, Skelton J, Squares S, Squares R, Sulton JE, Taylor K, Whitehead S, Barrell BG (1997) The complete genome sequence of the gastric pathogen *Helicobacter pylori*. Nature 388: 539–47; *12* Parkhill J, Barrell BG, Rajandream MA (1997) EMBL database; *13* Cole ST, Brosch R, Parkhill J, Garnier T, Churcher C, Harris D, Gordon SV, Eiglmeier K, Gas S, Barry III CE, Tekaia F, Badcock K, Basham D, Brown D, Chillingworth T, Connor R, Davies R, Devlin K, Feltwell T, Gentles S, Hamlin N, Holroyd S, Hornsby T, Jagels K, Krogh A, McLean J, Moule S, Murphy L, Oliver K, Osborne J, Quail MA, Rajandream MA, Rogers J, Rutter S, Seeger K, Skelton J, Squares S, Squares R, Sulston JE, Taylor K, Whitehead S, Barrell BG (1998). Deciphering the biology of *Mycobacterium tuberculosis* from the complete genome sequence. Nature 393: 537–44

5
Functions of PolyP

5.1
ATP Substitute and Energy Source

PPK converts polyP to ATP by catalyzing an ADP attack on the termini of the polyP chain (Fig. 1). An aggregate of polyP associated with this membrane-bound enzyme could generate significant amounts of ATP at that very spot. However, as a source of energy, polyP, even at a cellular concentration in *E. coli* ten times that of ATP, would, in view of the ATP turnover of a fraction of a second, sustain the cell for only a second or two.

Another source of ATP could come from an AMP attack on polyP (Fig. 1) by AMP-phosphotransferase to produce ADP [Eq. (2)], which is readily converted to ATP by coupling with PPK [Eq. (3)] or with adenylate kinase [Eq. (4)]:

$$polyP_n + AMP \rightarrow polyP_{n-1} + ADP \tag{2}$$

$$polyP_n + ADP \rightarrow polyP_{n-1} + ATP \tag{3}$$

$$2\ ADP \leftrightarrow ATP + AMP \tag{4}$$

AMP-phosphotransferase has been purified from *Acinetobacter* (Bonting et al. 1991) and has been identified in *E. coli* (Kim and Kornberg, unpubl. observ.) and *Myxococcus xanthus* (Shiba and Kornberg, unpubl. observ.); adenylate kinase is a potent and ubiquitous enzyme.

Through the action of these enzymes, polyP is a potential phosphagen in cells where and when its levels var exceed those of ATP. Compared to the usual cellular ATP levels of 5 to 10 mM, the massive vacuolar deposits in yeast, expressed on the basis of total cell volume, can exceed 200 mM; in *Myxobacteria*, in stationary phase, the granular aggregates of polyP can reach 50 mM (Voelz et al. 1966; Shiba and Kornberg, unpubl. observ.).

In view of its energy equivalence to ATP, polyP could qualify as an ATP substitute in all its kinase roles involving a variety of acceptors. In addition to the observed transfers to AMP and ADP, polyP, as already noted, can replace ATP in the phosphorylation of glucose in many bacteria. All these glucokinases use either ATP or polyP as donors; the more phylogenetically ancient species appear to show a preference for polyP over ATP (Hsieh et al. 1993). One might expect to find polyP kinases for other sugars, sugar derivatives (e.g., nucleosides and coenzyme precursors), proteins and carboxylic acids. Indeed, phosphorylation of a 40-kDa protein in the ribosomal fraction of the archaebacterium *Sulfolobus acidocaldarius* was observed with polyP as the donor (Skorko 1989).

An energy recycling mechanism operating in the efflux of organic end products (e.g., lactate in enteric bacteria) in symport with protons can generate a proton motive force (Van Veen et al. 1994). Such a mechanism may function in the utilization of polyP. The efflux of a protonated metal chelate of P_i released from polyP creates a proton motive force that may be coupled to the accumulation of amino acids from the medium or the synthesis of ATP.

5.2
A Reservoir for P_i

A stable level of P_i, essential for metabolism and growth, can be ensured by a reservoir in which polyP can be converted to P_i by associated exopolyphosphatases. The polymer, as an aggregate complexed with multivalent counterions, enjoys a clear osmotic advantage over free P_i. Regulation of the *ppk* operon, which encodes both the polyP kinase and exopolyphosphatase in *E. coli* (Akiyama et al. 1993), appears to be responsive to the *pho* regulon that controls more than twenty genes related to phosphate metabolism (Rao and Torriani 1990). Multiple exopolyphosphatases in *E. coli* (Akiyama et al. 1993; Keasling et al. 1993) and in yeast (Andreeva and Okorokov 1993; Wurst and Kornberg 1994; Wurst and Kornberg, unpubl. observ.) are potentially available to produce P_i in various cellular locations.

5.3
Chelator of Metal Ions

As expcted of a phosphate polyanion, polyP is a strong chelator of metal ions. *Lactobacillus plantarum*, unusual in lacking a superoxide dismutase, a metalloenzyme that catalyzes the removal of the damaging superoxide radical, has an inorganic catalyst instead, an extraordinarily high, 30-mM level of Mn^{2+} chelated to 60 mM polyP (Archibald and Fridovich 1982). With regard to chelation of Ca^{2+}, the regulation of cellular Ca^{2+} in yeast by vacuolar Ca^{2+} depends on its binding to polyP; the polyP acts as a Ca^{2+} sink within the vacuole lumen (Dunn et al. 1994). Chelation of Ca^{2+} and Mg^{2+}, structurally essential in the cell walls of gram-positive bacteria, is regarded as the basis for the antibacterial action of polyP (Lee et al. 1994). Chelation of other metals (e.g., Zn, Fe, Cu, Cd) may either reduce their toxicity or affect their functions.

5.4
Buffer Against Alkali

Algae, like yeast, accumulate polyP in their vacuoles. In the halotolerant green alga *Dunaliella*, deposits of polyP reach levels near 1 M in P_i equivalents. When stressed at alkaline pH, amines enter the algal vacuoles and are neutralized by protons released by the enzymatic hydrolysis of polyP (Pick and Weiss 1991). The specific polyphosphatase, presumably activated by the amines, produces $polyP_3$ by mechanisms that have yet to be determined. Thus, polyP, as a result of its hydrolysis, can provide a high-capacity buffering system that sustains compartmentation of amines in vacuoles and protects the cytoplasmic pH. This alga, cultivated in large outdoor ponds, is an important commercial source of β-carotene for health foods and for food-coloring.

5.5
Channel for DNA Entry

Transforming competent *E. coli* with DNA for the cloning and expression of genes is currently one of the world's favored indoor sports. Despite the widespread use of a Ca^{2+} recipe to induce competence, there is little understanding of the mechanism whereby the highly charged DNA molecule penetrates the lipid bilayer membranes surrounding the cell. The discovery of polyhydroxybutyrate (PHB) complexed with Ca^{2+} and polyP in the membranes of competent cells was a significant advance (Reusch and Sadoff 1988). In a proposed structure, Ca^{2+} is bonded by ion dipoles to the carbonyl ester groups of PHB and by ionic interactions with polyP. This complex produces profound physical changes in the competent-cell membranes – increased rigidity at ambient temperatures and biphasic melting (Reusch and Sadoff 1988; Castuma and Kornberg, unpubl. results; see Chap. 9). Whether and how these alterations facilitate DNA entry still remain unclear.

The PHB-Ca^{2+}-polyP complex has been reconstituted in large, unilamellar vesicles by adding PHB, Ca^{2+} and polyP to phospholipids (Castuma and Kornberg, unpubl. results). The capacity of these vesicles for the uptake of small and large molecules, charged and neutral, needs to be extended. Although mutants lacking the long-chain polyP can attain competence, their membranes still contain a short polyP chain of about 60 residues, synthesized by a presumed novel route during the development of competence (Castuma and Kornberg, unpubl. results). Of great importance is the demonstration that the complex can function as an ion-selective channel (Das et al. 1997).

5.6
Regulator for Stress and Survival

Regulatory roles for polyP, a phosphate polyanion with some resemblance to RNA and DNA, seem reasonable. PolyP readily interacts with basic proteins (e.g., histones), with basic domains of proteins, as in polymerases, and has been observed in association with nonhistone nuclear proteins (Offenbacher and Kline 1984). Such roles could affect gene functions in positive or negative ways. Inasmuch as polyP is present in several sizes and complex forms, is located in the nucleus and other cell compartments, and fluctuates in response to nutritional and other parameters, it seems possible that polyP might function in the network of responses to stresses and the many signals that govern stages in the cell cycle and development. In response to a variety of stringencies and in the dynamics of metabolic adjustments to stationary phase (Siegele and Kolter 1992), regulatory roles for polyP are found (Rao and Kronberg 1996; Rao et al. 1998) and further considered in Chapter 9.

5.7
Regulator of Development

Developmental changes in microorganisms – fruiting body and spore formation in *Myxobacterial* (e.g., *M. xanthus*), sporulation in bacteria (e.g., *Bacillus*) and fungi, and heterocyst formation in cyanobacteria (e.g., *Anabaena*) – occur in response to starvation of one or several nutrients. In view of the involvement of polyP in the stationary stage of *E. coli*, polyP may well participate in other instances of cellular adjustments to deprivation.

In *M. xanthus* upon the conclusion of vegetative growth, the levels of polyP and of polyP-AMP phosphotransferase activity increase more than 10-fold (Shiba and Kornberg, unpubl. observ.). At these levels (Voelz et al. 1966), polyP may contribute in various ways to the development of fruiting bodies and spores. Inasmuch as an increase in ppGpp precedes polyP formation and mutants that fail to produce ppGpp also fail to increase their polyP levels (Shiba and Kornberg, unpubl. observ.), it seems that ppGpp has a regulatory role in polyP formation (see Chap. 9).

5.8
Cell Capsule

PolyP is a component of the capsule (Tinsley et al. 1993) which is loosely attached to the surface of *Neisseria* species and represents about half of the cellular polyP. Whether the capsule contributes to the pathogenesis of infections, and, if so, by combating phagocytosis, by chelating metals needed in complement fixation, or by some other way, has yet to be discovered; *ppk* mutants lack the resistance to human serum displayed by wild type cells.

6
PolyP in Animal Cells and Tissues

Although the presence of polyP in fungi and algae has been widely noted, the distribution and abundance of polyP in more complex eukaryotic forms have remained uncertain. The very low levels of polyP in animal cells (Gabel and Thomas 1971) and subcellular compartments and the lack of definitive and sensitive methods have left its metabolic and functional roles entirely obscure.

Improved enzymatic assay methods have confirmed that polyP is present in a wide variety of cell cultures and animal tissues. The concentrations of polyP generally range from 10 to 100 μM (expressed in P_i equivalents) in sizes of 100 to nearly 1000 residues. Among the subcellular organelles, polyP has been identified in lysosomes (Pisoni and Lindley 1991) and in mitochondria (Liu and Kornberg, unpubl. observ.) and is relatively enriched in nuclei (Kumble and Kornberg 1995). In rat brain, polyP is present throughout the course of embryonic and postnatal development (Kumble and Kornberg 1995). Uptake of P_i into polyP has been observed in cultured mammalian cells (Cowling and Birnboim 1994; Kumble and Kornberg 1995). Synthesis of polyP from P_i supplied in the medium bypasses dilution in the intracellular pools of P_i and ATP suggesting the involvement of separate compartments or a novel membrane-based fixation of P_i into polyP.

A strikingly rapid turnover of polyP was seen in a confluent culture of PC 12 cells, a neuron-like cell derived from an adrenal pheochromocytoma. Although these cells have a generation time of 48 to 72 h, the turnover was nearly complete in 1 h (Kumble and Kornberg 1995). Studies of the dynamics of polyP formation and utilization in a variety of cells should reveal novel functions for the polymer in different stages of growth and metabolism.

7
Prebiotic Role of PolyP

RNA may have preceded DNA and proteins in prebiotic evolution, but it seems likely that polyP appeared on earth before any of these organic polymers. PolyP arises simply from the dehydration and condensation of P_i at elevated temperatures and is evident in volcanic condensates (Yamagata et al. 1991) and oceanic steam vents. The anhydride-bond energy and the phosphate of polyP are plausible sources for nucleoside triphosphates, the activated building blocks of RNA and DNA (Waehneldt and Fox 1967; Kulaev and Skryabin 1974); mixed carboxylic-phosphate anhydrides provide a route to chemical polypeptide synthesis starting with amino acids and polyP (Harada and Fox 1965).

Among the species of polyP, mention should be made of the very simplest, namely inorganic pyrophosphate (PP_i). Once regarded solely as a metabolic product of biosynthetic reactions (A. Kornberg 1957), and fated to be hydrolyzed by potent and ubiquitous inorganic pyrophosphatases to drive these pathways, PP_i was revealed by later studies to be a substitute for ATP (Wood 1985). A role for PP_i as well as for polyP in prebiotic events leading to the evolution of ATP deserves attention. Inasmuch as the synthesis of one ATP by extant de novo pathways requires the input of at least eight ATPs, one must wonder how ATP was made in the first place.

PolyP accumulations, also prominent in archaea, may well be the substrates for enzymatic attack by nucleoside to produce nucleosides mono-, di- or triphosphates. A systematic search among these ancient organisms might uncover enzymes that carry out such salvage reactions in the biosynthesis of nucleotides, coenzymes and other factors.

8
Industrial Applications of PolyP

Aspects of the chemistry and physical properties of polyP that emerge from its industrial uses are of general interest and in some instances may impinge directly on its place and functions in biological systems (Chap. 9).

8.1
Depollution of P_i in the Environment

P_i accumulation in wastewaters containing runoff of fertilizers and discharges of industrial agents is a global problem that results in destructive algal blooms in bays, lakes and waterways. The use of lime, alum or ferric chloride to remove P_i is expensive and inefficient. Currently, sanitary engineers employ a biological process in which aerobically activated bacteria take up the P_i and convert it to polyP which is then removed along with the bacteria as a sludge. Because the microbial process is slow and inadequate, genetically engineered improvements are needed in *Acinetobacter* and other gram-negative organisms that dominate the flora. A start has been made in this direction by showing that the *ppk* gene on a runaway plasmid introduced into a selected coliform species vastly increases the rate and extent of P_i removal from the medium (Hardoyo et al. 1994). When coupled with an improved *pst* uptake system for P_i, polyP accumulations have reached 38 to 48 % of the cell-dry weight.

8.2
Antibacterial Action

PolyP is a safe, biodegradable additive to meats, enhancing water binding, emulsification and color retention while retarding oxidative rancidity. In its use in virtually all processed meat, poultry and fish products, polyP also serves as an antibacterial agent (Lee et al. 1994).

8.3
PolyP Kinase as an Antimicrobial Target

The PPK sequence (Figs. 2–4) has a high degree of conservation among diverse bacterial species, including some of the pathogens of the major

Table 2. Microorganisms with high homology of PPK sequence (see figs. 2–4)

Escherichia coli
Campylobacter coli
Acinetobacter calcoaceticus
Neisseria meningitidis
Synechocystis sp.
Vibrio cholerae
Mycobacterium tuberculosis
Mycobacterium leprae
Klebsiella aerogenes
Helicobacter pylori
Deinococcus radiodurans
Streptomyces coelicolor
Pseudomonas aeruginosa
Salmonella typhimurium

infectious diseases (Table 2). In view of the essentiality of polyP for stationary phase responses in *E. coli*, a similar role for polyP appears plausible in the expression of virulence factors which also appear in the stationary phase of some pathogens and the production of antibiotics in *Streptomyces* sp. (Tzeng and Kornberg 1998).

Knockouts of *ppk* in each of these organisms may indicate the polyP-dependence of virulence factors, antibiotics and other phenotypic features. If polyP were to prove necessary for virulence, then PPK would become an attractive target for antimicrobial drugs. An effective inhibitor of PPK might enjoy a broad spectrum of microbial targets with little toxicity, inasmuch as homologies to the bacterial *ppk* gene have not been observed in mammalian cells.

8.4
Source of ATP (NTPs)

The cost of ATP for use as an enzymatic phosphorylating agent on an industrial scale may be prohibitive, as is the high cost of agents, such as creatine phosphate and phosphoenolpyruvate, that might be used in an enzymatic ATP-regenerating system. In their place, polyP has been employed to regenerate ATP using PPK immobilized on a column (Butler 1977). In this system, a commercial form of polyP costing 25 US cents per lb. can provide ATP equivalents that would cost over $2000. The nucleoside diphosphate kinase capacity of PPK has also been utilized in regenerating UTP in the production of oligosaccharides, by replacing phosphoenolpyruvate with polyP (Noguchi and Shiba 1998).

8.5
Insulating Fibers

Phosphate fibers form bones and teeth. Polyphosphates are added to cheese, meats, toothpaste and drinking water. A calcium polyphosphate fiber has been synthesized with all the properties of asbestos (Griffith 1992) and could be a health-safe substitute, but has been abandoned by its inventor, Monsanto Chemical Company, which cited its fear of litigation from lawyers trained to bring suits for injuries from mineral fibers. Unfortunately, the decision to abandon polyP fibers as an insulator also extends to its use for nonflammable infant sleepwear, hospital mattress covers and fabric for aircraft interiors.

9
Summary

Pursuit of the enzymes that make and degrade polyP has provided analytic reagents which confirm the ubiquity of polyP in microbes and animals and provide reliable means for measuring very low concentrations. Many distinctive functions appear likely for polyP depending on its abundance, chain length, biologic source and subcellular location: an energy supply and ATP substitute, a reservoir for P_i, a chelator of metals, a buffer against alkali, a channel for DNA entry, a cell capsule, and, of major interest, a regulator of responses to stresses and adjustments for survival in the stationary phase of culture growth and development. Whether microbe or human, we depend on adaptations in the stationary phase, a dynamic phase of life. Much attention has focused on the early and reproductive phases of organisms, rather brief intervals of rapid growth, but more concern needs to be given to the extensive period of maturity. Survival of microbial species depends on being able to manage in the stationary phase. In view of the universality and complexity of basic biochemical mechanisms, it would be surprising if some of the variety of polyP functions observed in microorganisms did not apply to aspects of human growth and development, to aging and to the aberrations of disease.

Of theoretical interest regarding polyP is its antiquity in prebiotic evolution, which, along with its high energy and phosphate content, make it a plausible precursor to RNA, DNA and proteins. Of practical interest is its many industrial applications, among which is its use in the microbial depollution of P_i in marine environments.

Acknowledgments. I want to thank my colleagues – Kyunghye Ahn, Dana Ault-Riché, Masahiro Akiyama, LeRoy Bertsch, Celina Castuma, Claudio Villar, Cres Fraley, Elliott Crooke, Hong-Yeoul Kim, Jay Keasling, Kwang-Seo Kim, K.D. Kumble, Akio Kuroda, Sheng-jiang Liu, Nobuo Owaga, S.-J. Park,

Pradeep Ramulu, M.H. Rashid, Narayana Rao, Rohini Vij, Toshikazu Shiba, C.-M. Tzeng and Helmut Wurst – for their courage in pursuing a largely forgotten polymer and for the contributions that made it more memorable. I am also grateful to the National Institutes of Health (USA) for the support of this research.

References

Ahn K, Kornberg A (1990) Polyphosphate kinase from *Escherichia coli*. J Biol Chem 265: 11734–11739

Akiyama M, Crooke E, Kornberg A (1992) The polyphosphate kinase gene of *Escherichia coli*. Isolation and sequence of the *ppk* gene and membrane location of the protein. J Biol Chem 267: 22556–22561

Akiyama M, Crooke E, Kornberg A (1993) An exopolyphosphatase of *Escherichia coli*. The enzyme and its *ppx* gene in a polyphosphate operon. J Biol Chem 268: 633–639

Andreeva NA, Okorokov LA (1993) Purification and characterization of highly active and stable polyphosphatase from *Saccharomyces cerevisiae* cell envelope. Yeast 9: 127–139

Andreeva NA, Kulakovskaya T, Sidorov I, Karpov A, Kulaev I (1998) Purification and properties of polyphosphatase from *Saccharomyces cerevisiae* cytosol. Yeast 14: 383–390

Archibald FS, Fridovich I (1982) Investigations of the state of the manganese in *Lactobacillus plantarum*. Arch Biochem Biophys 215: 589–596

Bonting CFC, Kortstee GJJ, Zehnder AJB (1991) Properties of polyphosphate: AMP phosphotransferase of *Acinetobacter* strain 210A. J Bacteriol 173: 6484–6488

Butler L (1977) A suggested approach to ATP regeneration for enzyme technology applications. Biotechnol Bioeng 19: 591–593

Cowling RT, Birnboim HC (1994) Incorporation of [^{32}P] orthophospate into inorganic polyphosphates by human granulocytes and other human cell types. J Biol Chem 269: 9480–9485

Crooke E, Akiyama M, Rao NN, Kornberg A (1994) Genetically altered levels of inorganic polyphosphate in *Escherichia coli*. J Biol Chem 269: 6290–6295

Das S, Lengweiler UD, Seebach D, Reusch RN (1997) Proof for a non-proteinaceous calcium-selective channel in *Escherichia coli* by total synthesis from (R)-3-hydroxybutanoic acid and inorganic polyphosphate. Proc Natl Acad Sci USA 94: 9075–9079

Dunn T, Gable K, Beeler T (1994) Regulation of cellular Ca^{2+} by yeast vacuoles. J Biol Chem 269: 7273–7278

Ebel JP (1949) 'Participation de lácide metaphosphorique à la constitution des bacteries et des tissus animaux'. C.R. Acad. Sci Paris 228: 1312

Gabel NW, Thomas V (1971) Evidence for the occurence and distribution of inorganic polyphospate in vertebrate tissues. J Neurochem 18: 1229–1242

Griffith EJ (1992) In search of a safe mineral fiber. Chemtech 22: 220–226

Harada K, Fox S (1965) Thermal polycondensation of free amino acids with polyphosphoric acid. In: Fox S (ed) Origins of prebiological systems. Academic Press, New York, pp 289–298

Hardoyo K, Yamada K, Shinjo H, Kato J, Ohtake H (1994) Production and release of polyphosphate by a genetically engineered strain of *Escherichia coli*. Appl Environ Microbiol 60: 3485–3490

Harold FM (1967) Inorganic polyphosphates in biology: structure, metabolism and function. Bacteriol Rev 30: 772–794

Hsieh P-C, Shenoy BC, Jentoft JE, Phillips NFB (1993) Purification of polyphosphate and ATP glucose phosphotransferase from *Mycobacterium tuberculosis* H$_{37}$Ra: evidence that poly (P) and ATP glucokinase activities are catalyzed by the same enzyme. Protein Expr Purif 4: 76–84

Keasling, JD, Bertsch L, Kornberg A (1993) Guanosine pentaphosphate phosphohydrolase of *Escherichia coli* is a long-chain exopolyphosphatase. Proc Natl Acad Sci USA 90: 7029–7033

Kornberg A (1957) Pyrophosphorylases and phosphorylases in biosynthetic reactions. Adv Enzymol 18: 191–240

Kornberg A (1993) Recollections. ATP and inorganic pyro- and polyphosphate. Protein Sci 2: 131–132

Kornberg A (1994) Inorganic polyphosphate: a molecular fossil come to life. In: Torriani-Gorini A, Silver S, Yagil E (eds) Phosphate in microorganisms: cellular and molecular biology. American Society for Microbiology, Washington, DC, pp 204–208

Kornberg A, Kornberg SR, Simmes ES (1956) Metaphosphate synthesis by an enzyme from *Escherichia coli*. Biochim Biophys Acta 20: 215–227

Kornberg SR (1957) Adensine triphosphate synthesis from polyphosphate by an enzyme from *Escherichia coli*. Biochim Biophys Acta 26: 294–300

Kulaev IS (1979) The biochemistry of inorganic polyphosphates. Wiley, New York

Kulaev IS, Skryabin KG (1974) Reactions of nonenzymic trans-phosphorylation performed by high-polymeric polyphosphates and role in abiogenesis. Zh Evol Biokhim Fiziol 10: 533

Kulaev IS, Vagabov VM, Shabalin YA (1987) New data in biosynthesis of polyphosphates in yeast. In: Torriani-Gorini A, Rothman FG, Silver S, Wright A, Yagil E (eds) Phosphate metabolism and cellular regulation in microorganisms. American Society for Microbiology, Washington, DC, pp 233–238

Kumble KD, Kornberg A (1995) Inorganic polyphosphate in mammalian cells and tissues. J Biol Chem 270: 5818–5822

Kumble KD, Kornberg A (1996) Endopolyphosphatases for long chain inorganic polyphosphate in yeast and mammals. J Biol Chem 271: 27146–27151

Kumble KD, Ahn K, Kornberg A (1996) Phosphohistidyl active sites in polyphosphate kinase of *Escherichia coli*. Proc Natl Acad Sci USA 93: 14391–14395

Kuroda A, Kornberg A (1997) Polyphosphate kinase as a nucleoside diphosphate kinase in *Escherichia coli* and *Pseudomonas aeruginosa*. Proct Natl Acad Sci USA 94: 439–442

Lee RM, Hartmann PA, Stahr HM, Olson DG, Williams FD (1994) Antibacterial mechanism of long-chain polyphosphate in *Staphylococcus aureus*. J Food Protect 57: 289–294

Liebermann L (1890) Detection of metaphosporic acid in the nuclein of yeast. Pflugers Arch 47: 155–160

Meyer A (1904) Orientierende Untersuchungen über Verbreitung, Morphologie, und Chemie des Volutins. Bot Z 62(1): 113–152

Noguchi T, Shiba T (1998) Use of *Escherichia coli* polyphosphate kinase for oligosaccharides synthesis. Biosci Biotechnol Biochem. 62: 1594–1596

Nyren P, Nore BF, Strid A (1991) Proton-pumping *N,N'*-dicyclohexylcarbodiimide-sensitive inorganic pyrophosphate synthase from *Rhodospirillum rubrum*: purification, characterization, and reconstitution. Bichemistry 30: 2883–2887

Offenbacher S, Kline ES (1984) Evidence for polyphosphate in phosphorylated nonhistone nuclear proteins. Arch Biochem Biophys 231: 114–123

Pick U, Weiss M (1991) Polyphosphate hydrolysis within acidic vacuoles in response to amine-induced alkaline stress in the halotolerant alga *Dunaliella salina*. Plant Physiol 97: 1234–1240

Pisoni RL, Lindley ER (1991) Incorporation of [^{32}P] orthophosphate into long chains of inorganic polyphosphate within lysosomes of human fibroblasts. J Biol Chem 267: 3626–3631

Rao NN, Kornberg A (1996) Inorganic polyphosphate supports resistance and survival of stationary-phase *Escherichia coli*. J Bacteriol 178: 1394–1400

Rao NN, Liu S, Kornberg A (1998) Inorganic polyphosphate in *Escherichia coli*: the phosphate regulon and the stringent response. J. Bacteriol 180: 2186–2193

Rao NN, Torriani A (1990) Molecular aspects of phosphate transport in *Escherichia coli*. Mol Microbiol 4: 1083–1090

Reusch RN, Sadoff HL (1988) Putative structure and functions of a poly-β-hydroxybutyrate/calcium polyphosphate channel in bacterial plasma membranes. Proc Natl Acad Sci USA 85: 4176–4180

Robinson NA, Clark JE, Wood HG (1987) Polyphosphate kinase from *Propionibacterium shermanii*. J Biol Chem 262: 5216–5222

Rodriguez RJ (1993) Polyphosphate present in DNA preparations from filamentous fungal species of *Colletotrichum* inhibits restriction endonucleases and other enzymes. Anal Biochem 209–291–297

Schmidt G (1951) The biochemistry of inorganic pyrophosphates and metaphosphates. In: McElroy WD, Glass B (eds) Phosphorus metabolism. Johns Hopkins Press, Baltimore, pp 443–475

Shabalin YA, Kulaev IS (1989) Solubilization and properties of yeast dolichylpyrophosphate: polyphosphate phosphotransferase. Biokhimiya 54: 68–74 (in Russian)

Siegele DA, Kolter R (1992) Minireview: life after log. J Bacteriol 174: 345–348

Skorko R (1989) Polyphosphate as a source of phosphoryl group in protein modification in the archaebacterium *Sulfolobus acidocaldarius*. Biochimie 71: 1089–1093

Tinsley CR, Manjula BN, Gotschlich EC (1993) Purification and characterization of polyphosphate kinase from *Neisseria meningitidis*. Infect Immun 61: 3703–3710

Tzeng C-M, Kornberg A (1998) Polyphosphate kinase is highly conserved in many bacterial pathogens. Mol Microbiol 29: 381–382

Van Veen HW, Abee T, Kleefsman AWF, Melgers B, Kortstee GJJ, Konings WN, Zehnder AJB (1994) Energetics of alanine, lysine and proline transport in cytoplasmic membranes of the polyphosphate-accumulating *Acinetobacter johnsonii* strain 210A. J Bacteriol 176: 2670–2676

Voelz H, Voelz U, Ortigoza RO (1966) The "polyphosphate overplus" phenomenon in *Myxococcus xanthus* and its influence on the architecture of the cell. Arch Mikrobiol 53: 371–388

Waehneldt TV, Fox S (1967) Phosphorylation of nucleosides with polyphosphoric acid. Biochim Biophys Acta 134: 9–16

Wiame J-M (1947) Yeast metaphosphate. Fed Proc 6: 302

Wood HG (1985) Inorganic pyrophosphate and polyphosphates as sources of energy. Curr Top Cell Regul 26: 355–369

Wood HG, Clark JE (1988) Biological aspects of inorganic polyphosphates. Annu Rev Biochem 57: 235: 260

Wurst H, Kornberg A (1994) A soluble exopolyphosphatase of *Saccharomyces cerevisiae*. J Biol Chem 269: 10996–11001

Yamagata Y, Watanabe H, Saitoh M, Namba T (1991) Volcanic production of polyphosphates and its relevance to prebiotec evolution. Nature 352: 516–519

Research on Inorganic Polyphosphates: The Beginning

P. Langen[1]

1
From the Last Decade of the Previous Century to the Middle of the Present One: the Discovery of Polyphosphates in Yeast and Other Microorganisms and the Elucidation of Their Structure

In 1928 Karl Lohmann published three papers on pyrophosphate in cells in the *Biochemische Zeitschrift* (Lohmann 1928a–c). At that time, the identification of phosphate-containing compounds in cells or cell extracts was still under way, though many representatives, e.g. the hexose-phosphates of the glycolytic pathway, had already been characterized. Soon after the publication of these three papers, Lohmann found that the pyrophosphate he had studied was not "inorganic" pyrophosphate but was bound to adenylic acid (Lohmann 1929; Rapoport 1978). Thus he discovered ATP, the universal coin in life's energy trade. We are now aware that pyrophosphate should not accumulate in cells, because this would seriously impair the equilibrium of biosynthetic processes, such as that of nucleic acids, giving rise to its formation. It has to be split immediately. Thus, the polyphosphate found in cells begins with the triphosphate.

The great progress achieved by Lohmann in the characterization of phosphate-containing compounds in cell extracts was based to a large extent on his introduction of a fractional hydrolysis under acid conditions which enabled him to assay the "acid-labile phosphate", including ATP and, later on, the inorganic polyphosphates. He described this in Lohmann (1928a). The samples were hydrolyzed in 1 M hydrochloric acid for 7 or 30 min at 100 °C and analyzed colorimetrically. The amount of phosphate set free in the first 7 min is the labile phosphate, only to be corrected by subtraction of a small amount of phosphate due to some hydrolyses of hexose-1,6-diphosphates or nucleic acids between 7 and 30 min. In my early work with Lohmann on polyphosphates I carried out thousands of these assays and can confirm they were simple and very useful.

[1] Max-Delbrück-Centrum für molekulare Medizin, FB Zellbiologie, Robert-Rössle Str. 10, 13125 Berlin, Germany.

Progress in Molecular and Subcellular Biology, Vol. 23
H. C. Schröder, W. E. G. Müller (Eds.)
© Springer-Verlag Berlin Heidelberg 1999

Greater general interest in inorganic polyphosphates in cells did not occur until after World War II. One should, however, mention the remarkable paper by Majorie Macfarlane in 1936, describing that "a complex containing iron, nucleic acid and metaphosphate", from which she could isolate "inorganic metaphosphate", accounted for 30 % of the total phosphate in yeast. To my knowledge this was the first report on the presence of polyphosphates in this organism. As she mentioned in her paper, the complex "is probably identical with the Plasminsäure described by Kossel (1892)".

Later, up until the early 1960s, the presence of polyphosphates was reported for many "lower" organisms. Examples are: *fungi* such as yeast, *Aspergillus, Penicillium, Neurospora, Phycomyces*; *Mycobacteria, Corynebacteria, bacteria* such als *Aerobacter*; red algae; green algae such as *Chlorella, Euglena*; *Acetabularia*; *Amoeba*. For more extensive information, the reviews by Kuhl (1960, in German) and Harold (1966) are recommended.

The reason for the new wave of reports between 1946 and 1961 was that modern methods, developed at that time, were available for studies of phosphate-containing compounds in cells. These new methods included the introduction of ^{32}P-phosphate in biochemical research, the development of paper chromatography and new methods of extracting and fractionating cells and tissues. The latter allowed a rough separation of the cellular phosphates by differential extraction in: (*1*) the acid-soluble phosphates, containing the inorganic monophosphate, also called "orthophosphate" [even this could be divided in to a "free" monophosphate being metabolically active, in contrast to the inactive "Struktur-Phosphat" (Holzer and Lynen 1950; Langen und Liss 1960)], the phosphate esters of sugars and nucleosides, and, as found at that time, the low-molecular polyphosphates up to a chain length of about 10 phosphate residues per molecule; (*2*) the phopholipid-phosphate; (*3*) the nucleic acids; and (*4*) the phosphate from phosphoproteins. Since these fractionation methods were developed for animal tissues mainly, they had to be adapted in a special way when using them on microorganisms, expecially if these contained polyphosphates. Greater problems arose from the separation of polyphosphates and nucleic acids, mainly RNA, as both types of compounds were present in large amounts, e.g. in yeast, and it was often proposed that they may occur in common complexes (e.g. Kulaev and Belozerskij 1958). These were found to contain calcium or magnesium bridges between both types of polymers (Stahl and Ebel 1963). Complexes may also form between proteins and polyphosphates. A biological relevance of such complexes, however, was never demonstrated unequivocally (see also below).

In this context it should be mentioned that the very first hints of "metaphosphate" in organic material occur in studies of the last century on "nuclein" and its separation into protein and nucleic acid. Kossel (1892) reported at a session of the Physiological Society of Berlin, October 14, that the composition of the nucleic acids from salmon (discovered by Miescher as early as 1874) and from yeast was different. From, the latter he could obtain the "Plasminsäure" (as cited by Macfarlane 1936 see above). Further chemical treatment of this acid yielded, among other compounds, a product with contained "less oxygen than phosphoric acid, resembling metaphosphoric acids in this respect". Liebermann (1890) obtained some evidence of the presence of a compound he defined as "monometaphosphoric acid" (a definition not accepted by Kossel). One cannot but admire what the scientists at that time managed to discover from the crude material they had and the simple methods available to them.

In the second part of the 1940s, the great time of paper chromatography began. Hanes and Isherwood (1949) devised the separation and detection of phosphate compounds with this method. Ebel (1952a) and Grunze and Thilo (1953) followed, describing the separation of low-molecular polyphosphates by paper chromatography. According to a nomenclatur introduced by Thilo (1959, 1962), the term "polyphosphate" had to be used for the chain-type compounds, while "metaphosphate" formerly often used for the whole group of condensed phosphates, should be used for cyclic phosphate only (e.g. tripolyphosphate vs trimetaphosphate). It was later found that metaphosphates do not occur in cells. However, since trimetaphosphate under special conditions form from polyphosphates, this is now used as a specific assay for the presence of polyphosphates in cells such as eukaryotes.

2
Polyphosphate Synthesis.
How and Where Does it Start?

There are many reports from this time on the synthesis and degradation of polyphosphates, mainly from experiments using differential cell fractionation in yeast. To my knowledge, the first biochemists (after Macfarlane 1936) to recognize the phosphate accumulating in yeast under certain conditions as polyphosphate (they still called it "metaphosphate") were Schmidt et al. (1946), followed by Wiame (1947), who analyzed its presence in the metachromatic "volutin" grana where they occur together with proteins, lipids and nucleic acids. Later, Liss and Langen (1962) described in more detail the accumulation of polyphosphate in phosphate-starved yeast during incubation in phosphate-containing medium. We found that in phosphat-starved yeast during incubation in phosphate-containing medium, the polyphosphate synthesis increased about 20-fold over that of normal yeast under the same conditions. Then, the polyphosphate content of the yeast can account for up to 25 % of the yeast's dry mass. The amount of monophosphate converted to polyphosphates during conditions of polyphosphate accumulation is about 10 % of the monophosphate bound by the glycolysis. Thus, polyphosphates could certainly serve as a store of energy-rich phosphates, as suggested by, e.g., Ebel (1952b) and Hoffmann-Ostenhof and Weigert (1952). The energy of the anhydro bond in polyphosphates was found to be 9700 cal and is therefore within the range of that of ATP (Yoshida 1955). The distribution of phosphate in yeast to different fractions during phosphate starvation and resynthesis was also described by Ehrenberg (1961). Juni et al. (1947) and Wiame (1949) were the first to describe two fractions of polyphosphates from yeast: an acid-soluble one and another one extracted with alkali. As they found with ^{32}-P-phosphate, it was in the acid-insoluble fraction where the radioactivity appeared first.

Lohmann took up his studies in this field in the mid 1950s when I joined his lab as a student. There, Eberhard Liss and I were able to fractionate the polyphosphates of yeast into four groups (Langen and Liss 1958a, b; Langen et al. 1962; see also review by Harold 1966) whose properties are given in Table 1. By means of ^{32}-P-phosphate we found that monophosphate was first incorporated into fraction 4, i.e. the fraction with the highest molecular weight. A similar conclusion could be drawn from Wiame's studies, since it could be anticipated that his acid-insoluble polyphosphate was the one with the higher molecular weight, which also follows form experiments with synthetic model compounds (Katchman and van Walzer 1954).

This means that the monophosphate does not climb up the polyphosphate ladder step by step but is propelled right to the top and then tumbles down the ladder. This can be nicely shown if one blocks the energy gain of the yeast cell by iodoacetate, thus inhibiting the synthesis of the energy-rich polyphosphates. The radioactivity then disappears form fraction 4 in an amount which corresponds exactly to that added to the lower molecular weight fractions ast the same time.

To my knowledge it is still not known to which site in the cell the new incoming phosphate is attached (probably via ATP; see below). Of special interest in this context is that our fraction 4 differs from all the other fractions in that its extraction under alkaline conditions follows a chemical reaction of first order and is temperature dependent (Fig. 1; Langen and Liss 1959; Langen et al. 1962). This suggests the splitting of a chemical bond. The removal of RNA by RNase prior to the extraction does not change the extraction course. This is not in favor of a binding to nucleic acids. Extraction of 75 % of this fraction with detergent instead of alkaline yielded a poly-

Table 1. Polyphosphates in yeast as obtained by differential extraction

Fraction no.	Extracted by	mg P per g dry weight of resting yeast	Percent of total phosphate	Average chain length (phosphate groups per molecule)
1	TCA	1.8	6	4
2	NaClO$_4$	3.44	21	20
3	PH 10	0.6	4	55
4	0.05 N NaOH	1.8	6	260

TCA, trichloroacetic acid.

Fig. 1. Time course of extraction of polyphosphate fraction 4 from yeast with 0.05 N sodium hydroxide. This extraction in contrast to that of fractions 1–3 (see Table 1) follows a chemical reaction of first order, pointing to splitting of covalent bonds (as within the polyphosphate molecules themselves or between polyphosphate and an organic substrate). This is of special interest because fraction 4 contains polyphosphates formed first in the course of polyphosphate synthesis. Thus, a study of this fraction might provide a better knowledge of the cellular site and possible primer of polyphosphate synthesis. P_a total amount of fraction 4; P_x amount already extracted

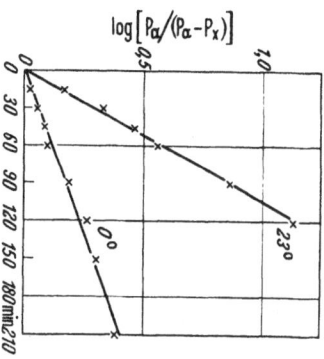

phosphate with 165 to 250 phosphate groups per molecule (Liss and Langen 1960; Langen et al. 1962). This is much lower than that found for *Aerobacter* by Harold (1963) with 600 phosphate groups per molecule.

The administration of calcium ions to the extraction medium abolished for fraction 4 the character of a chemical reaction of first order, which points to a chemical degradation of polyphosphates since calcium ions facilitate this process. Their presence during the extraction might speed it up to an extent that the chemical order is no longer recognizable. Then, is the primer a polyphosphate itself with a very high molecular weight? Is it to be found in the 25 % of fraction 4 not extracted with detergent? Of special interest in this context are experiments with isolated enzymes which reversibly transfer phosphate between polyphosphate and ATP/ADP, as found by the Kornbergs (Kornberg et al. 1956; Kornberg 1957) in *E. coli* and Muhammed (1961) in *Corynebacterium xerosis*. Both these enzymes did not require a polyphosphate for the cell-free synthesis of new polyphosphates. The *Escherichia* enzyme synthezised polyphosphate without the addition of primer polyphosphate to the test sample. For the enzyme from *Corynebacterium* the rate of polyphosphate synthesis did not change after the addition of polyphosphates of various structures. Thus, evidence of an inorganic polyphosphate primer for their synthesis is still lacking, renewing the discussion on a binding to an organic substrate. Is it possible that the purified enzyme preparation still contained bound polyphosphates in very small amounts, too small to be detected? One is reminded of the early statement by J.P. Ebel, a pioneer in polyphosphate research (Ebel 1952b): "pour être métaboliquement actifs, les polyphosphates devraient être liés". Bound to which partner was (is still?) the question.

At that time also the first enzymes involved in polyphosphate metabolism were described, such as polyphosphate-degrading enzymes (Mattenheimer 1956a–c; Muhammed et al. 1959). Szymona (1962), Szymona et al. (1962), Szymona and Szumilo (1966) and Dirheimer and Ebel (1962) found that phospate from polyphosphate can be transferred to glucose or fructose directly without the intermediate formation of ATP.

Growth and polyphosphate accumulation seem to be antagonistic processes. For cultures of *bacteria* and *fungi* it has been found that the polyphosphate content is low while the culture is still young, but increases with age and duration of the culture and decreasing growth. This might point to the fact that polyphosphate accumulation is a kind of luxury for the cell which it cannot afford during severe biosynthetic expenses. A function such as storage of energy and phosphate would also be compatible with the discovery by Harold and Harold (1963) that mutants, unable to form polyphosphates, survive happily in cultures, a finding which comes as a shock at least to those people who liked the idea that the compounds they worked on were essential for life. They might be consoled by the consideration that this was in a lab and that under the harsh environment in nature (e.g. yeast cells growing on grapes) the ability to form polyphosphates from phosphate and energy when available, as a storage for harder times, might be a decisive evolutionary advantage.

In 1961 we had a great symposium on "Acides Ribonucléiques et polyphosphates; structure, synthèse et fonctions" in Strasbourg, organized by J.P. Ebel for the "Centre National de la Recherche Scientifique", demonstrating that until that time both groups were still considered in some common context. From that time on, the fate for both groups was different: the interest in nucleic acids increasing with molecular biology, the research on polyphosphate stagnating (except for the work of Harold, with important findings, some cited above, and his comprehensive review on polyphosphate research from 1945 onwards).

In "The Beginning" we also considered the possibility that polyphosphates may occur in animal cells. My own research in this direction was only cursory and failed to yield any evidence of their presents. This might also have been due to the lack of proper methods. The fact that polyphosphates have now been found unambiguously in animal cells is exciting and further research is urgently needed on their structure, metabolism and function, with emphasis on the application of methods of molecular biology/genetics on the enyzmes involved. Our knowledge of the polyphosphates in microorganisms may be helpful, but I doubt if all the facts will be valid for animal cells. The future for polyphosphate research might now be brighter than it was 35 years ago. If I were now a young researcher, I think I would give work on these compounds a try.

References

Dirheimer G, Ebel JP (1962) Mise en évidence d'une polyphosphate-glucose-phospho-transferase dans *Corynebacterium xerosis*. Acad Sci 254: 2850–2852

Ebel JP (1952a) Recherches sur les polyphosphates contenue dans diverses cellules vivantes. I. Mise au point d'une méthode d'extraction. Bull Soc Chim Biol 34: 321–329

Ebel JP (1952b) Recherches sur les polyphosphates contenus dans diverses cellules vivantes. IV. Localisation cytologique et rôle physiologique des polyphosphates dans la cellule vivante. Bull Soc Chim Biol 34: 498–505

Ehrenberg M (1961) Der Phosphorstoffwechsel von *Saccharomyces cerevisiae* in Abhängigkeit von intra- und extrazellulärer Phosphatkonzentration. Arch Mikrobiol 40: 126–152

Grunze H, Thilo E (1955) Die Papierchromatographie der kondensierten Phosphate. Akademie-Verlag, Berlin

Hanes CD, Isherwood FA (1949) Separation of the phosphoric esters on the filter paper chromatogram. Nature 164: 1107–1112

Harold FM (1963) Inorganic polyphosphate of high molecular weight in *Aerobacter aerogenes*. J Bacteriol 86: 885–887

Harold FM (1966) Inorganic polyphosphates in biology: structure, metabolism, and function. Bacteriol Rev 30: 772–794

Harold RL, Harold FM (1963) Mutants of *Aerobacter aerogenes* blocked in the accumulation of inorganic polyphosphate. J Gen Microbiol 31: 241–246

Hoffmann-Ostenhof O, Weigert W (1952) Über die mögliche Funktion des polymeren Metaphosphats als Speicher energiereichen Phosphats in der Hefe. Naturwissenschaften 39: 303–304

Holzer H, Lynen F (1950) Über den aeroben Phosphatbedarf der Hefe. III. Labil an die Struktur gebundenes Phosphat in lebender Hefe. Liebigs Ann 569: 138

Juni E, Kamen MD, Siegelmann S, Wiame JM (1947) Physiological heterogeneity of metaphosphate in yeast. Nature 160: 717–718

Katchman A, Van Wazer JR (1954) The "soluble" and "insoluble" polyphosphates in yeast. Biochim Biophys Acta 1954: 445–446

Kornberg SR (1957) Adenosine triphosphate synthesis from polyphosphates by an enzyme from *Escherichia coli*. Biochim Biophys Acta 26: 294–300

Kornberg A, Kornberg SR, Simms ES (1956) Metaphosphate synthesis by an enzyme from *Escherichia coli*. Biochim Biophys Acta 26: 215–227

Kossel A (1892) Über die Nucleinsäure. Verhandlungen der physiologischen Gesellschaft zu Berlin, Jahrgang 1892–1893. I. Sitzung am 14. Okt. 1892. Arch Anat Physiol: 158–164

Kuhl A (1960) Die Biologie der kondensierten Phosphate. Ergeb Biol 23: 144–185

Kulaev IS, Belozerskij AN (1958) Electrophoretic studies on polyphosphate-ribonucleic acid complexes from *Aspergillus niger*. Proc Acad Sci USSR 120: 128–131

Langen P, Liss E (1958a) Über die Polyphosphate der Hefe. Naturwissenschaften 45: 191–192

Langen P, Liss E (1958b) Über Bildung und Umsatz der Polyphosphate in Hefe. Biochem Z 330: 455–466

Langen P, Liss E (1959) Zur Charakterisierung der Polyphosphate der Hefe. Naturwissenschaften 46: 151–152

Langen P, Liss E (1960) Differenzierung des Orthophosphates der Hefezelle. Biochem Z 392: 403–406

Langen P, Liss E, Lohmann K (1962) Art, Bildung und Umsatz der Polyphosphate der Hefe. In: Acides ribonucléiques et polyphosphates: structure, synthèse et fonctions. Colloques internationaux du Centre National de la Recherche Scientifique No 105. Editions du Centre National de la Recherche Scientifique, Paris, pp 604–614

Liebermann L (1890) Nachweis der Metaphosphosäure im Nuclein der Hefe. Pflugers Arch 47: 155–160

Liss E, Langen (1960) Über ein hochmolekulares Polyphosphat der Hefe. Biochem Z 333: 193–201

Liss E, Langen P (1962) Versuche zur Polyphosphat-Überkompensation in Hefezellen nach Phosphatverarmung. Arch Microbiol 41: 383–392

Lohmann K (1928a) Über das Vorkommen und den Umsatz von Pyrophosphat in Zellen. I. Mitteilung: Nachweis und Isolierung des Pyrophosphates. Biochem Z 202: 466–493

Lohmann K (1928b) Über das Vorkommen und den Umsatz von Pyrophosphat in Zellen. II. Mitteilung: Die Menge der leicht hydrolysierbaren P-Verbindung in tierischen und pflanzlichen Zellen. Biochem Z 208: 164–171

Lohmann K (1928c) Über das Vorkommen und den Umsatz von Pyrophosphat in Zellen. III. Mitteilung: Das physiologische Verhalten des Pyrophosphats. Biochem Z 203: 171–207

Lohmann K (1929) Über die Pyrophosphatfraktion im Muskel. Naturwissenschaften 17: 624–625

Macfarlane MG (1936) Phosphorylation in living yeast. Biochem J 30: 1369–1379

Mattenheimer H (1956a) Die Substratspezifität anorganischer Poly- und Metaphosphatasen. I. Optimale Wirkungsbedingungen für den Abbau von Poly- und Metaphosphaten. Z Physiol Chem. 303: 107–114

Mattenheimer H (1956b) Die Substratspezifität anorganischer Poly- und Metaphosphatasen. II. Trennung der Enzyme. Z Physiol Chem 303: 115–124

Mattenheimer H (1956c) Die Substratspezifität anorganischer Poly- und Metaphosphatasen. III. Papierchromatographische Untersuchungen beim enzymatischen Abbau von anorganischen Poly- und Metaphosphaten. Z Physiol Chem 303: 125–139

Muhammed A (1961) Studies on biosynthesis of polymetaphosphate by an enzyme from *Corynebacterium xerosis*. Biochim Biophys Acta 54: 121–132

Muhammed A, Rodgers A, Hughes DE (1959) Purification and properties of a polymetaphosphatase from *Corynebacterium xerosis*. J Gen Microbiol 20: 484–495

Rapoport SM (1978) Lohmann and ATP. TIBS 3: 184

Schmidt G, Hecht L, Thannhauser SJ (1946) The enzymatic formation and the accumulation of large amounts of a metaphosphate in baker's yeast under certain conditions. J Biol Chem 166: 775–776

Stahl AJC, Ebel JP (1963) Étude des complexes entre acides ribonucléiques at polyphosphates inorganiques dans la levure. III. Rôle des cations divalents dans la formation des complexes. Bull Soc Chim Biol 45: 887–900

Syzmona M (1962) Purification and properties of the new hexokinase utilizing inorganic polyphosphate. Acta Biochim Pol 9: 165–180

Szymona O, Szumilo T (1966) Adenosine triphosphate and inorganic polyphosphate fructokinases of *Mycobacterium phlei*. Acta Biochim Pol 17: 129–144

Szymona M, Szymona I, Kulesza S (1962) On the occurrence of inorganic polyphosphate hexokinase in some microorganisms. Acta Microbiol. Pol 11: 287–300

Thilo E (1959) Die kondensierten Phosphate. Naturwissenschaften 59: 367–373

Thilo E (1962) Condensed phosphates and arsenates. Adv Inorg Chem Radiochem 4: 1–77

Wiame JM (1947) Étude d'une substance polyphosphoreé basophile et metachromatique chez les levures. Biochim Biophys Acta 1: 234–255

Wiame JM (1949) The occurrence and physiological behavior of two metaphosphate fractions in yeast. J Biol Chem 178: 919–929

Yoshida A (1955) Studies on metaphosphate. II. Heat of hydrolysis of metaphosphate extracted from yeast cells. J Biochem 42: 162–168

Metabolism and Function of Polyphosphates in Bacteria and Yeast

I. S. Kulaev, T. V. Kulakovskaya, N. A. Andreeva and L. P. Lichko[1]

1
Introduction

Inorganic polyphosphate (polyPs) found in living organisms more than 100 years ago by Lieberman (1888) are linear polymers containing 2–1000 residues of orthophosphate linked by the energy-rich phosphoanhydride bond. They are widely spread in various microoganisms and are found in small amounts in the cells of animals and plants (Kulaev 1979; Kulaev and Vagabov 1983; Wood and Clark 1988). PolyPs perform varied biological functions. Recent investigations have shown the polyPs belong to biopolymers, the function of which changes when passing from prokaryotic cells to cells of lower and then higher eukaryotes. The comparative characteristics of the functions and metabolic ways of polyPs in different organisms are of interest from the viewpoint of the evolutionary physiology of the cell.

2
Polyphosphates and Enzymes of Their Metabolism in Prokaryotes

The study of polyP localization in prokaryotic cells offered a clearer view of some functions of these biopolymers. In prokaryotes, the compartmentation of biochemical processes is lowly developed. The main compartments of a bacterial cell are cytoplasm and cell envelope, both possessing polyPs.

PolyP-containing, or volutine, granules have been demonstrated in the cytoplasma of various bacteria (Kulaev and Vagabov 1983; Kornberg 1995). Bacteria are able to accumulate ultra-high amounts of polyPs under certain conditions. *Microlunatus phosphorovorus* isolated from activated sludge (Nakamura et al. 1995) and *Acinetobacter johnsonii* (Van Groenestijn et al. 1989) are examples of such bacteria. In *A. johnsonii*, this polymer can account for 30 % of dry biomass (Deinema et al. 1985). Therefore, one of the polyP functions in bacteria is that of phosphate storage. The localization of

[1] Institute of Biochemistry and Physiology of Microorganisms, Russian Academy of Sciences, Pushchino, Moscow Region, 142292, Russia.

Progress in Molecular and Subcellular Biology, Vol. 23
H. C. Schröder, W. E. G. Müller (Eds.)
© Springer-Verlag Berlin Heidelberg 1999

polyP granules in the vicinity of the bacterial nucleoid suggests their possible involvement in the regulation of gene activity (Kulaev and Vagabov 1983).

Some quantities of polyP are localized in the periplasmic region outside the cytoplasmic membrane (Kulaev 1979; Kulaev and Vagabov 1983).

A complex consisting of polyPs, Ca^{2+}, and poly-β-hydroxybutyrate has been isolated from *Escherichia coli* membranes (Reusch and Sadoff 1988). The complex is a double helix where the outher chain is represented by poly-β-hydroxybutyrate, and the inner chain, by polyP, and they are linked to each other by Ca^{2+} ions. The amount of this complex rises in transformation-competent cells whose membranes are easily permeable for DNA (Huang and Reusch 1995). Increase in the content of such complexes in the membranes is followed by appearance of rigid domains and disruption of a lipid bilayer conformation (Castuma et al. 1995). The authors assume that these complexes are involved in DNA transport into the cell. Now, new data have evolved that the complexes serve as voltage-activated calcium channels (Huang and Reusch 1996).

PolyP is a component of the capsule, which is loosely attached to the surface of *Neisseria* species and represents about half cellular polyP of *Neisseria* (Tinsley et al. 1993). PolyP pool-reduced mutant of *Neisseria* has a lower pathogeneity than the wild-type strain (Tinsley and Gotschlich 1995).

Investigation of polyP functions is closely associated with the study of enzymes of their metabolism. The most important enzymes involved in biosynthesis and degradation of polyP in bacteria are 1,3-diphosphoglycerate: polyphosphate phosphotransferase (EC 2.7.4.17) (Kulaev 1979), polyphosphate: adenosine monophosphate phosphotransferase (Dirheimer and Ebel 1965), polyphosphate: D-glucose 6-phosphotransferase (EC 2.7.1.63) (Szymona 1957; Szymona and Ostrowski 1964), polyphosphate: ADP phosphotransferase (polyphosphate kinase, EC 2.7.4.1) (Kornberg et al. 1956), exopolyphosphatase (EC 3.6.1.11) (Akiyama et al. 1993), and tripolyphosphatase (EC 3.6.1.25) (Van Alebeek et al. 1994).

Polyphosphate kinase revealed in bacteria by Kornberg et al. (1956) catalyzes the reactions of reverse transfer of the energy-rich phosphate residue from ATP to polyphosphates and from polyphosphates to ADP, thus linking the energy-rich pools. This enzyme, whose gene *ppk* has been recently cloned at Kornberg's laboratory (Akiyma et al. 1992), is a membrane-bound homotetramer with subunit molecular mass of 80 kDa. It is responsible for the processive synthesis of long polyP chains (ca. 750 residues) in vivo. Polyphosphate kinase plays an important role in the survival of *E. coli* under stress and starvation, which was established in mutants deficient in this gene (Kornberg 1995). Alteration of the conserved His-441 and His-460 to either glutamine or alanine by site-specific mutagenesis rendered the kinase protein incapable of enzymatic activity. The cells expressing such mutant proteins had virtually nor polyP (Kumble et al. 1996). The mutant cells show no phenotypic changes in the exponential phase of growth (Crooke et al. 1994; Kornberg 1995). In the stationary phase these mutants survive poorly, and are less resistant to heat, oxidants, and osmotic challenge (Crooke et al. 1994; Kornberg 1995; Rao and Kornberg 1996).

The amounts of polyP in the *ppk* mutants of *Neisseria gonorrhoeae* and *N. meningitidis* were reduced to 2–10 % of wild-type levels. The mutants grew less vigorously than the wild-type cells in vitro and showed a striking increase in sensitivity to killing by human serum (Tinsley and Gotschlich 1995).

Taken together, these observations point to the fact that polyphosphate kinase is the main enzyme providing polyP synthesis in bacteria, and one of the most important functions of polyPs is their involvement in the regulation of biochemical processes under stress. Bacteria, however, possess some other polyP-synthesizing enzymes, such as 1,3-diphosphoglycerate : polyphosphate phosphotransferase (Kulaev 1979). Poly-β-hydroxybutyrate calcium polyP membrane complexes are present in *ppk*-mutants of *E. coli*, which lack polyP (Castuma et al. 1995). Therefore, the polyP in these complexes synthesized without polyphosphate kinase. There is an additional example of polyP involvement in regulation. The archaebacterium *Sulfolobus acidocaldarius* has been shown to contain a protein kinase, which uses not ATP but high-polymer polyphosphates for protein phosphorylation (Skorko 1989).

In prokaryotes, the level of intracellular phosphate is regulated by exopolyphosphatase, the enzyme which splits orthophosphate off the end of the polyphosphate chain. It was purified from *E. coli* (Akiyama et al. 1993) and *A. johnsonii* (Bonting et al. 1993). In bacteria, this enzyme hydrolyzes mostly high-molecular polyP. As for polyPs with the chain length of 3–4 phosphate residues, they are hydrolyzed, but only sligthly. A specific tripolyphosphatase is purified from *Thermobacterium thermoautotrophicum*. This enzyme of 22 kDA hydrolyzes tripolyphosphate with five-fold higher activity than polyP$_{15}$ (Van Alebeek et al. 1994).

At Kornberg's laboratory, another enzyme possesing exopolyphosphatase activity was revealed in *E. coli* (Keasling et al. 1993). One of its substrates turned out to be guanosine pentaphosphate which yields guanosine tetraphosphate in hydrolysis. This is the so-called stringent factor, which is a certain regulator of ribosome function and protein biogenesis in bacteria under conditions of amino acid starvation.

The function of reservation of phosphate groups is closely associated with that of the energy-rich compound. In bacteria, polyP functions are connected with energy metabolism. Belozersky suggested that inorganic polyP in very primitive organisms could perform the functions of energy-rich compounds, which subsequently passed to ATP (Belozersky 1958). The involvement of polyP in energy metabolism in relation to the adenyl system has been confirmed recently by new data. Polyphosphate: AMP-phosphotransferase was purified from *Acinetobacter* strain 210A (Bonting et al. 1991). PolyPs can replace ATP in phosphorylation of glucose in many bacteria. The polyP- and ATP-dependent glucokinase activities are catalyzed by the same enzyme purified from *Propionibacterium shermanii* (Phillips et al. 1993) and *Mycobacterium tuberculosis* (Hsieh et al. 1993).

Thus, the functions of polyphosphates in prokaryotes are as follows:

1. reservation of phosphate;
2. participation in energy metabolism being closely linked to the adenyl system;
3. involvement in the control of biochemical processes:
   (a) participation in the survival of bacteria under stress,
   (b) formation of the complex with poly-β-hydroxybutyrate, which affects the membrane permeability,
   (c) serving as donors of protein phosphorylation.

3
Polyphosphates and Enzymes of Their Metabolism in Lower Eukaryotes

One of the principal properties distinguishing eukaryotes from prokaryotes is a pronounced compartmentation of biochemical processes when some functions are performed by specialized cell organelles. The metabolism peculiarities in each individual organelle are substantial enough to stimulate the theory of endosymbiotic origin of the eukaryotic cell; this theory is confirmed at present by many facts (Woese et al. 1990; Margulis 1993).

Eukaryotes possess polyP pools localized in different cell compartments (Kulaev 1979; Kulaev and Vagabov 1983). Yeast and other eukaryotic microorganisms reveal high-molecular polyPs in such compartments as cytosol, cell envelope (Kulaev 1979, 1994), and vacuoles (Urech et al. 1978; Lichko et al. 1982). PolyPs were found in the nuclei of *Neurospora crassa* (Kulaev 1979) and *Physarum polycephalum* (Piletus et al. 1989) and also in yeast mitochondria (Beauvoit et al. 1989). At present, the data available suggest that polyP functions in these compartments are different in many respects.

In vacuoles, the main polyP function is phosphate storage and chelation of cations. In vacuoles, which are storage compartments for a number of cations in the yeast cell (Wiemken and Durr 1974; Okorokov et al. 1980), the orthophosphate residues may confine different cations in an osmotically inert state. Arginine accumulated in vacuoles is demonstrated to form a complex with polyP (Durr et al. 1979). Accumulation of K^+, Mn^{2+}, and Mg^{2+} ions in vacuoles is in good correlation with the increase of the polyP content; therefore, it is possible to form appropriate complexes in this case again (Lichko et al. 1982). The complex-forming function is likely to be very important for the yeast cell, since even under phosphate starvation polyP decreases but little (Lichko et al. 1982). Under certain cultivation conditions, the vacuoles can contain the major part of the polyP pool of the yeast cell (Urech et al. 1978). At the same time, *Candida utilis* was reported to have no vacuolar polyP under phosphate starvation (Bourne 1991). Presumably, the polyP synthesis in vacuoles is accounted for by the work polyphosphate kinase revealed in these organelles (Shabalin et al. 1977). Recently, the occurrence of this enzyme in yeast has been questioned (Booth and Guidotti 1995). However, a different way of polyP synthesis in vacuoles is not ruled out. The vacuoles have a functional analogy with lysosomes of animal cells and are able to accumulate orthophosphate both in vivo and in vitro. It is well known that orthophosphate transported into isolated lysosomes is quickly incorporated into the fraction of the high-molecular polyphosphates (Pisoni and Lindley 1992). It is not improbable that the polyphosphatase of these organelles is able not only to hydrolyze but also to synthesize polyPs.

PolyPs with small chain lengths are revealed in yeast mitochondria (Beauvoit et al. 1989). Their formation is connected with oxidative phosphorylation and probably involves ATP (Beauvoit et al. 1989). Therefore, their function in mitochondria is believed to be similar to that in bacteria and is connected with bioenergetics.

The polyphosphate function in the nuclei is likely to be connected with their participation in the regulation of gene activity. It is well known that their biosynthesis occurs simultaneously with the total RNA synthesis (Kulaev 1979). Recent data show that in the fungus *Physarum polycephalum* the high-molecular polyPs of nuclei degrade to low-molecular fragments during sporulation (Piletus et al. 1989). Degradation of the high-molecular polyPs to low-molecular fragments is also found in the fungus *Agaricus bisporus* (Kulaev 1979). Recently, new evidence for participation of nuclear polyP in the regulation of gene activity has been obtained. PolyP$_{60}$ present in DNA preparation from filamentous fungal species of *Colleotrichum* inhibits restriction endonucleases and other enzymes (Rodriguez 1993). *Neurospora crassa* nuclei are revealed to contain a polyphosphate depolymerase (Kulaev 1979), and *Saccharomyces cerevisiae* nuclei – a polyphosphatase (Lichko et al. 1995). A body of compelling evidence of polyP involvement in the regulation of gene activity was obtained for animal cells as well (Offenbacher and Kline 1984; Schroder et al. 1996).

Cell envelope polyPs are of great importance for the formation of a negative charge on the cell surface of fungi (Vagabov et al. 1990; Ivanov et al. 1996). In the same compartment, the polyphosphate synthesis is connected with the glycoprotein one (Kulaev et al. 1987). At our laboratory, a new enzyme localized in membranes of transport vesicles of the yeast was revealed. It is dolichyl pyrophosphate:polyphosphate phosphotransferase, which catalyzes the transfer of phosphate from dolichyl pyrophosphate involved in the biosynthesis of the yeast cell wall mannoproteins to high-polymer polyphosphates (Kulaev et al. 1987).

PolyP synthesis in the yeasts at the exit of K^+ ions from cells suggests also the participation of these polymers in the regulation of transport processes (Okorokov et al. 1983). The polyphosphate complexes with poly-β-hydroxybutyrate were revealed in eukaryotic membranes (Reusch 1989). They are similar to those found in bacteria and seem to perform the same functions associated with the regulation of membrane permeability. Taken together, these data suggest that the role of polyP in the structure of membranes and cell envelope as well as in transport processes in eukaryotes is substantial.

The polyP functions in the yeast cytosol have not been specially analyzed yet; however, the cytosol polyphosphatase may be up to 60% of the total polyphosphatase activity in the yeast cell and should play an important role. The polyP metabolism in the yeast might be directly related to that of such second messengers as diadenosinetetra-, penta-, and hexapolyphosphates. Diadenosinetetra-, penta- and hexapolyphosphates are involved in the regulation of intracellular Ca^{2+} level (Tepel et al. 1996). Recent data show that all alleged yeast polyphosphate kinase is actually diadenosine-5',5'''-P^1,P^4-tetraphosphate α,β-phosphorylase (Booth and Guidotti 1995). The authors attempted to repurify the putative yeast polyphosphate kinase following the protocol of Felter and Stahl (1973) and using the same assay for the activity, namely the production of ^{32}P-labeled ATP in the presence of ^{32}P-labeled polyP and ADP. They found the activity thus assayed and purified was not due to a polyphosphate kinase but rather to the enzyme diadenosine-5',5'''-P^1,P^4-tetraphosphate α,β-phosphorylase (AP_4 phosphorylase), acting in concert with one or more yeast polyphosphatases. The enzyme mentioned was isolated from soluble yeast extract. However, one cannot rule out a possibility that the yeast possess polyphosphate kinase in certain membrane fractions, say, in the vacuolar membrane where it was detected earlier (Shabalin et al. 1977).

Adenosine-5'-tetraphosphate, guanosine-5'-tetraphosphate, and dinucleoside polyphosphates were found in the yeast (Baltzinger et al. 1986). Recently, we have found that the cytosol exopolyphosphatase of S. cerevisiae is capable of hydrolyzing adenosine-5'-tetraphosphate and guanosine-5'-tetraphosphate with activities which are 1.5–2 time higher than that with polyP$_{15}$ (Kulakovskaya et al. 1997). The apparent Km values for hydrolysis of adenosine-5'-tetraphosphate and guanosine-5'-tetraphosphate are 100 and 80 μM, respectively. The high specific activity and affinity of the cytosol polyphosphatase with respect to adenosine-5'-tetraphosphate and guanosine-5'-tetraphosphate allow us to consider them as physiologically important substrates of this enzyme in the yeast cell (Kulakovskaya et al. 1997).

In the process of evolution from prokaryotes to eukaryotes, the polyP functions greatly changed and were distributed among the cell compartments. The most essential functions of polyP in the lower eukaryotes are as follows:

1. phosphate storage in vacuoles;
2. retention of positively charged substance pools in vacuoles in an osmotically inert form;
3. participation in bioenergetic processes in mitochondria;
4. participation in the regulation of transport processes in membranes including formation of polyphosphate-poly-β-hydroxybutyrate complexes;
5. participation in the structure and regulation of the synthesis of some cell wall components;
6. participation in the control of gene activity.

Several exopolyphosphatases have been purified from yeast, catalyze the hydrolysis of inorganic polyP with splitting of inorganic phosphate from the chain end (Andreeva and Okorokov 1993; Lorenz et al. 1994; Wurst and Kornberg 1994), as well as endopolyphosphatase splitting long polyP molecules into shorter ones (Kumble and Kornberg 1996). Endopolyphosphatase is a dimer of 35 kDa subunits. Its activity requires divalent metal cations. Mn^{2+} is more active than Mg^{2+} with the optimum of about 2.5 mM. This enzyme hydrolyzes $polyP_{750}$ to shorter chains and even to $polyP_3$ (Kumble and Kornberg 1996). The authors suggest that this enzyme is localized in vacuoles. Endopolyphosphatases escaped detection in prokaryotes. The presence of these enzymes in eukaryotes is possibly attributed to compartmentation and the necessity to transport polyP from one compartment to other. A detailed description of the properties of yeast exopolyphosphatases of different localization is given below.

4
Exopolyphosphatases of Yeast Cells: Compartmentation and Properties

The difference in functions and metabolic peculiarities as well as possible different evolutionary origin of organelles suggest that the same reaction in different organelles of eukaryotic cells may be often catalyzed by the enzyme specific for a given organelle and different from the corresponding enzymes from other organelles in structure, genetic determinants, properties and functions. Examples of such enzymes are H^+-ATPases belonging to different classes [F-ATPase of mitochondria and chloroplasts, P-ATPase of plasma membrane, and V-ATPase of vacuolar membrane (Nelson 1989, 1992)], Ca^{2+}-ATPases whose different forms are revealed in different membrane fractions (Okorokov 1994), isocitrate dehydrogenases that have both mitochondrial and cytosolic forms (Popova 1993), protein kinases (Hunter 1987), and pyrophosphatases (Rea and Pool 1993).

Some data obtained recently indicate that each compartment of the yeast cell contains not only their own specific pool of polyP but also their own polyphosphatase that differs by properties from the corresponding enzymes of other compartments. We purified polyphosphatases from *S. cerevisiae* cell envelope (Andreeva et al. 1990; Andreeva and Okorokov 1993) and the cytosol of the same yeast (Andreeva et al. 1996) and characterized the polyphosphatase activities of vacuoles (Andreeva et al. 1993), nuclei (Lichko et al. 1995, 1996a), and mitochondria (Lichko et al. 1996b) isolated from the same yeast strain.

All *S. cerevisiae* polyphosphatases hydrolyzed polyPs with different chain lengths and were optimal at neutral pH. The purified enzymes from the cell envelope and cytosol did not hydrolyze pyrophosphate, p-nitrophenylphosphate, and nucleoside triphosphates (Andreeva and Okorokov 1993), just as the polyphosphatases from other cell compartments under study (Andreeva et al. 1993, 1994; Lichko et al. 1996a,b). Other phosphohydrolases of these cell compartments, such as ATPases, PPases, and nonspecific phosphatases, failed to hydrolyze polyPs (Andreeva et al. 1993, 1994; Lichko et al. 1996a,b)

The polyphosphatase activities were nearly the same with substrates ranging from polyP$_9$ to polyP$_{208}$ for the purified cell envelope and cytosol polyphosphatases (Andreeva and Okorokov 1993; Andreeva et al. 1996), as well as for those of nuclei (Lichko et al. 1996a) and soluble mitochondrial preparation (Lichko et al. 1997). On the contrary, the vacuolar and membrane-bound polyphosphatases of mitochondria hydrolyzed polyP$_{208}$ better than polyP$_{15}$ (Table 1).

The purified cell envelope and cytosol polyphosphatases hydrolyzed polyP$_3$ with a higher rate than the substrates of greater chain lengths (Table 1). The following results provide evidence of the ability of *S. cerevisiae*

Table 1. Apparent K_m values and specific activities of polyphosphatases of different compartments of *S. cerevisiae*

	Cell envelope[a]	Cytosol[a]	Vacuolar sap[b]	Nuclei[b]	Mitochondria	
					Membrane-bound[b]	Soluble[b]
Substrate	Activities (U/mg protein)					
PolyP$_3$	320	420	0.28	0.087	0.04	0.2
PolyP$_9$	220	320	0.46	0.056	0.045	0.065
PolyP$_{15}$	180	300	0.62	0.055	0.085	0.080
PolyP$_{208}$	180	270	0.67	0.047	0.110	0.075
	K_m (mM)					
PolyP$_{15}$	15	11	110	5	23	4
PolyP$_{188-208}$	0.9	1,2	6	4	1	0.06

[a] Purified enzymes.
[b] Isolated subcellular fractions. All analysis were made in the presence of 2.5 mM MgSO$_4$.

cell envelope and cytosol polyphosphatases to hydrolyze $polyP_3$: co-purification of both activities during chromatography steps, detection of both activities in the same band after PAGE under non-denaturing conditions, the same temperature- and pH-optima, and sensitivity to inhibitors including heparine and antibodies against the purified cell envelope polyphosphatase (Andreeva and Okorokov 1993; Andreeva et al. 1996).

In mitochondrial matrix and nuclei, the tripolyphosphatase activities were also higher compared with the activities with the long-chain polyPs. In vacuoles and mitochondrial membranes, the tripolyphosphatase activities were somewhat lower than that in the case of $polyP_{15}$ (Table 1). However, we cannot rule out the possible presence of specific tripolyphosphatases in these subcellular fractions.

The cell envelope and cytosol polyphosphatases were similar in their K_m values to polyPs with different chain lengths. An increase of the affinity to polyPs of about ten-fold with the extension of polyP chain length was observed both for the cell envelope polyphosphatase and for those of cytosol, vacuoles, and mitochondria, while the nuclear polyphosphatase was found to have practically identical affinity to $polyP_{15}$ and $polyP_{208}$.

Thus, only polyphosphatases of cell envelope and cytosol of S. cerevisiae were found to be similar in their substrate specificity to polyPs with different chain lengths. Polyphosphatase activities of vacuoles, nuclei, and membrane-bound and soluble polyphosphatases of mitochondria as well were unique in substrate specificity and differed both from the above two enzymes and from one another. Of particular interest are the soluble mitochondrial polyphosphatases having a very high affinity to long chain polyPs and the nuclear one having the same K_m values with polyPs of different chain lengths.

The cell envelope and vacuolar polyphosphatases were stimulated by addition of 200 mM K^+ in the presence of 50 mM Tris-HCl and 2.5 mM Mg^{2+} by approximately 20 %, the cytosol enzyme was almost insensitive to K^+, while the polyphosphatase activity of nuclei was inhibited with 200 mM K^+ by 30 %.

In the presence of 50 mM Tris-HCl and 2.5 mM Mg^{2+}, monovalent cations (100–200 mM K^+, Na^+, and NH_4^+) did not actually affect the soluble polyphosphatase activity of mitochondria while stimulating the polyphosphatase activity in the mitchondrial membranes 1.5- to 2-fold. NH_4^+ was the best stimulator.

The polyphosphatase activities of cell envelope, cytosol, vacuoles, nuclei, and mitochondrial matrix increased in the presence of divalent metal cations. The degree of stimulation of these polyphosphatases depended differently on the nature and concentration of the cation (Table 2). The cell envelope and cytosol polyphosphatases revealed a high similarity with respect to cation stimulation. A difference in the degree of stimulation was due to different activities of these two enzymes in the absence of divalent cations. The

Table 2. Effect of divalent cations on polyphosphatases of different compartments of *S. cerevisiae*

Cation	Polyphosphatase activity with polyP$_{15}$ (activities without cations taken as a unit)					
	Cell envelope[a]	Cytosol[a]	Vacuolar sap[b]	Nuclei[b]	Mitochondria	
					Membrane-bound[b]	Soluble[b]
2.5 mM Mg^{2+}	10	39	3.6	1.8	0.95	4.0
0.1 mM Co^{2+}	14	66	2.6	3.2	0.7	4.0
0.1 mM Zn^{2+}	7.5	30	2.0	1.7	0.6	2.5

[a] Purified enzymes.
[b] Isolated subcellular fractions.

activity of cell envelope polyphosphatase was 18 U/mg protein without divalent cations and 200 U/mg protein in the presence of 2.5 mM Mg^{2+} with polyP$_{15}$ as a substrate. The corresponding values for the cytosol enzyme were 7 and 280 U/mg protein.

The requirement for divalent metal cations of the nuclear, vacuolar, and soluble mitochondrial polyphosphatases were lower than those of cytosol and cell-envelope ones. Stimulation of the nuclear polyphosphatase activity by divalent cations was only two- to three-fold. It should be pointed out that the endogenous ions were removed from all preparations used in these experiments by dialysis as described (Andreeva and Okorokov 1993).

Addition of divalent cations to the membrane preparation of mitochondria resulted in inhibition of the polyphosphatase activity, which also depended on the nature and concentration of cations (Table 2). Based on the degree of inhibition of the membrane-bound polyphosphatase activity, the tested cations were arranged in the following order: Co^{2+} = Zn^{2+} > Mn^{2+} > Mg^{2+}.

The effect of divalent metal cations on polyphosphatase activities in various yeast cell compartments showed significant differences between the corresponding enzymes. Only the cytosolic and cell envelope enzymes showed similar dependence on divalent cations.

Little is known about the inhibitors of polyphosphatases. We found only one inhibitor, which suppressed polyphosphatase activities in all compartments tested (Table 3). It is heparine, a concurrent inhibitor, probably acting as a substrate analog (Andreeva et al. 1994; Lichko et al. 1994).

Several SH-reagents (*N*-ethylmaleimide, iodacetamide) had little or no effect on the polyphosphatase activities of the yeast (Andreeva and Okorokov 1993; Andreeva et al. 1993, 1994; Lichko et al. 1996a,b). Such commonly used inhibitors of phosphohydrolases as orthovanadate and molybdate were ineffective. The sensitivity to molibdate reported earlier (Andreeva and Okorokov 1993; Andreeva et al. 1994; Lichko et al. 1996a) was due to acidification of the incubation medium under molibdate addition. This inhibition

Table 3. Effect of some reagents on polyphosphatase activities of different cell compartments of *S. cerevisiae*

Reagent	Activity with polyP$_{15}$ (% to control)					
	Cell envelope[a]	Cytosol[a]	Vacuolar sap[b]	Nuclei[b]	Mitochondria	
					Membrane-bound[b]	Soluble[b]
Without effector	100	100	100	100	100	100
EDTA (1 mM)	145	143	68	70	100	100
Heparin (20 µg/ml)	6	8	26	11	0	0
Antibodies against cell envelope poly-phosphatase	10	20	100	100	100	70

[a] Purified enzymes.
[b] Isolated subcellular fractions. All analyses were made in the presence of 2.5 mM MgSO$_4$.

was unspecific and, when pH was adjusted to 7.2 after addition of molibdate, all polyphosphatase activities under study demonstrated insensitivity to this effector.

EDTA had a varied effect on the polyphosphatase activities under study. While decreasing the polyphosphatase activities of vacuoles and nuclei by approximately 30 % and having a low effect on the mitochondrial activities, EDTA increased those of purified cytosol and cell envelope enzymes (Table 3). Different EDTA effects on the polyphosphatase activities under study are possibly due to their different sensitivity to Me^{2+}-influence both in the presence and absence of Mg^{2+} (Andreeva and Okorokov 1993; Andreeva et al. 1996). Thus, the cell envelope and cytosol polyphosphatases are closely related in their sensitivity to EDTA in contrast to those of nuclei, vacuoles, and mitochondria.

Antibodies against the purified cell envelope polyphosphatase were effective inhibitors of polyphosphatases from cell envelope and cytosol but did not affect the vacuolar, nuclear, and mitochondrial membrane-bound polyphosphatase activities (Table 3). These antibodies inhibited the soluble mitochondrial polyphosphatase by approximately 30 % in contrast to 80–90 % inhibition of polyphosphatases of cell envelope and cytosol. This again indicated the similarity of cell envelope and cytosol polyphosphatases and their distinction from vacuolar, nuclear, and mitochondrial enzymes.

We revealed the polypeptides reacting with the antibodies against the cell envelope polyphosphatase in different fractions of cell organelles with the aim of primary identification of polyphosphatases. Only one polypeptide band with a molecular mass of about 40 kDa was recognized by these antibodies both in the cell envelope homogenous enzyme and in partially purified preparation of cytosol polyphosphatase containing some additional polypeptides. These molecular masses corresponded to those determined by gel filtration and SDS-PAGE methods for cell envelope polyphosphatase (Andreeva and Okorokov 1993) and for cytosol polyphosphatase (Andreeva et al. 1994, 1996).

Immunoblotting of the vacuolar preparation revealed two main polypeptides bound to antibodies and having molecular masses of 72 and 40 kDa, and some minor bands as well. The intensity of the 40-kDa band varied from experiment to experiment. This polypeptide might be a product of proteolysis of the 72-kDa one, since the vacuoles possess a diversity of proteases (Wiemken et al. 1979). The probability that the polypeptide of 40 kDa has cytosolic origins is little, since otherwise at least partial inhibition of the vacuolar polyphosphatase activity by antibodies would be detected. The molecular mass of vacuolar polyphosphatase determined by gel filtration was 280 kDa. This enzyme is supposed to be an oligomer.

Antibodies against the purified cell envelope polyphosphatase reacted with polypeptides of about 115 and 78 kDa in the mitochondrial membrane preparation, and 37 kDa in mitochondrial matrix. The molecular masses determined by gel filtration were of 120 and 76 kDa for the membrane-bound enzyme and 36 kDa for the soluble one (Lichko et al. 1997). The two membrane polyphosphatase may differ in the non-protein components.

In the nuclear preparation, two main polypeptides of about 64 and 32 kDa bound with the antibodies against cell envelope polyphosphatase were revealed. No protein bands were stained in immunoblots of the tested preparations when using serum of a control rabbit.

Thus, the immune properties both of cell envelope and cytosol polyphosphatases of *S. cerevisiae* turned out to be identical. Antibodies against the purified cell envelope polyphosphatase suppressed their acitivity and interacted with polypeptides of the same molecular mass of 40 kDa in both preparations. Vacuolar, nuclear, and mitochondrial soluble and membrane-bound polyphosphatases differ in their molecular masses from both cell envelope and cytosol polyphosphatases and from each other. All yeast polyphosphatases under study probably have certain similar structure sites, and therefore the antibodies against cell envelope polyphosphatase are able to bind polyphosphatases of vacuoles, nuclei, and mitochondria of the same yeast strain.

The biochemical and immune properties of cell envelope and cytosol polyphosphatases are similar. Besides, they have the same molecular masses. These enzymes differ only in some chromatographic properties which is due to different methods of their purification.

Polyphosphatases detected in the preparations of isolated vacuoles, nuclei, and mitochondria differ in their kinetic and immune properties, substrate specificity, requirements in divalent cations and in some effector actions both from the preceding two and from each other. We consider that the yeast cell possesses at least five distinct polyphosphatases, such as cell envelope and cytosol ones, the polyphosphatases of vacuoles and nuclei, and membrane-bound and soluble enzymes of mitochondria.

Two works concerning polyphosphatase purification from *S. cerevisiae* homogenate have appeared recently (Lorenz et al. 1994; Wurst and Kornberg 1994). The first one deals with the purified polyphosphatase, which seems to be similar to both purified cell envelope (Andreeva and Okorokov 1993) and cytosol polyphosphatases (Andreeva et al. 1996) in a number of properties including the molecular mass. The enzyme appears to be localized in the yeast cytosol. The gene of this polyphosphatase is cloned and sequenced (Wurst et al. 1995). The other polyphosphatase from *S.cerevisiae* homogenate (Lorenz et al. 1994) is similar to the above polyphosphatases in its kinetic properties, substrate specificity, pH-optimum, and dependence on divalent cations, but its molecular mass is 28 kDa. Its location in the cell is still unknown.

All exopolyphosphatases purified until now from the yeast are monomeric proteins. These enzymes have neutral pH optima, similar kinetic properties and substrate specificity and require divalent cations for the maximal activity, preferably Mg^{2+} or Co^{2+}. Their activity with tripolyphosphate is nearly 1.5-fold higher than that with long chain polyPs. All purified yeast polyphosphatases seem to be products of one gene *ppx* with possible post-transcriptional modifications. Our study of polyphosphatases in purified fractions of cell organelles indicates that the yeast cell possesses other enzyme forms differing from those described above in their localization, properties, and probably gene determinants.

Yeast exopolyphosphatases purified until now differed significantly from bacterial enzymes. Exopolyphosphatase from *E. coli* was a dimer with the subunit molecular mass of about 58 kDa. Its affinity to high molecular weight polyphosphates was nearly 100-fold higher than that of yeast polyphosphatases (K_m = 9 nM polyP$_{500}$ as polymer). This enzyme had a high requirement for K^+ (21-fold stimulation by 175 mM K^+) (Akiyama et al. 1993). Exopolyphosphatase from *A. johnsonii* was monomeric protein of 55 kDa. The K_m value for polyphosphate with the average chain length of 64 phosphate residues was 5.9 µM. The acitivity was maximal in the presence of 2.5 mM Mg^{2+} and 0.1 mM K^+. No activity was observed in the absence of cations or in the presence of Mg^{2+} or K^+ alone (Bonting et al. 1993).

Thus, bacterial exopolyphosphatases differ from each other in properties to a great extent. Their similar features are low activity with tripolyphosphate and requirement of K^+ for the maximal activity. Purified yeast exopolyphosphatases (Andreeva and Okorokov 1993; Lorenz et al. 1994; Wurst and Kornberg 1994) hydrolyzed tripolyphosphate better than polyPs with longer chains, and were little stimulated by K^+. At present, this is the most appreciable difference between yeast and bacterial polyphosphatases. It is possible that polyphosphatases from other yeast cell compartments might have a certain similarity with bacterial enzymes.

Thus, the yeast cell possesses several forms of polyphosphatases, such as the polyphosphatases of cytosol and cell envelope and those of vacuoles, nuclei, and mitochondria. The detection of various polyphosphatase forms in cell organelles might be one more argument in favor of the endosymbiotic theory (Kulaev 1994). Conceivably, the existence of these forms might be determined not only by the organelle origin but also because their polyphosphate functions are different, which in turn requires special ways of regulating the level of these biopolymers in each compartment. Physiological significance of the occurrence of individual polyphosphatases in different organelles of the yeast cell is a goal for future investigations.

References

Akiyama M, Crooke E, Kornberg A (1992) The polyphosphate kinase gene of *Escherichia coli*. Isolation and sequence of the ppk gene and membrane location of the protein. J Biol Chem 267: 22556–22561

Akiyama M, Crooke E, Kornberg A (1993) An exopolyphosphatase of *Escherichia coli*. The enzyme and its ppx gene in a polyphosphate operon. J Biol Chem 268: 633–639

Andreeva NA, Okorokov LA (1993) Purification and characterization of highly active and stable polyphosphatase from *Saccharomyces cerevisiae* cell envelope. Yeast 9: 127–139

Andreeva NA, Okorokov LA, Kulaev IS (1990) Purification and certain properties of cell envelope polyphosphatase of the yeast *Saccharomyces carlsbergansis*. Biochemistry (Moscow) 55: 819–826

Andreeva NA, Lichko LP, Kulakovskaya TV, Okorokov LA (1993) Characterization of polyphosphatase activity of vacuoles of the yeast *Saccharomyces cerevisiae*. Biochemistry (Moscow) 58: 737–744

Andreeva NA, Kulakovskaya TV, Kulaev IS (1994) Characteristics of the cytosol polyphosphatase activity of the yeast *Saccharomyces cerevisiae*. Biochemistry (Moscow) 59: 1411–1417

Andreeva NA, Kulakovskaya TV, Kulaev IS (1996) Purification and characterization of polyphosphatase from *Saccharomyces cerevisiae* cytosol. Biochemistry (Moscow) 61: 1213–1220

Baltzinger M, Ebel JP, Remy P (1986) Accumulation of dinucleoside polyphosphates in *Saccharomyces cerevisiae* under stress conditions. High levels are associated with cell death. Biochimie 68: 1231–1236

Beauvoit B, Rigonlet M, Guerin B, Canioni P (1989) Polyphosphates as a source of high energy phosphates in yeast mitochondria: a P-NMR study. FEBS Lett 252: 17–22

Belozersky AN (1958) The formation and function of polyphosphates in the development processes of some lower organisms. In: Communications and Reports of the 4th International Biochemistry Congress, Vienna, p 3

Bonting CF, Korstee GJ, Zehnder AJ (1991) Properties of polyphosphate: AMP phosphotransferase of *Acinetobacter* strain 210A. J Bacteriol 173: 6484–6488

Bonting CF, Kortstee GJ, Zehnder JA (1993) Properties of polyphosphatase of *Acinetobacter johnsonii* 210A. Antonie van Leeuwenhoek 64: 75–81

Booth JW, Guidotti G (1995) An alleged yeast polyphosphate kinase is actually diadenosine-5',5'''-P^1,P^4-tetraphosphate α,β-phosphorylase. J Biol Chem 270: 19377–19382

Bourne RM (1991) Net phosphate-transport in phosphate-starved *Candida utilis* – relationship with pH and K^+. Biochim Biophys Acta 1067: 81–88

Castuma CE, Huang R, Kornberg A, Reusch RN (1995) Inorganic polyphosphates in the acquisition of competence in *Escherichia coli*. J. Biol Chem 270: 12980–12983

Crooke E, Akiyama M, Rao NN, Kornberg A (1994) Genetically altered levels of inorganic polyphosphate in *Escherichia coli*. J Biol Chem 269: 6290–6295

Deinema MH, Van Loosdrecht M, Scholten A (1985) Some physiological characteristics of *Acinetobacter* spp. accumulating large amounts of phosphate. Water Sci Technol 17: 119–125

Dirheimer G, Ebel JP (1965) Charactérisation d'une polyphosphate AMP-phosphotransferase dans *Corynebacterium xerosis*. R C Acad Sci 260: 3787–3790

Durr M, Urech K, Boller T, Wiemken A, Schwencke J, Nagy M (1979) Sequestration of arginine by polyphosphate in vacuoles of yeast *Saccharomyces cerevisiae*. Arch Microbiol 121: 169–175

Felter S, Stahl AJC (1973) Enzymes du métabolisme des polyphosphates dans la levure. III. Purification et propriétés de la polyphosphate-ADP-phosphotransférase. Biochimie 55: 245–251

Hsieh PC, Shenoy BC, Jentoft JE, Philipps NFB (1993) Purification of polyphosphate and ATP glucose phosphotransferase from *Mycobacterium tuberculosis* H_{27}Ra: evidence that poly(P) and ATP glucokinase activities are catalyzed by the same enzyme. Protein Expr Purif 4: 76–84

Huang RP, Reusch RN (1995) Genetic competence in *Escherichia coli* requires poly-beta-hydroxbutirate calcium polyphosphate membrane complex and certain divalent cations. J Bacteriol 177: 586–490

Huang RP, Reusch RN (1996) Poly (3-hydroxybutirate) is associated with specific proteins in the cytoplasm and membranes of *Escherichia coli*. J Biol Chem 271: 22196–22202

Hunter TA (1987) Thousand and one protein kinases. Cell 50: 823–829

Ivanov AJ, Vagabov VM, Fomchenkov VN, Kulaev IS (1996) Study of the influence of polyphosphates of cell envelope on the sensitivity of yeast *Saccharomyces carlsbergensis* to the cytyl-3-methylammonium bromide. Microbiologia (Moscow) 65: 611–616

Keasling JD, Bortish LR, Kornberg A (1993) Guanosine pentaphosphate phosphohydrolase of *Escherichia coli* is a long-chain exopolyphosphatase. Proc Natl Acad Sci USA 90: 7029–7033

Kornberg A (1995) Inorganic polyphosphate: toward making a forgotten polymer unforgettable. J Bacteriol 177: 491–496

Kornberg A, Kornberg S, Simms E (1956) Methaphosphate synthesis by enzyme from *Escherichia coli*. Biochim Biophys Acta 20: 215–227

Kulaev IS (1979) Biochemistry of inorganic polyphosphates. Wiley, Chichester

Kulaev IS (1994) Inorganic polyphosphate function at various stages of cell evolution. J Biol Phys 20: 255–273

Kulaev IS, Vagabov VM (1983) Polyphosphate metabolism in microorganisms. Adv Microbiol Physiol 24: 83–171

Kulaev IS, Vagabov VM, Shabalin YA (1987) New data on biosynthesis of polyphosphates in yeasts. In: Torrlani-Gorini A, Rothman FG, Silver S et al. (eds) Phosphate metabolism and cellular regulation in microorganisms. American Society for Microbiology, Washington, DC, pp 233–238

Kulakovskaya TV, Andreeva NA, Kulaev IS (1997) Adenosine 5'-tetraphosphate and guanosine-5'-tetraphosphate – new substrates of the cytosol exopolyphosphatase of *Saccharomyces cerevisiae*. Biochemistry (Moscow) 62: 1225–1227

Kumble KD, Kornberg A (1996) Endopolyphosphatases for long chain polyphosphate in yeast and mammals. J Biol Chem 271: 27146–27151

Kumble KD, Ahn K, Kornberg A (1996) Phosphohistidyl active sites in polyphosphate kinase of *Escherichia coli*. Proc Natl Acad Sci USA 93: 14391–14395

Lichko LP, Okorokov LA, Kulaev IS (1982) Participation of vacuoles in regulation of K^+, Mg^{2+} and orthophosphate ions in cytoplasm of the yeast *Saccharomyces carlsbergensis*. Arch Microbiol 132: 289–293

Lichko LP, Kulakovskaya TV, Kulaev IS (1994) Effects of platelet-activating factor, sphingosine and heparin on some phosphohydrolase and transport activities of yeast vacuoles. Biochemistry (Moscow) 59: 815–821

Lichko LP, Kulakovskaya TV, Dmitriev VV, Kulaev IS (1995) *Saccharomyces cerevisiae* nuclei possess polyphosphatase activity. Biochemistry (Moscow) 60: 1465–1468

Lichko LP, Kulakovskaya TV, Kulaev IS (1996a) Characterization of the nuclear polyphosphatase activity in *Saccharomyces cerevisiae*. Bichemistry (Moscow) 61: 361–366

Lichko LP, Kulakovskaya TV, Kulaev IS (1996b) Characterization of the polyphosphatase activity in isolated *Saccharomyces cerevisiae* mitochondria. Biochemistry (Moscow) 61: 1664–1671

Lichko LP, Kulakovskaya TV, Kulaev IS (1997) Detection and some properties of membrane-bound and soluble polyphosphatases in mitochondria of yeast *Saccharomyces cerevisiae*. Biochemistry (Moscow) 62: 1139–1145

Lieberman L (1888) Über das Nuclein der Hefe und künstliche Darstellung eines Nucleus-Eiweiss und Metaphosphatsäure. Ber Chem Ges 21: 598–607

Lorenz B, Muller WEG, Kulaev IS, Schroder HCJ (1994) Purification and characterization of an exopolyphosphatase from *Saccharomyces cerevisiae*. J Biol Chem 269: 22198–22204

Margulis L (1993) Symbiosis in cell evolution. Freeman, San Francisco

Nakamura K, Hiraishi A, Yoshimi Y, Kawaharasaki M, Masuda K, Kamagata Y (1995) *Microlunatus phosphorovorus* gen.nov. sp.nov., a new gram-positive polyphosphate-accumulation bacterium isolated from activated sludge. Int J Syst Bacteriol 45: 17–22

Nelson N (1989) Structure, molecular genetics and evolution of vacuolar H^+-ATPases. J Bioenerg Biomembr 21: 553–571

Nelson N (1992) Evolution of organellar proton-ATPases. Biochim Biophys Acta 1100: 109–124

Offenbacher S, Kline H (1984) Evidence for polyphosphate in phosphorylated non-histone nuclear proteins. Arch Biochem Biophys 231: 114–123

Okorokov LA (1994) Several compartments of *Saccharomyces cerevisiae* are equipped with Ca^{2+}-ATPase(s). FEMS Microbiol Lett 117: 311–318

Okorokov LA, Lichko LP, Kulaev IS (1980) Vacuoles: the main compartment of potassium, magnesium and phosphate ions in *Saccharomyces carlsbergensis* cells. J Bacteriol 144: 661–665

Okorokov LA, Lichko LP, Andreeva NA (1983) Changes of ATP, polyphosphate and K^+ contents in *Saccharomyces carlsbergensis* during uptake of Mn^{2+} and glucose. Biochem Int 6: 481–488

Phillips NF, Horn PJ, Wood HG (1993) The polyphosphate and ATP dependent glucokinase from *Propionibacterium shermanii*: both activities are catalyzed by the same protein. Arch Biochem Biophys 300: 309–319

Piletus U, Meyer A, Hildebrandt A (1989) Nuclear polyphosphate as a possible source of energy during the sporulation of *Physarum polycephalum*. Arch Biochem Biophys 275: 215–223

Pisoni RL, Lindley ER (1992) Incorporation of [32-P] orthophosphate into long chain of inorganic polyphosphate within lysosomes of human fibroblasts. J Biol Chem 267: 3626–3631

Popova TN (1993) Isocitratedehydrogenases: forms, localization, properties, and regulation. Biochemistry (Moscow) 58: 1861–1879

Rao NN, Kornberg A (1996) Inorganic polyphosphate support resistance and survival of stationary-phase *Escherichia coli*. J Bacteriol 178: 1394–1400

Rea PA, Poole RJ (1993) Vacuolar H^+-translocating pyrophosphatase. Annu Rev Plant Physiol Plant Mol Biol 44: 157–180

Reusch RN (1989) Poly-beta-hydroxybutirate/calcium polyphosphate complexes in eukaryotic membranes. R Soc Exp Biol Med 191: 377–381

Reusch RN, Sadoff HL (1988) Putative structure and functions of poly-beta-hydroxybutirate/calcium polyphosphate channel in bacterial plasma membranes. Proc Natl Acad Sci USA 85: 4176–4180

Rodriguez RJ (1993) Polyphosphate present in DNA preparation from filamentous fungal species of *Colleotrichum* inhibits restriction endonucleases and other enzymes. Anal Biochem 209: 291–297

Schroder HC, Oliveira M, Tully G, Leitao JM, Muller WEG (1996) Inorganic polyphosphate – age-related synthesis of a forgotten macromolecule. Z Gerontol Geriatr 29: 85

Shabalin YA, Vagabov VM, Tsiomenko AB, Zemlianuhina OA, Kulaev IS (1977) Study of polyphosphate kinase activity in the yeast vacuoles. Biokhimiya 42: 1642–1648

Skorko K (1989) Polyphosphate as a source of phosphoryl group in protein modification in archaebacterium *Sulfolobus acidocaldarius*. Biochimie 71: 9–10

Szymona M (1957) Utilization of inorganic polyphosphates for phosphorilation of glucose in *Micobacterium phlei*. Bull Acad Pol Sci Ser Sci Biol 5: 379–382

Szymona M, Ostrowsky W (1964) Inorganic polyphosphate glucokinase of *Micobacterium phlei*. Biochim Biophys Acta 85: 283–295

Tepel M, Bachmann J, Schluter H, Zidek W (1996) Diadenosine polyphosphates, increase cytosolic calcium and attenuate angiotensin-II-induced changes of calcium in vascular smooth-muscle cells. J Vasc Res 33: 132–138

Tinsley CR, Gotschlich EC (1995) Cloning and characterization of the meningococcal polyphosphate kinase gene: production of polyphosphate synthesis mutant. Infect Immun 63: 1624–1630

Tinsley CR, Manjula BN, Gotschlich EC (1993) Purification and characterization of polyphosphate kinase from *Neisseria meningitis*. Infect Immun 61: 3703–1710

Urech K, Durr M, Boller T, Wiemken A (1978) Localization of polyphosphate in vacuoles of *Saccharomyces cerevisiae*. Arch Microbiol 116: 274–278

Vagabov VM, Chemodanova OV, Kulaev IS (1990) Effect of inorganic polyphosphates on negative charge of yeast cell wall. Dokl Akad Nauk SSSR 313: 989–992

Van Alebeek GJWM, Keitjens JT, Van der Drift C (1994) Tripolyphosphatase from *Methanobacterium thermoautotrophicum* (strain DH). FEMS Microbiol Lett 117: 263–268

Van Groenestijn JW, Bentvelzen MMA, Deinema MH, Zehnder AJB (1989) Polyphosphate degrading enzymes in *Acinetobacter* spp. and activated sludge. Appl Environ Microbiol 55: 219–223

Wiemken A, Durr M (1974) Characterization of amino acid pools in the vacuolar compartment of *Saccharomyces cerevisiae*. Arch Microbiol 101: 45–57

Wiemken A, Schellenberg M, Urech K (1979) Vacuoles: the sole compartments of digestive enzymes in yeast (*Saccharomyces cerevisiae*). Arch Microbiol 123: 23–35

Woese CR, Kandler O, Wheelis ML (1990) Towards a natural system of organisms. Proposal for the domenus Archaea bacteria, Eucaria. Proc Natl Acad Sci USA 87: 4576–4579

Wood HG, Clark JE (1988) Biological aspects of inorganic polyphosphates. Annu Rev Biochem 57: 235–260

Wurst H, Kornberg A (1994) A soluble exopolyphosphatase of *Saccharomyces cerevisiae*. J Biol Chem 269: 10966–11001

Wurst H, Shiba T, Kornberg A (1995) The gene for a major exopolyphosphatase of *Saccharomyces cerevisiae*. J Bacteriol 177: 898–906

Inorganic Polyphosphate in Eukaryotes: Enzymes, Metabolism and Function

H. C. Schröder[1], B. Lorenz[2], L. Kurz[1], and W. E. G. Müller[1]

1
Introduction

Inorganic polyphosphates (polyP) are linear polymers of orthophosphate (P_i) residues linked by high-energy phosphoanhydride bonds. These polymers are widely distributed in nature, from archaebacteria, eubacteria, fungi, algae, and protozoa to higher plants and animals (for reviews, see Kulaev 1979; Wood and Clark 1988; Kornberg 1994, 1995). PolyP molecules are stable in neutral aqueous solutions, but are hydrolyzed by heat-treatment, and under acidic or alkaline conditions. The chain length of polyP may range from 3 to more than 1000 P_i residues; it can be analyzed on urea/polyacrylamide gels (Clark and Wood 1987; Lorenz et al. 1994a). In contrast to the linear polymer, branched inorganic polyP has not yet been detected in living organisms (Kulaev 1979); for the occurrence of cyclic metaphosphates, see Chapter 5. Until recently, the occurrence and metabolism of polyP have been studied mainly in microorganisms (Kulaev 1979; Kulaev and Vagabov 1983), although significant amounts of these polymers are also present in mammalian cells and tissues (Kumble and Kornberg 1995; Lorenz et al. 1995, 1997a–c). They could be even detected in human cells (Pisoni and Lindley 1992; Kumble and Kornberg 1995; Lorenz et al. 1997a–c). In general, the polyP content in higher eukaryotes is much lower than in bacteria and yeast. Moreover, the amount and chain length of polyP may vary within the same species depending on several factors, including nutrition and growth phase. Some organisms (e.g., yeasts) are able to accumulate large amounts of polyP, up to 10% of the dry cell mass and more (Kulaev 1979). In view of the ubiquity of polyP, it has been concluded that these polymers have a multiplicity of biological functions.

2
Occurrence of PolyP in Eukaryotic Cells

The occurrence of inorganic polyP has been extensively studied in lower eukaryotes (yeast and algae; for reviews, see Kulaev 1979; Kulaev and Vaga-

[1] Institut für Physiologische Chemie, Universität, Duesbergweg 6, 55099 Mainz, Germany.
[2] Institut für Biochemie, Universität, Leipziger Straße 44, 39120 Magdeburg, Germany.

Progress in Molecular and Subcellular Biology, Vol. 23
H. C. Schröder, W. E. G. Müller (Eds.)
© Springer-Verlag Berlin Heidelberg 1999

bov 1983), but there are only relatively few studies on polyP in higher animals or man. After a first report more than 30 years ago (Grossmann and Lang 1962), these polymers were ignored for a long time. One of the reasons may be the fact that the quantities of polyP in animal and human cells and tissues are rather small. Now it has been definitely shown that inorganic polyP is actually present in these organisms (Kumble and Kornberg 1995; Lorenz et al. 1997a–c; Leyhausen et al. 1998). The presence of polyP in mammalian cells and tissues has been demonstrated using different techniques, including highly specific enzymatic methods (Kumble and Kornberg 1995; see also Chap. 12); these methods make use of the recombinant *Escherichia coli* polyphosphate kinase (conversion of polyP to ATP) or of the recombinant *Saccharomyces cerevisiae* exopolyphosphatase (hydrolysis of polyP to P_i). Using these methods, the intracellular concentrations of polyP in organs from rodents (brain, liver, heart, kidneys, and lungs) were found to range from 25 to 120 µM (in terms of P_i residues). The polyP chain lengths were between 50 and 800 residues; in brain, the majority of polyP was approximately 800 residues long.

Another novel method used for determination of polyP is based on the abolition of the Mn^{2+}-induced quenching of fura-2 fluorescence caused by polyP (Lorenz et al. 1997c). As summarized in Table 1, determinations of polyP content by fura-2 method revealed that polyP content in rat brain is higher than in rat liver, where polyP is concentrated in the nucleus. Higher amounts of polyP in rat brain compared to rat liver were also reported by Gabel and Thomas (1971) and Kumble and Kornberg (1995). Similar results were obtained by toluidine blue method (metachromatic reaction). The values for step 2 and step 3 extracts (long-chain polyP fractions extracted according to the procedure described by Clark and Wood 1987) determined by fura-2 method are lower than those found by metachromatic reaction. By contrast, the amount of polyP in step 1 extracts (short-chain polyP) is higher when assayed with fura-2 compared to toluidine blue method. The possible reasons for these differences (e.g., weak metachromatic reaction of short-chain polyP) have been discussed (Lorenz et al. 1997c).

More recently, we (Lorenz et al. 1997a; Leyhausen et al. 1998) and others (Kumble and Kornberg 1995) showed that significant amounts of polyP are present also in human cells. The amount of polyP in human peripheral blood mononuclear cells (PBMC) and erythrocytes is in the range of 90 to 120 µM (expressed in P_i; Lorenz et al. 1997a; Table 1). This concentration of polyP is only two- to three-fold lower than that displaying antiviral activity (see Sect. 5.9). Furthermore, polyP was also found extracellularly in human blood plasma and serum (Table 1; Lorenz et al. 1997a). However, the content of insoluble long-chain polyP in plasma (and serum) is much lower than in erythrocytes and PBMC. It is unclear whether the plasma/serum polyP is a genuine component of these body fluids or is due to the lysis of erythrocytes,

Table 1. PolyP concentration in various cells and tissues[a]

Cells or tissue or blood fraction	PolyP fraction (step)	PolyP concentration[b]	
		Metachromatic reaction (μM [P_i])	Fura-2 method (μM [P_i])
Rat brain	1	4	10
	2	22	18
	3	64	37
Rat liver	1	2	10
	2	22	8
	3	38	19
Rat liver nuclei	1	1	5
	2	20	18
	3	88	34
HeLa cells	1	2	9
	2	18	15
	3	79	31
Human osteoblasts[c]	1	n.d.	
	2	394	
	3	134	
Human gingival cells	1	n.d.	
	2	141	
	3	15	
Human PBMC	1	11	
	2	51–56	
	3	29–30	
Human erythrocytes	1	9	
	2	71–76	
	3	28–32	
Human plasma	1	14	
	2	49	
	3	3	

n.d., Not determined; PBMC, peripheral blood mononuclear cells.

[a] PolyP was determined by either fura-2 method or metachromatic reaction after successive extractions; step 1 (short-chain polyP), step 2 (soluble long-chain polyP; 10–50 residues), and step 3 extracts (insoluble long-chain polyP; > 50 residues). PolyP$_{35}$ was used as standard. Concentrations are based either on volume of pelleted cells or on plasma volume.

[b] Data from Leyhausen et al. 1998 (metachromatic reaction), Lorenz et al. 1997a (metachromatic reaction), and Lorenz et al. 1997c (metachromatic reaction and fura-2 method).

[c] Unstimulated osteoblasts.

as suggested by its lower size. The highest amounts of polyP in human cells were found in bone-forming osteoblasts (Table 1; Leyhausen et al. 1998).

In cells and tissues of higher eukaryotes (HeLa cells, HL60 cells, rat brain and liver) different populations of polyP with different chain lengths exist; e.g., HL60 cells were found to contain two size classes with ~150 residues and 25 to 45 residues (Lorenz et al. 1997b), and Jurkat-, NIH3T3- and PC12 cells, with 500 to 800 residues and 5 to 15 residues (Kumble and Kornberg 1995).

Comparably high amounts of polyP have been detected in the lowest metazoa, the marine sponges (Lorenz et al. 1995; Imsiecke et al. 1996). Determinations of polyP in the marine sponge *Tethya lyncurium* revealed a polyP content of 30 µg/g wet mass (Lorenz et al. 1995). The occurrence of polyP has also been demonstrated in the protozoa *Leishmania major* (in the vacuoles; Blum 1989; LeFurgey et al. 1990) and *Entamoeba* (in the cytoplasm and nuclear membrane of trophozoites; Lopez-Revilla and Gomez-Dominguez 1985).

2.1
Variations in Cellular PolyP Content

The amount and size of polyP may vary considerably not only between different species, but also within the same species or even in the same individual; the polyP content may depend on nutrition and the period of growth during which the polyP was isolated. This has been studied in particular in yeast (Vagabov et al. 1998). Yeast cells accumulate up to 10% of their dry weight as polyP when they are transferred from a phosphate-depleted medium to a phosphate-rich medium (Langen and Liss 1958; Kulaev 1979; Schuddemat et al. 1989). A transient accumulation of large amounts of polyP (up to 20 mM P_i residues) was also found in bacteria (*E. coli*) following nutritional stress (Rao et al. 1998). In addition, the polyP content may change during development and ageing; see Section 6. Dramatic changes in polyP size were also found following induction of apoptosis; see Section 7.

It has been shown that in several mammalian cell lines, polyP is rapidly synthesized and displays a high turnover rate. For example, polyP turnover in confluent adrenal pheochromocytoma PC12 cells was found to be complete in an hour (Kumble and Kornberg 1995). On the other hand, only little turnover of polyP (4 h) was observed in fibroblasts.

Metabolic labeling studies revealed that $^{32}P_i$ taken up by human leukemic HL60 cells appears first in the lower molecular polyP fraction (25 to 45 residues) before it is incorporated into high-molecular polyP (around 150 residues; Lorenz et al. 1997b). This result indicates that the incorporation of exogenous P_i into polyP in human cells may be different from microorganisms. In microorganisms, exogenous P_i has been reported to enter first in the most highly polymerized polyP fraction (Kulaev 1979). From experiments in yeast, it was assumed that the most highly polymerized polyP is formed first and the less highly polymerized polyP fractions secondarily by degradation of the long-chain polyP through polyphosphatases (Langen and Liss 1958). However, later it was shown that in yeast cells different polyP pools exist in different cellular compartments. Therefore, individual polyP fractions may be synthesized independently (for a discussion, see Kulaev 1979). Other results suggested that formation of long-chain polyP in yeast (Schuddemat et al. 1989) and *E. coli* (Rao et al. 1985) may also occur via initial synthesis of shorter polyP chains.

2.2
Subcellular Localization

PolyP is nonrandomly distributed within the cell with an enrichment in the nucleus, the mitochondria, lysosomes and the plasma membrane (Reusch 1989; Pisoni and Lindley 1992; Kumble and Kornberg 1995; Lorenz et al. 1997b). Determination of polyP in nuclei from rat brain and liver revealed that the polyP is predominantly localized in these organelles (Griffin et al. 1965; Penniall and Griffin 1984; Kumble and Kornberg 1995; Lorenz et al. 1997b). The concentration of polyP in nuclei from rat brain amounts to 164 µM and that in nuclei from rat liver, 107 µM; a polyP concentration of 89 µM was found in rat liver nuclei using enzyme methods (Kumble and Kornberg 1995). The nuclear polyP mainly belongs to the long-chain polyP fraction. At the concentrations found in nuclei, polyP is able to destabilize chromatin and therefore could play a role in regulation of gene activity; see Section 5.5. Surprisingly, erythrocytes which lack a nucleus contain the highest amounts of polyP among all blood constituents tested (Table 1; Lorenz et al. 1997 a). Therefore, a relatively high polyP content of a cell is not necessarily associated with the presence of a nucleus.

Also in yeast, polyP ist nonrandomly distributed; highest amounts are found in the vacuoles (Urech et al. 1978). The vacuolar polyP ist present as Ca^{2+}-polyP complex (Ohsumi et al. 1988). PolyP in yeast is also found in mitochondria, in the cell nucleus, on the membranes of the endoplasmic reticulum, and on the surface of the plasma membrane (Kulaev 1979; Vorisek et al. 1982). PolyP localized outside the yeast plasma membrane can be released by osmotic shock (Tijssen et al. 1983). In many cellular membranes, Ca^{2+}-polyP may form complexes with poly(3-hydroxybutyrate) (Reusch and Sadoff 1988; Reusch 1989); see Section 5.7.

3
PolyP-Dependent Enzymes

Several enzymes involved in the synthesis and degradation of polyP have been identified. However, only a few of these enzymes have been isolated from both prokaryotic and eukaryotic sources. The enzymes purified from eukaryotic cells are mainly engaged in polyP breakdown; the enzymatic basis of polyP formation in yeast and animal cells is still an enigma. The polyP-degrading enzymes which have been studied as yet in eukaryotic cells include several endo- and exopolyphosphatases. However, it cannot be excluded that further enzymes shown to be involved in polyP breakdown in bacteria may also participate in the degradation of polyP in eukaryotes. Therefore, these enzymes will be mentioned in the following, too.

3.1
PolyP Synthesis: Polyphosphate Kinase

In bacteria, the synthesis of polyP from ATP is catalyzed by the enzyme polyphosphate kinase (ATP-polyphosphate phosphotransferase). The enzyme reaction is reversible and may therefore result in either biosynthesis or degradation of polyP (Yoshida 1955; Kornberg et al. 1956; Kornberg 1957):

$$polyP_n + ATP \leftrightarrow polyP_{n+1} + ADP,$$

where $polyP_n$ is inorganic polyphosphate with a chain length of n residues. An enzyme similar to bacterial polyphosphate kinase most likely does not exist in eukaryotic cells. A putative polyphosphate kinase purified to homogeneity from *S. cerevisiae* (Felter and Stahl 1973, 1975) was shown to be actually a diadenosine-5',5'''-P^1,P^4-tetraphosphate α,β-phosphorylase (Booth and Guidotti 1995). Therefore, it is assumed that eukaryotic cells use a different mechanism for polyP biosynthesis. Experiments with cultured mammalian cells revealed that $^{32}P_i$ taken up by the cells and incorporation into intracellular polyP most likely does not pass the intracellular P_i and ATP pools (Kumble and Kornberg 1995). Similar results were obtained with cell organelles and broken cell preparations (Pisoni and Lindley 1992; Cowling and Birnboim 1994). Incubation of lysosomes isolated from human fibroblast with $^{32}P_i$ resulted in a rapid transport of the P_i into the lysosomes and incorporation into long polyP chains (100 to 600 residues; Pisoni and Lindley 1992). A rapid incorporation of $^{32}P_i$ into polyP has also been reported for isolated rat liver nuclei (Penniall and Griffin 1984). Therefore it was concluded that synthesis of polyP from P_i in animal cells may occur without prior formation of ATP. However, the underlying mechanism(s) is still unkown.

Recently, it has been shown that the Ca^{2+}-ATPase (Ca^{2+} pump) of the human erythrocyte plasma membrane, which is involved in the maintenance of cytosolic Ca^{2+} concentration, contains polyP and poly(3-hydroxybutyrate) (Reusch et al. 1997). Interestingly, the plasma membrane Ca^{2+}-ATPase was found to act as "polyphosphate kinase". The enzyme is able to transfer both P_i from ATP to polyP and, the reverse reaction, P_i from polyP to ADP (Reusch et al. 1997).

An alternative route for biosynthesis of polyP might be possible in yeast. It has been proposed that polyP synthesis in *Saccharomyces carlsbergensis* is associated with the synthesis of mannan (Shabalin and Kulaev 1989). GDP-mannose, which serves as donor of mannosyl groups during mannoprotein biosynthesis, may also act as donor of phosphate in polyP synthesis. The transfer of P_i from GDP-mannose to polyP may occur via formation of dolichol-pyrophosphate mannose as follows:

1. GDP-mannose + dolichol phosphate $\rightarrow$ GMP + dolichol-pyrophosphate mannose;
2. Dolichol-pyrophosphate mannose + (mannan)$_n$ $\rightarrow$ dolichol-pyrophosphate + (mannan)$_{n+1}$;
3. Dolichol-pyrophosphate + polyP$_n$ $\rightarrow$ dolichol phosphate + polyP$_{n+1}$.

Other mechanisms for formation of polyP might be possible, too; e.g., formation of long-chain polyP via cyclic low molecular polyP intermediates (Bental et al. 1991) or association with RNA biosynthesis by intermediate formation of PP$_i$ (inorganic pyrophosphate; Kulaev et al. 1973).

Compared to eukaryotes, there is much more knowledge about polyP synthesis in bacteria. A polyphosphate kinase has been isolated from a number of eubacteria (e.g., *E. coli*, Ahn and Kornberg 1990; *Propionibacterium shermanii*, Robinson and Wood 1986; *Neisseria meningitidis*, Tinsley et al. 1993) and from archaebacteria (*Sulfolobus acidocaldarius*, Skorko et al. 1989). The polyphosphate kinase gene has been cloned from some bacteria, including *E. coli* (Akiyama et al. 1992) and *Klebsiella aerogenes* (Kato et al. 1993). The *E. coli* enzyme is a tretramer of 69-kDa subunits and attached to the cell outer membrane (Akiyama et al. 1992). The formation of polyP by this enzyme does not require a polyP primer and occurs via a phosphoenzyme intermediate (Ahn and Kornberg 1990; Kumble et al. 1996). The polyphosphate kinase in *E. coli* and *Pseudomonas aeruginosa* possesses also nucleoside diphosphate kinase activity (transfer of P$_i$ form polyP to NDPs, Kuroda and Kornberg 1997). The enzyme also catalyzes the transfer of a pyrophosphate (PP$_i$) group to GDP under formation of guanosine 5' tetraphosphate.

Recently, the *E. coli* polyphosphate kinase has been identified as a component of the bacterial RNA degradosome (Blum et al. 1997). RNA degradation by the degradosome is inhibited by polyP. This effect is abolished by addition of ADP, the substrate for polyP degradation by the polyphosphate kinase. Therefore it was proposed that the polyphosphate kinase in the degradosome removes the inhibitory polyP and regenerates ATP.

3.2
PolyP Degradation: Polyphosphatases

Polyphosphatases catalyze the hydrolysis of polyP to P$_i$. Endopolyphosphatases (polyphosphorylases) cleave polyP within the polyP chain:

$$polyP_n + H_2O \rightarrow polyP_{n-x} + polyP_x.$$

Exopolyphosphatases (polyphosphate-phosphohydrolases) hydrolyze polyP at the ends of the polymer under the formation of P$_i$:

$$polyP_n + H_2O \rightarrow polyP_{n-1} + P_i.$$

Several endo- and exopolyphosphatases, which differ in substrate specificity and product, have been isolated and purified (Kritskii et al. 1972; Andreeva and Okorokov 1993; Akiyama et al. 1993; Kowalczyk and Phillips 1993; Wurst and Kornberg 1994; Lorenz et al. 1994b, 1995; Kumble and Kornberg 1996). The acitivities of these enzymes have been shown to change depending on the availability of phosphate, the period of growth cycle (Kulaev 1979), and during development (Imsiecke et al. 1996) and ageing (Lorenz et al. 1997b). Hydrolysis of short-chain polyPs and metaphosphates may be catalyzed by distinct enzymes (Mattenheimer 1956a–c). It has been reported that in nuclei of animal cells, specific enzymes for hydrolysis of polyP$_3$, polyP$_4$ and tri- and tetrametaphosphate exist (Grossmann and Lang 1962).

3.2.1
Exopolyphosphatases

In contrast to polyP synthesis, there is much more knowledge on the breakdown of polyP in eukaryotic cells and tissues. In addition to a number of polyphosphatases (both endo- and exo-enzymes) from yeast, a few enzymes from animals have been described; some of them have been purified or purification is in progress.

3.2.1.1
Exopolyphosphatases from *Saccharomyces cerevisiae*

At least three exopolyphosphatases which cleave inorganic polyP to P$_i$ have been purified to homogeneity from *S. cerevisiae* and characterized in more detail (Andreeva and Okorokov 1993; Lorenz et al. 1994b; Wurst and Kornberg 1994; Andreeva et al. 1996).

   The 28-kDa exopolyphosphatase from *S. cerevisiae* is a monomeric protein (Lorenz et al. 1994). The enzyme has a pH optimum of 7.5 and requires Mg^{2+} ions for maximal activity. Co^{2+} is a further potent activator of the enzyme; other divalent cations are less effective. Immobilized polyP (polyP-modified zirconia) can be used as an effective affinity matrix for purification of the enzyme; in the absence of divalent cations, the exopolyphosphatase binds to polyP but does not degrade the polymer. The enzyme was found to be extremely unstable in solution, but can be stabilized by addition of Triton X-100. The activation energy with polyP$_{10}$ as substrate is 58 kJ/mol. o-Vanadate, Cu^{2+}, and Ca^{2+} are inhibitors of the exopolyphosphatase. The enzyme preferentially hydrolyzes linear polyP; PP$_i$ as well as cyclic tri- and tetrametaphosphate are degraded only very slowly, while ATP is not split by the exopolyphosphatase. The time kinetics of hydrolysis of short-chain [^{32}P]polyP$_{\leq 18}$ by the purified exopolyphosphatase is shown in Fig. 1. The mode of polyP degradation by the exopolyphosphatase is apparently non-processive. Therefore, the mechanism of polyP degradation by the 28-kDa

Fig. 1. Time kinetics of degradation of [^{32}P]polyP by yeast (*Saccharomyces cerevisiae*) exopolyphosphatase. [^{32}P]PolyP was incubated with purified exopolyphosphatase for the indicated time periods (*lanes a–j*). Lane k [^{32}P]orthophosphate marker. Analysis of size of degradation products was performed by electrophoresis on a 7-M urea/16.5% polyacrylamide gel. The autoradiogram is shown. P_i orthophosphate; *TM* trimetaphosphate. Chain lengths of polyPs (in number of residues) can be determined by counting bands on the gel (from *bottom* to *top*) starting with P_i

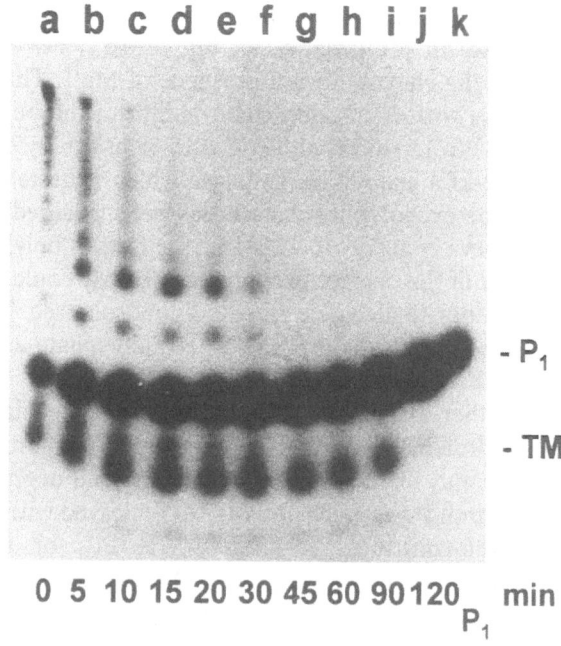

exopolyphosphatase differs from that of the 40-kDa exopolyphosphatase from *S. cerevisiae* described by Wurst and Kornberg (1994) which hydrolyzes polyP in strictly processive manner. A processive mechanism is defined as a process which occurs without repetitive dissocation of the polymer from the enzyme.

The K_m value for polyP of the 28-kDa exopolyphosphatase was found to decrease markedly with increasing chain length of the substrate (polyP$_3$ to polyP$_{33-36}$) from 410 to 1.3 µM (as polymer), while the decrease in the value of the catalytic rate constant, k_{cat}, from 143 to 81 s^{-1}, is comparatively low; as a result, the k_{cat}/K_m value, a criterion for enzyme specificity, increases. The k_{cat}/K_m for polyP$_{33-36}$ is 6.2 × 10^7 M^{-1} s^{-1}; this value is close to the diffusion-controlled limit of a second order rate constant. This result shows that, with increasing chain lengths, the affinity of exopolyphosphatase to polyP becomes higher, while the reaction rate slows down.

A 40-kDa exopolyphosphatase from the cell envelope of *S. cerevisiae* has been purified and characterized by Andreeva et al. (1990) and Andreeva and Okorokov (1990, 1993). This enzyme displays a broad pH optimum at 6.0–7.5 and is inhibited by *N*-ethylmaleimide and iodoacetamide. The cell envelope exopolyphosphatase is unable to split PP$_i$, in contrast to the 28-kDa exopolyphosphatase. In addition, this enzyme is maximally active at a 1:1 ratio of polyP/Mg^{2+} (Andreeva and Okorokov 1993), while the 28-kDa exopolyphosphatase requires excess amounts of Mg^{2+} for optimal activity. The ability of divalent cations to activate the 40-kDa exopolyphosphatase decreases in the order Co^{2+}, Mg^{2+}, Mn^{2+}, Fe^{2+}, and Zn^{2+}; Cu^{2+} and Zn^{2+} inhibit the enzyme in the presence of Mg^{2+}. The K_m values of the enzyme decrease with increasing chain length of the polyP substrate from 170 µM ($polyP_3$) to 150 nM ($polyP_9$) and to 88 pM ($polyP_{208}$).

A 40-kDa exopolyphosphatase which is quite similar to the purified cell envelope exopolyphosphatase has been purified from *S. cerevisiae* cytosol (Andreeva et al. 1996). This enzyme cleaves $polyP_n$ ($n \geq 3$) but not PP_i. The activity of the cytosol exopolyphosphatase could be inhibited by antibodies against purified cell envelope polyphosphatase. The enzyme has the ability to hydrolyze also adenosine-5'-tetraphosphate and guanosine-5'-tetraphosphate (Kulakovskaya et al. 1997).

Another exopolyphosphatase with a molecular mass of 40 kDa (monomeric protein) has been purified to apparent homogeneity from *S. cerevisiae* (Wurst and Kornberg 1994). This enzyme degrades polyP in a processive mode until PP_i is reached. $PolyP_{250}$ is cleaved with a k_{cat}/K_m near the limit for diffusion-controlled reactions. ATP, PP_i, or trimetaphosphate are not degraded by the enyzme. This 40-kDa exopolyphosphatase requires Mg^{2+}, Mn^{2+}, or Co^{2+} and a pH between 6.8 and 8.8 for optimal activity. The activation energy is 63 kJ/mol. The enzyme has a high turnover number of 500 P_i residues s^{-1} (37 °C). The gene encoding this exopolyphosphatase has been cloned from *S. cerevisiae*; the molecular mass deduced from the DNA sequence is 44 922 Da (Wurst et al. 1995). The enzyme is located in the cytosol.

A further polyphosphatase has been purified to homogeneity from the yeast *Neurospora crassa* (Umnov et al. 1975). This enzyme prefers long-chain polyP and requires the presence of divalent cations (Mg^{2+}; Co^{2+}, Mn^{2+} and Fe^{2+} are less effective). The molecular mass of the enzyme is 50 kDa; the K_m for long-chain polyP is 0.7 mM.

Studies of the compartmentation of exopolyphosphatases in yeast revealed that apparently each subcellular fraction (nucleus, mitochondria, cytosol, cell envelope, and vacuoles) of *S. cerevisiae* has its own enzymes for polyP degradation (Kulaev et al. 1995, 1997; Kulakovskaya et al. 1995; Lichko et al. 1996a,b; see also Chap. 3). These enzymes differ in their kinetic properties, substrate specificity, requirements of divalent cations, and sensitivity to inhibitors. The nuclear polyphosphatase activity of *S. cerevisiae* has a pH optimum of 7.5, depends on divalent metal cations (Co^{2+}, Mg^{2+}, Zn^{2+}, Mn^{2+}, in decreasing order) and hydrolyzes $polyP_3$, $polyP_{15}$, and $polyP_{208}$ with K_m's of 100, 5, and 4.1 µM, respectively (Lichko et al. 1995, 1996a). The mitochondrial enzyme activity is maximal at neutral pH, stimulated by divalent cations (as the nuclear enzyme) and degrades all polyP chain lengths from 9 to 208 with nearly the same efficiency; EDTA, heparin and different monovalent cations are inhibitory (Lichko et al. 1996b). Both membrane-bound and soluble mitochondrial polyphosphatases with different molecular masses could be detected (Lichko et al. 1997). The exopolyphosphatase present in *S. cerevisiae* vacuoles displays different properties, exhibits some activity with polyP$_3$ and adenosine-5'-tetraphosphate, but does not hydrolyze PP$_i$ (Andreeva et al. 1998). Exopolyphosphatases from cytosol, cell envelope, and vacuoles have been purified to homogeneity (see above).

A number of exopolyphosphatases has been purified also from bacteria (e.g., *E. coli*, Akiyama et al. 1993; *Corynebacterium xerosis*, Muhammed et al. 1959; *Acinetobacter johnsonii* 210A, Bonting et al. 1993). The gene encoding an *E. coli* exopolyphosphatase (*ppx*) has been cloned; it was found downstream of the polyphosphate kinase gene (*ppk*), in the same operon (Akiyama et al. 1993). The enzyme purified to homogeneity from overproducing cells is a dimer of 58-kDa subunits, processively degrades long-chain polyP and is stimulated by Mg^{2+}. A second exopolyphosphatase activity (molecular mass 100 kDa) purified to homogeneity from an *E. coli* mutant strain lacking *ppx* gene has been identified as guanosine pentaphosphate phosphohydrolase (Keasling et al. 1993). This enzyme, which cleaves long-chain polyP, also hydrolyzes the 5'-γ-phosphate of guanosine 5'-triphosphate 3'-diphosphate (pppGpp) under formation of guanosine 5'-diphosphate 3'-diphosphate (ppGpp). The K_m for long-chain polyP (0.5 nM) is much lower than that for pppGpp (0.13 mM) as a substrate.

3.2.1.2
Exopolyphosphatase I and II from *Tethya lyncurium*

The first polyP-dependent enzymes purified from an animal were two exopolyphosphatases (exopolyphosphatase I and exopolyphosphatase II) from the marine sponge *T. lyncurium*, a member of the most simplest metazoans (Lorenz et al. 1995). Both enzymes which release P$_i$ from inorganic polyP are monomeric proteins. Exopolyphosphatase I has a molecular mass of 45 kDa, a pH optimum of 5.0 and does not require divalent cations for activity, while exopolyphosphatase II has a molecular mass of 70 kDa, a pH optimum of 7.5 and displays optimal activity in the presence of Mg^{2+} ions. Final purification of the enzymes could be achieved by affinity chromatography on polyP-modified zirconia. The mode of action of both enzymes was found to be processive; analysis of the degradation products of polyP on urea/polyacrylamide gels did not reveal a transient increase of polyP chains of intermediate size in the course of exopolyphosphatase I and exopolyphosphatase II reaction. A typical time kinetics of degradation of [^{32}P]polyP by exopolyphosphatase I is shown in Fig. 2, lanes a–e. The size of a [^{32}P]polyP preparation used reached up to about 20 and more phosphate residues (lane e). P$_i$ is the final product of hydrolysis of polyP by exopolyphosphatase I. Incubation of polyP without enzyme did not result in degradation of the polymer (lane h). In contrast to the apparently processive degradation of polyP by exopolyphosphatase I, during hydrolysis of polyP under acidic conditions

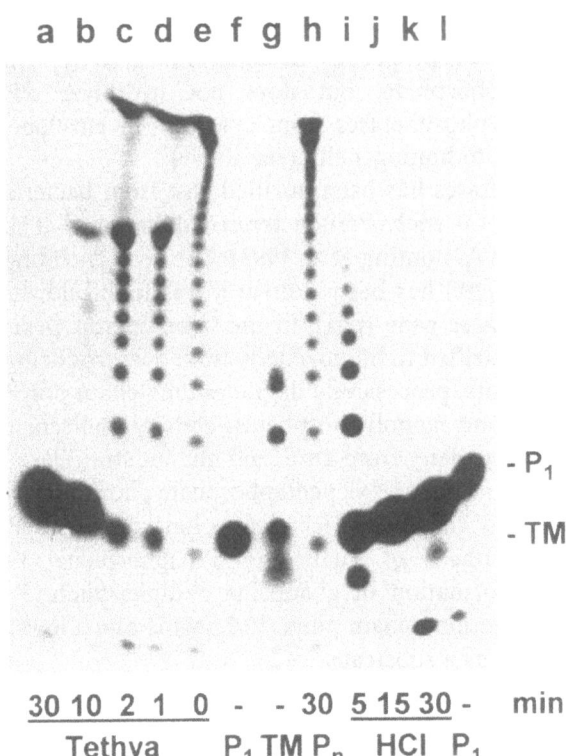

a b c d e f g h i j k l

- P$_1$

- TM

<u>30 10 2 1 0</u> - - 30 <u>5 15 30</u> -   min
    Tethya   P$_1$ TM P$_n$ HCl P$_1$

Fig. 2. Kinetics of degradation of polyP by exopolyphosphatase I (*Tethya lyncurium*). [^{32}P]PolyP was incubated with purified exopolyphosphatase I for the indicated time periods (*lanes a–e*). *Lanes f* and *l* [^{32}P]orthophosphate standard; *lane g* [^{32}P]trimetaphosphate; *lane h* [^{32}P]polyP incubated for 30 min without enzyme; *lanes i–k* products obtained by chemical hydrolysis of [^{32}P]polyP. For further details, see legend to Fig. 1

(10 mM HCl; 70 °C), the amount of shorter polyP molecules transiently increases in the course of the reaction (lanes i–k). In addition, formation of trimetaphosphate can be observed under these conditions (lane i).

The time kinetics of degradation of [^{32}P]polyP by exopolyphosphatase II is shown in Fig. 3, lanes a–i. Degradation of polyP by this enzyme also occurs by a processive mechanism, in contrast to hydrolysis of [^{32}P]polyP by the exopolyphosphatase activity present in extracts from the sponge *Geodia cydonium* (lane k). PolyP is degraded by exopolyphosphatase II until PP$_i$ is reached as end product, which is hydrolyzed, if at all, only very slowly. By contrast, exopolyphosphatase I totally cleaves the polymer to P$_i$ (Fig. 2). Cyclic trimetaphosphate present in the [^{32}P]polyP preparation (Fig. 3, lane a) is cleaved by exopolyphosphatase II very rapidly (lane b).

Both enzymes differ from the 28-kDa exopolyphosphatase from *S. cerevisiae* which hydrolyzes polyP in a non-processive manner (see above). A processive mechanism for polyP degradation has also been demonstrated for the 40-kDa exopolyphosphatase from yeast (see above) and the exopolyphosphatases from *E. coli* (Akiyama et al. 1993; Keasling et al. 1993) and *P. shermanii* (Robinson and Wood 1986; Robinson et al. 1987).

Fig. 3. Analysis of chain lengths of degradation products of [^{32}P]polyP produced by exopolyphosphatase II (*Tethya lyncurium*). [^{32}P]PolyP was incubated with purified exopolyphosphatase II for the indicated time periods (*lanes a–i*). *Lane j* [^{32}P]polyP incubated for 90 min in assay mixture without enzyme; *lane k* [^{32}P]polyP incubated for 30 min in assay mixture with *Geodia cydonium* extract. P_i orthophosphate; P_2 pyrophosphate; *TM* trimetaphosphate. For further details, see legend to Fig. 1

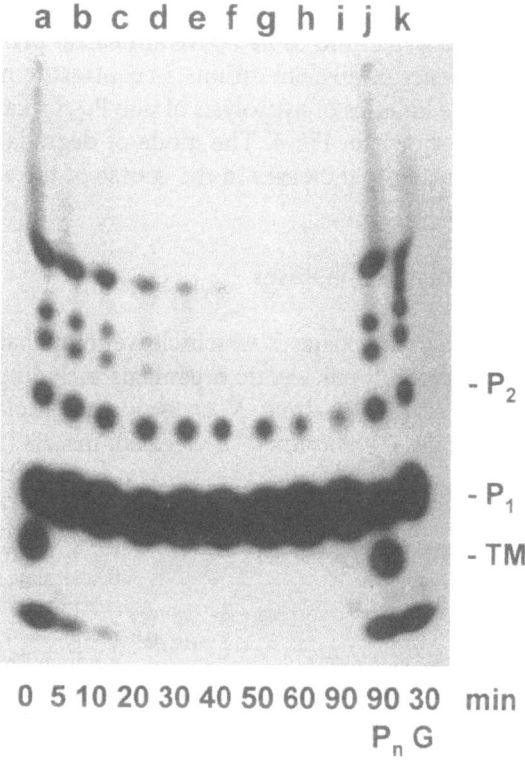

The K_m values for polyP$_{10}$ as substrate of exopolyphosphatase I and exopolyphosphatase II are 24 and 9 μM, respectively (based on polymer concentration); the activation energies for polyP$_{10}$ as substrate are 56 and 60 kJ/mol, respectively.

3.2.1.3
Exopolyphosphatase Activities in Mammalian Cells and Tissues

Extracts from many mammalian cell lines and tissues, including rat brain and liver, and human blood cells were found to contain a polyP-degrading exopolyphosphatase activity (Lorenz et al. 1997a–c). Relatively high amounts of exopolyphosphatase activity are present in human osteoblasts (Leyhausen et al. 1998). Exopolyphosphatase activity was also detected extracellularly in human blood plasma and serum (Lorenz et al. 1997a). The exopolyphosphatase from rat liver could be partially purified. The liver enzyme degrades long-chain polyP to P_i, depends on the presence of divalent metal ions and is optimal at neutral pH.

Recently, we demonstrated that highly purified preparations of mammalian alkaline phosphatase are able to degrade long-chain polyP, in addition to PP_i, one of the known substrates (Lorenz et al., unpublished results). The enzyme was found to be active at neutral pH. The enzyme is also active in the absence of divalent cations, in contrast to most other exopolyphosphatases. The kinetics of hydrolysis of $polyP_{750}$ by calf intestinal alkaline phosphatase is shown in Fig. 4. The mode of degradation is apparently processive; PP_i transiently increases in the course of the reaction.

3.2.2
Endopolyphosphatases

Endopolyphosphatases which cleave long-chain polyP have been identified in a variety of eukaryotic organisms including protozoa (*Giardia duodenalis*), yeast (*S. cerevisiae, N. crassa*), slime mold (*Dictyostelium discoideum*), nematodes (*Caenorhabditis elegans*), insects (*Drosophila melanogaster*) and

Fig. 4. Degradation of long-chain polyP by purified alkaline phosphatase (calf intestine). The substrate ($[^{32}P]polyP_{750}$) was synthesized by recombinant *E. coli* polyphosphate kinase (kind gift from Dr. A. Kornberg). Reaction products obtained after incubation for different time periods (0 to 60 min; lanes *a–h*) were analysed on urea/polyacrylamide gel

in different tissues from mammals, but only in some bacteria (Kritskii et al. 1972; Wiemken et al. 1979; Kumble and Kornberg 1996). The enzyme from *S. cerevisiae* has been purified to homogeneity (Kumble and Kornberg 1996). The yeast endopolyphosphatase is a dimer of 35-kDa subunits. The enzyme catalyzes the non-processive cleavage of long-chain polyP. $PolyP_{750}$ is degraded under formation of shorter chains up to a chain length of about 60 residues ($polyP_{60}$). In addition, $polyP_3$ is produced in a ratio of $polyP_{\geq 60}$ to $polyP_3$ of about 3 (in terms of polymer concentration). The K_m of the enzyme for $polyP_{750}$ as substrate is 185 nM. The enzyme has a pH optimum of 7.5 and requires divalent metal ions (Mg^{2+} or Mn^{2+}); EDTA, Ca^{2+} and Zn^{2+} are inhibitory. An endopolyphosphatase activity has also been detected in rat and bovine brain (Kumble and Kornberg 1996); the enzyme activity in brains is much higher ($\sim$10 times) than in other tissues. On the other hand, yeast, where the enzyme is localized in vacuoles, contains 10 times higher activity than brain.

3.3
Further PolyP-Dependent Enzymes

The enzymes described in this section have been purified from lower organisms. Their presence at least in higher eukaryotes is uncertain or could not be confirmed. The polyphosphate glucokinase (polyphosphate:D-glucose-6-phosphate-phosphotransferase) catalyzes the phosphorylation of glucose by polyP (or ATP):

$$polyP_n + glucose \rightarrow polyP_{n-1} + glucose\text{-}6\text{-phosphate}.$$

This enzyme has been identified in several bacteria, but not in yeast, animals or other eukaryotes (Szymona and Ostrowski 1964; Szymona et al. 1977; Szymona and Szymona 1979; Pepin and Wood 1986, 1987; Phillips et al. 1993; Kowalczyk et al. 1996). The polyphosphate glucokinase is also present in some bacteria which are pathogenic for humans (see Sect. 9), and might therefore represent a potential target for chemotherapy.

The enzyme from the important human pathogen *Mycobacterium tuberculosis* has been purified to homogeneity (Hsieh et al. 1993b). The gene encoding the enzyme has been cloned and expressed in *E. coli* (Hsieh et al. 1996). The enzyme, a dimer of 33 kDa, utilizes preferentially longer polyP chains by a non-processive mechanism; tryptophan at the active site of the enzyme is involved in the reaction (Hsieh et al. 1993a). The two phosphate donors of the enzyme, polyP and ATP, have different binding sites; the K_m for $polyP_{32}$ is 18 μM and that for ATP is 1.5 mM (Hsieh et al. 1993b). It has been speculated that the polyphosphate glucokinases found in present-day bacteria, which use both polyP and ATP as substrates, may represent an intermediate in evolution of an ancestor enzyme which solely used polyP to phosphorylate glucose, and the ATP glucokinase of higher animals, which has lost the ability to use polyP as phosphate donor (Phillips et al. 1993).

Another enzyme which may participate in polyP degradation is the polyphosphate:AMP phosphotransferase which catalyzes the transfer of phosphate from polyP to AMP to produce ADP:

$$polyP_n + AMP \rightarrow polyP_{n-1} + ADP.$$

This enzyme has not been described in eukaryotic cells, but has been found in some bacteria (Winder and Denneny 1957; Dirheimer and Ebel 1965; Bonting et al. 1991).

Both degradation of biosynthesis of polyP may be mediated by a 1,3-diphosphoglycerate:polyphosphate phosphotransferase:

$$polyP_n + 1,3\text{-diphosphoglycerate} \leftrightarrow polyP_{n+1} + 3\text{-phosphoglycerate}.$$

Crude extracts from *N. crassa* were found to contain low activities of this enzyme (Kulaev and Bobyk 1971; Kulaev and Konoshenko 1971). The physiological significance of this enzyme activity is, however, uncertain.

The polyP-dependent NAD-kinase catalyzes the following reaction:

$$polyP_n + NAD \rightarrow polyP_{n-1} + NADP.$$

This enzyme has been isolated only from bacterial cells (Murata et al. 1979), but not from eukaryotes.

4
PolyP-Binding Proteins

Inorganic polyP is capable of forming complexes with other polymers such as basic proteins or nucleic acids (Kulaev 1979). Histones are precipitated by increasing concentrations of polyP (Stros et al. 1984). We could detect the presence of a variety of such polyP-binding proteins in crude cell extracts from different organisms (yeast, invertebrate and vertebrate cells) by using an novel filter-binding technique or affinity chromatography on polyP zirconia (Lorenz et al. 1994a). Inorganic polyP has also been identified as an integral part of alkaline phosphatase preparations (Gabel und Thomas 1976).

5
Biological Functions of PolyP

The multiple biological functions of polyP have now become a focal point of attention, although a direct proof of these functions is often difficult. They may be different in different species and different compartments of the cell. Inorganic polyP has been proposed to serve as:

1. An energy source
2. A phosphate reserve and phosphate donor for certain sugar kinases
3. A regulator of cellular enzymes
4. A regulator of levels of intracellular adenylate nucleotides
5. A regulator of DNA–histone interaction and gene activity
6. A component of membrane channels with poly(3-hydroxybutyrate) and calcium
7. A chelator for calcium and toxic metal ions, and counterion for basic molecules
8. A regulator of cellular stress response, and
9. An antibacterial and antiviral agent

These functions will be discussed in the following Sections (5.1–5.9).

5.1
Energy Source

The presence of high-energy phosphoanhydride bonds in polyP suggests some role in energy metabolism. It has been proposed that polyP as storage molecule for energy-rich phosphate might act as "phosphagen" involved in the regulation of the intracellular concentrations of ATP and other nucleotides (Kulaev and Vagabov 1983). This might be true in yeast but seems to be unlikely in higher eukaryotic cells, where the concentration of polyP is smaller than that of ATP. For example, in mammalian cells, intracellular concentrations of ATP range from 5 to 8 mM, depending on metabolic conditions (Regen et al. 1964), compared to polyP concentrations varying, with some exceptions (osteoblasts), from 25 to 120 µM (in terms of P_i residues; see Sect. 2); in yeast cells, ATP concentrations are around 3 mM (den Hollander et al. 1986), but polyP can reach concentrations of up to 100 mM (although it is not totally present in soluble form). Nevertheless, it cannot be excluded that in certain circumstances polyP could act as a (solid-state) energy source at a very spot in place of diffusible ATP in higher eukaryotes.

In yeast, polyP has been proposed to be utilized as a energy source for antibiotic production, e.g., during chlortetracycline biosynthesis by *Streptomyces aureofaciens* (Hostalek et al. 1976; Kulaev et al. 1976) and valinomycin biosynthesis by *Streptomyces cyaneofuscatus* (Telesnina et al. 1986). PolyP has also been proposed to be hydrolyzed as a source of energy during repair of radiation damage following exposure of yeast to ^{60}Co γ-radiation (Holahan et al. 1988).

5.2
Phosphate Reserve and Phosphate Donor for Kinases

Inorganic polyP may serve as an osmotically advantageous and an easily available phosphate reserve that can be mobilized under conditions of phosphate starvation (Kornberg 1957; Kulaev 1979; Gillies et al. 1981; Kulaev and Vagabov 1983). This phosphate reserve might be important in biotopes where phosphate is growth limiting. It is known that the amount of polyP decreases when yeast cells are deprived of phosphate. On the other hand, these cells accumulate large amounts of polyP when they are transferred to a phosphate-rich medium after phosphate starvation ("hypercompensation effect") (Kulaev and Vagabov 1983). The vacuolar polyP in yeast may provide an active pool for phosphate, which is mobilized to the cytosol during phosphate starvation (Shirahama et al. 1996).

Furthermore, polyP is a substrate for certain sugar kinases (polyphosphate glucokinases) from bacteria (see Sect. 3.3), but not from yeast and animal cells. Therefore, this function seems to be restricted to prokaryotic organisms. PolyP may also act as phosphate donor in protein phosphorylation, as shown in archaebacteria (Skorko 1989).

5.3
Regulator of Cellular Enzymes

The activity of several enzymes has been reported to be influenced by polyP. Inorganic polyP was found to act as inhibitor of AMP deaminase which is involved in the control of glycolysis and adenylate energy charge in yeast (Yoshino and Murakami 1988). In addition, polyP was shown to act as an inhibitor of phospholipase A_2 from pig pancreas, snake (*Naja naja*) and bee venom, and of phospholipase B from the yeast *Torulaspora delbrueckii* (Sultana et al. 1995). Furthermore, polyP has been reported to inhibit cyclic 3',5'-AMP phosphodiesterase activity of *S. carlsbergensis* (Speziali and Van Wijk 1971). Cathepsin D from horse spleen is inhibited by polyP (Ducastaing et al. 1976), while the activity of the enzyme from bovine spleen is increased by polyP (Watabe et al. 1979). PolyP associated with the membrane of mammalian lysosomes could have a protective function by preventing degradation of membrane components by lysosomal hydrolases (Pisoni and Lindley 1992). Moreover, polyP present in DNA preparations from filamentous fungi was found to inhibit restriction endonucleases, *Taq*I DNA polymerase and T4 DNA ligase (Rodriguez 1993).

5.4
Regulator of Levels of Intracellular Adenylate Nucleotides

Inorganic polyP was found to inhibit the formation of ATP by adenylate kinase in yeast and animal cells (Lorenz et al. 1995). This enzyme plays an important role in regulation of the energy state of a cell by catalyzing the transformation of the three adenosine phosphates, AMP, ADP and ATP, in each other. $PolyP_4$ and longer-chain polyPs inhibit the adenylate kinase in a competitive manner. The inhibitory effect of the longer polymers was higher than that of shorter polyPs. Based on these results, it has been proposed that polyP may regulate the intracellular concentrations of adenylate nucleotides (Lorenz et al. 1995), as follows. If high amounts of ATP are available in the cell, polyP is synthesized. Thereby the surplus energy is stored and can be made available for other metabolic reactions. If the concentration of long-chain polyP increases, adenylate kinase will be inhibited, and consequently the formation of ATP and AMP and the synthesis of further polyP will be suppressed. In addition, ATP formation decreases because the rate of glycolysis is reduced in the presence of low levels of AMP as a result of inhibition of adenylate kinase reaction. If the concentration and size of polyP decline again due to consumption of the energy-rich phosphate residues of the polymer, the inhibition of adenylate kinase becomes weaker and ATP can be formed by the enzyme, now being available for the synthesis of new polyP.

5.5
Regulator of DNA–Histone Interaction and Other Functions in the Cell Nucleus

In mammalian cells, polyP is present predominantly in the nucleus (see Sect. 2.2). It has been demonstrated that at the concentrations found in nuclei (i.e., ~1 % of DNA), polyP is able to destabilize chromatin, suggesting that this polyanion may play a role in regulating chromatin template activity (Ansevin et al. 1975; Mansurova et al. 1975; Weinstein and Li 1976). The destabilization of chromatin by polyP may be caused by a dissociation of histones (Ansevin et al. 1975; Weinstein and Li 1976) or an interaction with non-histone nuclear proteins (Offenbacher and Kline 1984).

In gel-mobility-shift assays we could demonstrate that polyP affects the binding of histones to DNA in vitro. If DIG-labeled DNA was incubated with purified histones (calf thymus) and then separated from free DNA by electrophoresis on polyacrylamide gel, a marked mobility-shift of the DNA–histone complexes formed was observed (Fig. 5, lane a). Incubation of the complexes with increasing concentrations of polyP$_{35}$ and analysis by mobility-shift gel electrophoresis revealed an increase in the intensity of the free DNA band which is correlated with the increasing amounts of polyP (Fig. 5, lanes b–f). This result shows that polyP$_{35}$ is indeed able to compete with the DNA for the DNA binding site at the histones. The competition depends on the chain length of polyP. In order to determine whether polyP also interferes with DNA–histone interaction in isolated chromatin (HL60 cells), co-incubation experiments of chromatin with micrococcal nuclease and different concentrations of polyP were performed. It is known that chromatin is cleaved by micrococcal nuclease at the linker DNA, whereas the core DNA is protected against digestion. Therefore, after nuclease treatment and subsequent extraction of the DNA, a stepladder-like pattern of DNA bands representing multiplicates of 150–200 bp (= length of core DNA)

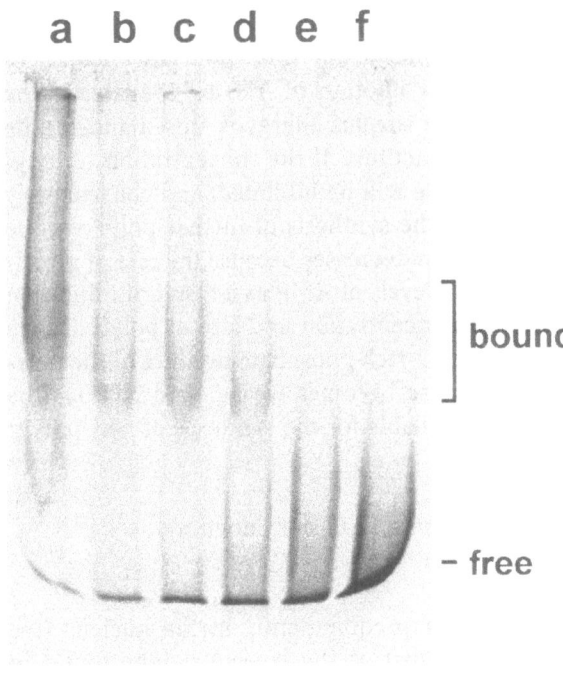

Fig. 5. Effect of increasing concentrations of polyP$_{35}$ on integrity of DNA–histone complexes in mobility-shift assay. DNA (1 ng 154-bp DNA restriction fragment 3-end labeled with DIG-11-dUTP) was incubated with 200 ng histones (calf thymus) for 45 min at 25 °C. Then increasing amounts of polyP$_{35}$ (*lane a* 0; *lane b* 500 pg; *lane c* 5 ng; *lane d* 50 ng; *lane e* 500 ng; *lane f* 5 µg) were added and incubation was continued for 30 min. The samples were then separated on a 5 % polyacrylamide gel and blot-transferred to PVDF membrane, followed by immuno-detection of DNA. Positions of histone-bound DNA and free DNA are indicated

should be expected in agarose gel. If polyP is able to compete with the DNA for the binding sites on the core complexes, a part of the core DNA should dissociate from chromatin; digestion with micrococcal nuclease should then result in formation of DNA fragments smaller than 150 bp. Indeed, we found that smaller fragments of DNA are formed after preincubation with polyP (Lorenz et al., unpubl. results). The activity of micrococcal nuclease was not influenced by polyP. These results indicate that polyP may also interact with DNA–histone binding in chromatin. Moreover, binding of polyP has been shown to inhibit the activity of a series of nuclear enzymes including topo-isomerases (Lorenz et al., unpubl. results).

In *E. coli*, binding of polyP to RNA polymerase from stationary growth phase cells was found to be associated with a reduced transcription of the genes expressed in exponentially growing cells, suggesting that polyP may play a role in the control of promoter selectivity of RNA polymerase in *E. coli* (Kusano and Ishihama 1997). It seems to be possible that polyP could play a similar role in regulation of gene activity in eukaryotes.

5.6
Formation of Calcium Channels Across Membranes

Inorganic polyP has been shown to be capable of forming channels through cell membranes together with poly(3-hydroxybutyrate) and calcium (see

Chap. 8). These channels consisting of a complex of polyP, Ca^{2+}, and poly(3-hydroxybutyrate) are found both in bacterial plasma membranes (Reusch and Sadoff 1988) and in cellular membranes of a variety of animals and plants (Reusch 1989). In bovine liver, the poly(3-hydroxybutyrate)/Ca^{2+}-polyP complexes are primarily localized in the mitochondria and microsomes; smaller amounts are found in the plasma membrane (Reusch 1989). The complexes in eukaryotic membranes consist of polyP with an estimated size of 170–220 residues and poly(3-hydroxybutyrate) with an estimated chain length of 120–200 subunits. The molar ratio of polyP:poly(3-hydroxybutyrate):calcium of these complexes is approximately 1:1:0.5. The poly(3-hydroxybutyrate) is assumed to form an outer transmembrane helix which is wound around an inner helix of Ca^{2+}-polyP; the Ca^{2+} ions in this structure bridge the two polymers. These poly(3-hydroxybutyrate)/Ca^{2+}-polyP complexes have been shown to act as non-proteinaceous voltage-activated Ca^{2+} channels in the plasma membranes of E. coli (Reusch et al. 1995) or in synthetic membranes (Das et al. 1997). They may also act as Ca^{2+} channels in eukaryotic cell membranes, e.g., in human erythrocytes; see Section 3.1. In addition, they have been proposed to play a role in the transport of phosphate and uptake of DNA across the plasma membranes of transformed bacteria (Castuma et al. 1995; Huang and Reusch 1995).

5.7
Chelator for Divalent Cations and Counterion for Basic Molecules

As a polyanionic ion exchanger, polyP is capable of chelating Ca^{2+} or other divalent cations and thus may act as metabolic regulator. In yeast, vacuolar polyP has been proposed to participate in regulation of cellular Ca^{2+} concentration (Dunn et al. 1994). PolyP seems to be involved in the control of the cellular tolerance to heavy metals such as Cd^{2+}, Zn^{2+}, Hg^{2+}, Pb^{2+}, Cu^{2+}, and Co^{2+}. This function has been demonstrated especially in microorganisms; e.g., polyP has been shown to affect cadmium tolerance of E. coli (Keasling and Hupf 1996). It has been assumed that this effect is due to the sequestering of toxic metals by polyP and the resulting reduction of their effective concentration in the cell. However, recent results indicate that hydrolysis of polyP to P_i by polyphosphatase and transport of the metal phosphates out of the cell may be important for heavy metal tolerance (Keasling 1997).

Furthermore, polyP can serve as counterion for basic amino acids (arginine, ornithine, histidine, and lysine) in vacuoles or lysosomes. In the yeast vacuole, the molar ratio of basic amino acids and polyP phosphorus is approximately 1 (Cramer et al. 1980; Cramer and Davis 1984).

5.8
Regulator of Cellular Stress Response

The response of cells to osmotic stress and pH stress has been shown to be associated with changes in polyP metabolism. The effects of these stress conditions on polyP metabolism in yeast have been studied mainly by ^{31}P NMR spectroscopy. A rapid hydrolysis of polyP with a concomitant increase in cytoplasmic phosphate, cytoplasmic pH, and vacuolar pH was found in *N. crassa* following hypoosmotic shock (Yang et al. 1993). It has been proposed that polyP serves as pH buffer in *S. cerevisiae* after alkalinization (Castro et al. 1995). In mammalian cells, inorganic polyP could also act as a buffer to maintain the intralysosomal acidic pH, leveling out variations in the activity of lysosomal proton pump (Pisoni and Lindley 1992). Mammalian lysosomes contain enriched amounts of polyP (Kumble and Kornberg 1995); the total content of phosphate (including P_i) in rat liver lysosomes even amounts to 0.13 μmol/mg protein (approximately 30 mM phosphate; Schneider 1983).

Like yeast, algae are able to accumulate large amounts of polyP (Langen 1958, 1961; Kulaev 1979). Cytoplasmic alkalinization of the unicellular green alga *Dunaliella salina*, induced by ammonium ions, resulted in degradation of long-chain polyP to polyP$_3$. The hydrolysis of polyP was found to be correlated with the recovery of cytoplasmic pH (Pick et al. 1990). Therefore, hydrolysis of polyP may provide a pH-stat mechanism to counterbalance alkaline stress.

The alga *Phaeodactylum tricornutum* was found to respond to hyperosmotic stress with a marked elongation of polyP molecules and a decrease in the total amount of cellular polyP, while exposure to hypoosmotic stress resulted in an increase in shorter polyP molecules and a rise in total polyP content (Leitão et al. 1995). The reaction to hypoosmotic shock was even much stronger, when the algal cultures were adapted to higher salt concentrations before. Thus, variations in concentration as well as in size of polyP might allow to adjust the intracellular osmotic pressure to the extracellular osmotic pressure. Moreover, changes in polyP following osmotic stress could help to adjust the intracellular ATP level to normal, since ATP can be formed or consumed by polyP-dependent reactions; e.g., ATP production via polyphosphate kinase for energy-consuming adaptation to osmotic stress. The level of ATP has been shown to decrease following hyperosmotic shock and to increase following hypoosmotic shock (Bental et al. 1990).

Also in bacteria, polyP is involved in stress response. PolyP has been demonstrated to promote survival of stationary-phase *E. coli* cells (Rao and Kornberg 1996). *Escherichia coli* mutant lacking polyphosphate kinase showed a greater sensitivity to heat, to H_2O_2, to menadione (a redox-cycling agent), and to hyperosmotic stress. Reduction of polyP levels in *E. coli* by expression of a plasmid carrying the gene for yeast exopolyphosphatase diminished the resistance to H_2O_2 (Shiba et al. 1997). PolyP is accumulated in bacteria subjected to osmotic or nutritional stress or to nitrogen exhaustion (Ault-Riche et al. 1998). The accumulation of polyP in response to amino acid starvation in *E. coli* was found to be promoted by an increase in guanosine tetraphosphate (ppGpp) and guanosine pentaphosphate (pppGpp) (Kuroda et al. 1997).

5.9
Antibacterial and Antiviral Agent

Inorganic polyP has been shown to possess antibacterial and antiviral activity. Because of its antibacterial activity, polyP is added as multifunctional ingredient to many foods (Matsuoka et al. 1995; Palumbo et al. 1995). Lorenz et al. (1997a) found that polyP at levels only three- to six-fold higher than in human blood constituents (see Sect. 2) displays a marked antiviral activity on cells infected with the human immunodeficiency virus HIV-1. PolyP molecules consisting of more than four P_i residues inhibit HIV-1 infection of cells in vitro already at concentrations of $\geq 300\,\mu M$ (in terms of P_i residues); $polyP_3$ is ineffective. The longer polymers ($polyP_{15}$ and $polyP_{34}$) but not $polyP_3$ and $polyP_4$ also abolish HIV-1-induced syncytium formation. Binding studies revealed that polyP reduces the binding of virus particles to the host cell (Lorenz et al. 1997a). Furthermore, polyP has been shown to protect cells against the cytopathogenic effect of *Clostridium difficile* toxin B (Florin and Thelestam 1984).

6
PolyP Metabolism During Ageing and Development

The polyP content and exopolyphosphatase activity in rat tissue were found to change in the course of ageing and development (Lorenz et al. 1997b; Fig. 6). The level of polyP in rat brain strongly (by six-fold) increases after birth. The polyP content in embryonic tissue is low compared to adult tissue; the polyP in rat embryo and newborns consists mainly of soluble polyP molecules. The increase in polyP in rat brain after birth is mainly caused by an increase in insoluble long-chain polyP. Maximal levels of polyP were found in 12-month-old animals. In old rat brain the amount of total polyP markedly decreased to about 50 %. This decrease is mainly due to a decrease in the amount of insoluble long-chain polyP; the amount of soluble long-chain polyP does not change significantly with age. In rat liver, the age-dependent changes are less pronounced. The changes in polyP content in rat brain and liver are accompanied by changes in exopolyphosphatase activity; highest enzyme activities were found when the polyP level was low (Lorenz et al. 1997b). Similarly, endopolyphosphatase activity was found to be highest in prenatal rat brain and decreases after birth (Kumble and Kornberg 1996).

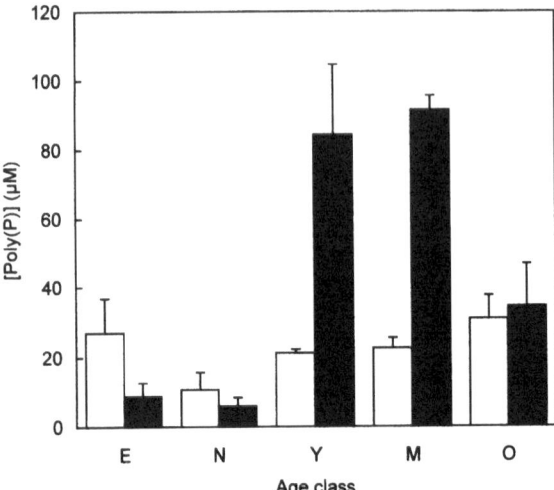

Fig. 6. Changes in polyP content of rat brain in the course of ageing and development. Both soluble long-chain polyP (*open bars*) and insoluble long-chain polyP (*solid bars*) were determined. Each value represents the mean ± SD of the assays performed on five or four animals (*E* embryo; *N* newborn; *Y* young adult; *M* middle-aged adult; *O* old rats)

Dramatic changes in polyP metabolism were also found during gemmule germination of the freshwater sponge *Ephydatia muelleri*. Gemmules are dormant forms playing a role in survival of the sponge under adverse conditions. *Ephydatia muelleri* contains relatively high amounts of polyP (about 55 μg/g wet mass), particularly in its gemmules (260 μg/g; Imsiecke et al. 1996). During gemmule germination a rapid rise in exopolyphosphatase activity and a strong decrease (by 94% at day 2) in polyP level occur (Imsiecke et al. 1996). Germination does not require nutrients as an exogenous energy source, suggesting that this process occurs by consumption of material already present in the gemmules. The long-chain polyP present in gemmule might serve as a phosphate reservoir for the synthesis of DNA and RNA occurring during germination. On the other hand, induction of gemmulation by theophylline causes an increase in the amount of long-chain polyP (Imsiecke et al. 1996).

Degradation of long polyP chains to shorter polymers has also been observed during sporulation of *Physarum polycephalum* (Pilatus et al. 1989). Since hydrolysis of polyP even occurred in the presence of a large pool of phosphate as demonstrated by ^{31}P NMR spectroscopy, the authors concluded that the polyP is utilized as an energy source rather than a phosphate reserve. On the other hand, synthesis of high-molecular polyP has been observed during the germination of fungal spores of *Aspergillus niger* (Kulaev and Belozersky 1957). Thereby, the formation of highly polymerized polyP may occur by conversion of shorter polyP molecules into longer polymers.

7
Changes in PolyP Metabolism During Apoptosis

PolyP displays characteristic changes in chain length in the course of programmed cell death (apoptosis). Induction of apoptosis in human leukemic
HL60 cells by actinomycin D causes a degradation of long-chain polyP to
shorter polyP molecules (Lorenz et al. 1997b). Both proliferating (Fig. 7,
lanes a and b) and non-proliferating HL60 cells (lanes c and d) contain two
distinct classes of polyP: a higher-molecular (long-chain) polyP class of

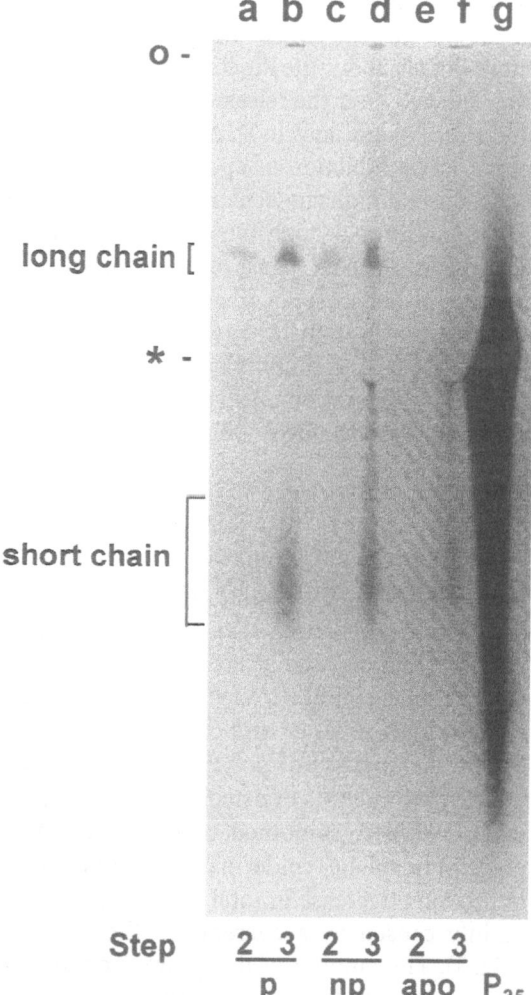

Fig. 7. Size of polyP extracted from non-apoptotic and apoptotic HL60 cells. PolyP chain lengths were determined in both *step 2* (soluble long-chain polyP) and *step 3* extracts (insoluble long-chain polyP) from proliferating (*p*) and non-proliferating (*np*) HL60 cells and from apoptotic (*apo*) HL60 cells, 19 h following induction of apoptosis by actinomycin D. Electrophoresis was performed on a 7-M urea/16.5 % polyacrylamide gel. The gel was stained with toluidine blue. *Lane a* step 2 extract from proliferating cells; *lane b* step 3 extract from proliferating cells; *lane c* step 2 extract from non-proliferating cells; *lane d* step 3 extract from non-proliferating cells; *lane e* step 2 extract from apoptotic cells; *lane f* step 3 extract from apoptotic cells; *lane g* commercial polyP$_{35}$ which actually consists of a heterogeneous population of polyP chains ranging from 3 to about 180 residues with mean size 34. Migration positions of short-chain and long-chain polyP are indicated. *o* Orgin; * aberrations due to a salt region

around 150 residues and lower-molecular (short-chain) polyP classes of 25–45 residues. In apoptotic cells (19 h after induction of apoptosis) the long-chain polyP class disappeared (Fig. 7, lanes e and f). Changes in polyP chain length were observed as soon as 2 h following induction of apoptosis, simultaneous with the onset of DNA fragmentation. This was shown by metabolic labeling of the polyP in HL60 cells with $^{32}P_i$ (Lorenz et al. 1997b). The incorporation of radioactivity into different polyP classes occurred in a time-dependent manner, whereby labeling of higher-molecular polyP (around 150 residues) required longer incubation periods (>1 h). Induction of apoptosis by actinomycin D in metabolically labeled (incubation period: 2 h) HL60 cells resulted in the appearance of a ^{32}P-labeled low-molecular polyP fraction (around 20 residues) not visible in untreated cells. This fraction is assumed to be formed by degradation of the longer polymers.

Induction of apoptosis in synchronized HL60 cells allowed us to observe more details about the kinetics of polyP changes during apoptosis. Apoptosis was induced after the release of the cells from the second thymidine block by addition of actinomycin D. Analysis of the time course of the degradation of polyP after induction of apoptosis in urea/polyacrylamide gels revealed, in addition to a high-molecular polyP fraction (~150 residues), the occurrence of an additional high-molecular, but somewhat shorter polyP fraction already in an early phase of apoptosis (3 h following induction), simultaneous with the onset of DNA fragmentation (Lorenz et al., unpubl. results).

These findings, together with results showing that polyP binds to histones (see Sect. 5.5) and inhibits nuclease activities (Rodriguez 1993), suggest that nuclear polyP may be involved in the process of apoptosis either by affecting the integrity of the DNA–histone complex or by regulating nuclease activity.

8
PolyP in Human Bone Formation

Comparably high amounts of polyP are present in human bone-forming osteoblasts (Leyhausen et al. 1998). As summarized in Table 1, the amount of polyP in human osteoblasts is markedly higher than those in human gingival fibroblasts, PBMC, erythrocytes, and blood plasma. Osteoblasts also contain high levels of exopolyphosphatase activity (Leyhausen et al. 1998).

The metabolism of polyP in human osteoblasts was found to be modulated by stimulators of osteoblast proliferation or differentiation (Leyhausen et al. 1998). A combined treatment of the cells with dexamethasone, β-glycerophosphate, epidermal growth factor (EGF), and ascorbic acid results in a strong decrease in total polyP content; this decrease in polyP content is mainly caused by a decrease in the amount of soluble, long-chain polyP (Fig. 8). The amount of this polyP fraction but not the amount of insoluble long-chain polyP further decreases after additional treatment of the cells with 1α,25-dihydroxyvitamin D_3 [1,25(OH)$_2$D$_3$] (Fig. 8). The decrease in

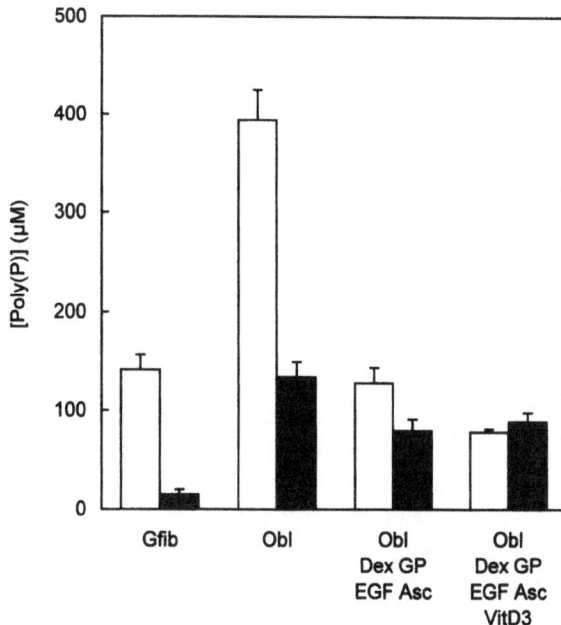

Fig. 8. Changes in polyP content of human osteoblasts after treatment with stimulators of osteoblast proliferation and differentiation. Amount of soluble long-chain polyP (*open bars*) and insoluble long-chain polyP (*solid bars*) was determined in unstimulated osteoblasts (*Obl*) and in osteoblasts subjected to combined treatment with dexamethasone (*Dex*), β-glycerophosphate (*GP*), epidermal growth factor (*EGF*), and ascorbic acid (*Asc*) for 6 days, or to combined treatment with the same additives for 3 days in the absence and then for 3 days in the presence of 1,25(OH)$_2$D$_3$ (*VitD3*). For comparison, the polyP content of gingival fibroblasts (*Gfib*) is shown. Results are means ± SD of three independent experiments

polyP content during treatment with dexamethasone, β-glycerophosphate, EGF, and ascorbic acid is accompanied by a decrease in exopolyphosphatase activity. However, additional treatment with 1,25(OH)$_2$D$_3$ results in an increase in the enzyme activity (Leyhausen et al. 1998). 1,25(OH)$_2$D$_3$ is known to increase the expression of genes which are associated with the mineralization process such as osteocalcin in mature osteoblasts, resulting in an enhanced mineralization.

Similar effects of dexamethasone and 1,25(OH)$_2$D$_3$ were observed for osteoblast inorganic pyrophosphatase and alkaline phosphatase activity (Leyhausen et al. 1998) and ecto-nucleoside triphosphate pyrophosphatase activity (Oyajobi et al. 1994). The latter enzyme, which is located on the cell surface, is assumed to serve in the generation of extracellular PP$_i$ in bone (Caswell and Russell 1988).

It is reasonable to assume that polyP may be involved in modulation of the mineralization process in bone tissue, in addition to inorganic PP$_i$. Extracellular PP$_i$ is a calcification inhibitor; it is assumed that such inhibitors must be removed from the site of mineralization before calcification can occur (Fleisch and Neuman 1961). PP_i can be hydrolyzed by alkaline phosphatase and inorganic pyrophosphatase activities. These enzymes can locally hydrolyze PP_i in bone, thus allowing local formation of hydroxyapatite crystals. The increase in both enzyme activities induced by $1,25(OH)_2D_3$ may result in an increased production of P_i and a decrease in PP_i concentration, thus modulating mineralization during bone formation. It has been shown that long-chain polyP inhibits hydroxyapatite precipitation more strongly than PP_i (Fleisch and Neuman 1961). Therefore, polyP might act as a further regulator of calcification and decalcification in bone tissue. The presence of an exopolyphosphatase, which may locally destroy polyP, might be necessary for bone tissue to mineralize, in addition to PP_i-degrading pyrophosphatase and alkaline phosphatase activities (Leyhausen et al. 1998).

The osteoblast exopolyphosphatase was found to be inhibited by bisphosphonates used in the therapy of bone diseases (Leyhausen et al. 1998). Bisphosphonates are synthetic analogs of PP_i containing a non-hydrolyzable P-C-P bond instead of P-O-P bond (Fleisch 1991; see Chap. 10). They are effective inhibitors of bone resorption. Bisphosphonates are clinically used in the treatment of disorders characterized by excessive bone loss, including Paget's disease, tumor-induced hypercalcemia, metastatic bone disease, and osteoporosis (Fleisch 1991; Singer and Minoofar 1995). These compounds strongly bind to hydroxyapatite crystals and inhibit, like PP_i and polyP, both precipitation and dissolution of calcium phosphate (Fleisch et al. 1969; Francis 1969; Russell et al. 1970). The exopolyphosphatase activity in human osteoblasts is inhibited by the bisphosphonates etidronate and, to a lesser extent, clodronate and pamidronate; methylene-1,1-bisphosphonate has only a very weak effect (Leyhausen et al. 1998). Etidronate inhibits the degradation of both longer and shorter polyP chains by the exopolyphosphatase when analyzed by electrophoresis on urea/polyacrylamide gels (Fig. 9). These results indicate that polyP metabolism in bone tissue may be a further target for these drugs.

9
PolyP and Disease

PolyP is also present in human pathogens, e.g., *Neisseria meningitidis* and *N. gonorrhoeae* (Noegel and Gotschlich 1983; Tinsley et al. 1993). These organisms incorporate P_i into long-chain polyP on the exterior of the cells, suggesting a function as a protective, capsule-like coating. The polyphosphate kinase from *N. meningitidis* has been purified (Tinsley et al. 1993) and its gene has been cloned and sequenced (Tinsley and Gotschlich 1995); highly homologous sequences can be found in a number of other pathogenic bacteria (see Chap. 1). PolyP has also been identified in *Mycobacterium tuberculosis* H37Ra (Winder and Denneny 1957) and in *Helicobacter pylori*, a risk factor for the development of gastric cancer (Bode et al. 1993). The polyP granules detected in *H. pylori* may serve as energy and phosphorus stores. In

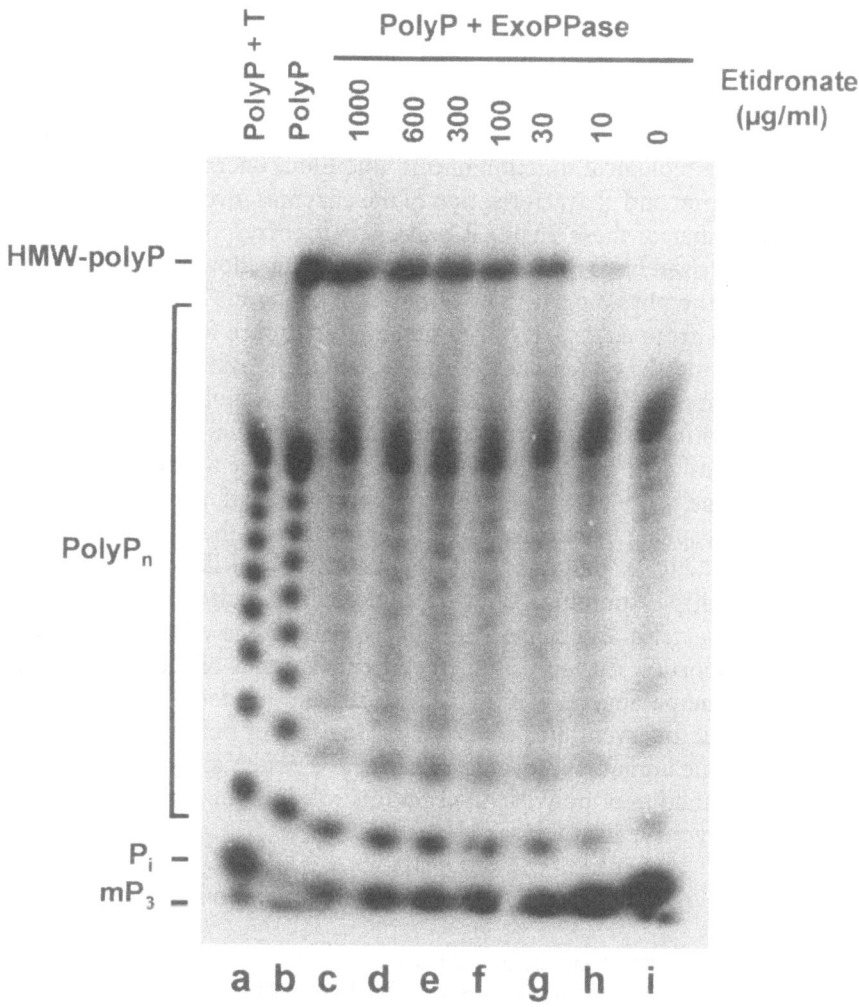

Fig. 9. Inhibition of exopolyphosphatase-mediated hydrolysis of polyP by etidronate. The [^{32}P]polyP preparation (*lane b*) was incubated with enzyme for 10 min in the presence of 1000 (*lane c*), 600 (*lane d*), 300 (*lane e*), 100 (*lane f*), 30 (*lane g*), 10 (*lane h*), or 0 (*lane i*) μg/ml etidronate. Chain lengths of the degradation products of [^{32}P]polyP were analyzed by gel electrophoresis on a urea/polyacrylamide gel. *Lane a* thermal degradation of [^{32}P]polyP (10 mM HCl, 70 °C, 3 min). Migration positions of high molecular weight (*HMW*, approximate size, 35–40 residues) polyP, medium-sized polyP$_n$, P$_i$, and trimetaphosphate (*mP$_3$*) are indicated

Propionibacterium acnes polyP is partly located outside the cell membrane and responds to UV irradiation (Kjeldstad et al. 1991). Intralysosomal polyphosphatase activities in mammalian phagocytic cells may play an important role in the defense against polyP-containing bacteria and yeast taken up by these cells via phagocytosis (Pisoni and Lindley 1992).

10
Concluding Remarks

In the last decade, inorganic polyP, which had been neglected for a long time, has become a rapidly expanding area of research. The basis for the elucidation of the biological function of this ubiquitous energy-rich polymer is the identification and characterization of the enzymes involved in its metabolism. A number of these enzymes and of their corresponding genes have been characterized from prokaryotic organisms. This allowed us to gain first insights into the physiological role of polyP in bacteria. Only very recently, have enzymes from eukaroytes been obtained in a pure form and only one of these enzymes has been cloned and sequenced (yeast exopolyphosphatase). It appears that the enzymatic basis of polyP metabolism in eukaryotes is somewhat different from prokaryotes. Yeast and animal cells seem to lack an enzyme similar to polyphosphate kinase responsible for polyP synthesis in bacteria, and some enzymes involved in polyP breakdown in prokaryotes (e.g., polyphosphate glucokinase) are not found in these cells. A more detailed knowledge of polyP metabolism in eukaryotes will help to prove the proposed multiple functions of these molecules, which are likely not all identical in bacteria and higher organisms. Knowledge of these functions might be important not only from the scientific point of view but also to understand the pathogenesis of some diseases and to identify novel targets for therapeutic intervention. In particular, the role of polyP, present in the capsule of some human pathogenic bacteria, as a potential factor of infectivity, and the function of polyP in bone and its role in bone diseases might be important topics for future research.

References

Ahn K, Kornberg A (1990) Polyphosphate kinase from *Escherichia coli*. Purification and demonstration of a phosphoenzyme intermediate. J Biol Chem 265: 11734–11739

Akiyama M, Crooke E, Kornberg A (1992) The polyphosphate kinase gene of *Escherichia coli*. Isolation and sequence of the *ppk* gene and membrane location of the protein. J Biol Chem 267: 22556–22561

Akiyama M, Crooke E, Kornberg A (1993) An exopolyphosphatase of *Escherichia coli*. The enzyme and its *ppx* gene in a polyphosphate operon. J Biol Chem 268: 633–639

Andreeva NA, Okorokov LA (1990) Some properties of highly purified polyphosphatase of the yeast *Saccharomyces carlsbergensis* cell envelope. Biokhimiya 55: 2286–2293 (in Russian)

Andreeva NA, Okorokov LA (1993) Purification and characterization of highly active and stable polyphosphatase from *Saccharomyces cerevisiae* cell envelope. Yeast 9: 127–139

Andreeva NA, Okorokov LA, Kulaev IS (1990) Purification and certain properties of cell envelope polyphosphatase of the yeast *Saccharomyces carlsbergensis*. Biochemistry (Moscow) 55: 819–826

Andreeva NA, Kulakovskaya TV, Kulaev IS (1996) Purification and characterization of polyphosphatase from *Saccharomyces cerevisiae* cytosol. Biochemistry (Moscow) 61: 1213–1220

Andreeva NA, Kulakovskaya TV, Kulaev IS (1998) Purification and properties of exopolyphosphatase isolated from *Saccharomyces cerevisiae* vacuoles. FEBS Lett 429: 194–196

Ansevin AT, MacDonald KK, Smith CE, Hnilica LS (1975) Mechanis of chromatin template activation. Physical evidence for destabilization of nucleoproteins by polyanions. J Biol Chem 250: 281–289

Ault-Riche D, Fraley CD, Tzeng CM, Kornberg A (1998) Novel assay reveals multiple pathways regulating stress-induced accumulations of inorganic polyphosphate in *Escherichia coli*. J Bacteriol 180: 1841–1847

Bental M, Pick U, Avron M, Degani H (1990) Metabolic studies with NMR spectroscopy of a alga *Dunaliella salina* trapped with agarose beads. Eur J Biochem 188: 111–116

Bental M, Pick U, Avron M, Degani H (1991) Polyphosphate metabolism in the alga *Dunaliella salina* studied by ^{31}P-NMR. Biochim Biophys Acta 1092: 21–28

Blum E, Py B, Carpousis AJ, Higgins CF (1997) Polyphosphate kinase is a component of the *Escherichia coli* RNA degradosome. Mol Microbiol 26: 387–398

Blum JJ (1989) Changes in orthophosphate, pyrophosphate and long chain polyphosphate levels in *Leishmania major* promastigotes incubated with and without glucose. J Protozool 36: 254–257

Bode G, Mauch F, Ditschuneit H, Malfertheiner P (1993) Identification of structures containing polyphosphate in *Helicobacter pylori*. J Gen Microbiol 139: 3029–3033

Bonting CF, Kortstee GJ, Zehnder AJ (1991) Properties of polyphosphate:AMP phosphotransferase of *Acinetobacter* strain 210A. J Bacteriol 173: 6484–6488

Bonting CF, Kortstee GJ, Zehnder AJ (1993) Properties of polyphosphatase of *Acinetobacter johnsonii* 210A. Antonie Van Leeuwenhoek 64: 75–81

Booth JW, Guidotti G (1995) An alleged yeast polyphosphate kinase is actually diadenosine-5',5'''-P^1,P^4-tetraphosphate α,β-phosporylase. J Biol Chem 270: 19377–19382

Castro CD, Meehan AJ, Koretsky AP, Domach MM (1995) In situ ^{31}P nuclear magnetic resonance for observation of polyphosphate and catabolite responses of chemostat-cultivated *Saccharomyces cerevisiae* after alkalinization. Appl Environ Microbiol 61: 4448–4453

Castuma CE, Huang R, Kornberg A, Reusch RN (1995) Inorganic polyphosphates in the acquisition of competence in *Escherichia coli*. J Biol Chem 270: 12980–12983

Caswell AM, Russell RG (1988) Evidence that ecto-nucleoside-triphosphate pyrophosphatase serves in the generation of extracellular inorganic pyrophosphate in human bone and articular cartilage. Biochim Biophys Acta 966: 310–317

Clark JE, Wood HG (1987) Preparation of standards and determination of sizes of long-chain polyphosphates by gel electrophoresis. Anal Biochem 161: 280–290

Cowling RT, Birnboim HC (1994) Incorporation of [^{32}P]orthophosphate into inorganic polyphosphates by human granulocytes and other human cell types. J Biol Chem 269: 9480–9485

Cramer CL, Davis RH (1984) Polyphosphate-cation interaction in the amino-acid-containing vacuole of *Neurospora crassa*. J Biol Chem 259: 5152–5157

Cramer CL, Vaughn LE, Davis RH (1980) Basic amino acids and inorganic polyphosphates in *Neurospora crassa*: independent regulation of vacuolar pools. J Bacteriol 142: 945–952

Das S, Lengweiler UD, Seebach D, Reusch RN (1997) Proof for a nonproteinaceous calcium-selective channel in *Escherichia coli* by total synthesis from (R)-3-hydroxybutanoic acid and inorganic polyphosphate. Proc Natl Acad Sci USA 94: 9075–9079

Den Hollander JA, Ugurbil K, Shulman RG (1986) ^{31}P and ^{13}C NMR studies of intermediates of aerobic and anaerobic glycolysis in *Saccharomyces cerevisiae*. Biochemistry 25: 212–219

Dirheimer G, Ebel JP (1965) Caractérisation d'une polyphosphate-AMP-phosphotransférase dans *Corynebacterium cerosis*. C R Acad Sci Paris 260: 3787–3790

Ducastaing A, Azanza JL, Raymond J, Robin JM, Creac'h O (1976) Cathepsin D from horse spleen. II. Study of certain enzymatic properties. Biochimie 58: 783–791

Dunn T, Gable K, Beeler T (1994) Regulation of cellular Ca^{2+} by yeast vacuoles. J Biol Chem 269: 7273–7278

Felter S, Stahl AJC (1973) Enzymes du métabolisme des polyphosphates dans la levure. III. Purification et propriétés de la polyphosphate-ADP-phosphotransférase. Biochimie 55: 245–251

Felter MS, Stahl A (1975) The polyphosphate synthetase of *Saccharomyces cerevisiae*. C R Acad Sci [D] 280: 1903–1906

Fleisch H (1991) Bisphosphonates: pharmacology and use in the treatment of tumour-induced hypercalcaemia and metastatic bone disease. Drugs 42: 919–944

Fleisch H, Neuman WF (1961) Mechanisms of calcification: role of collagen, polyphosphates, and phosphatase. Am J Physiol 200: 1296–1300

Fleisch H, Russell RGG, Francis MD (1969) Diphosphonates inhibit hydroxyapatite dissolution in vitro and bone resorption in tissue culture and in vivo. Science 165: 1262–1264

Florin I, Thelestam M (1984) Polyphosphate-mediated protection from cellular intoxication with *Clostridium difficile* toxin B. Biochim Biophys Acta 805: 131–136

Francis MD (1969) The inhibition of calcium hydroxyapatite crystal growth by polyphosphonates and polyphosphates. Calcif Tissue Res 3: 151–162

Gabel NW, Thomas V (1971) Evidence for the occurrence and distribution of inorganic polyphosphates in vertebrate tissues. J Neurochem 18: 1229–1242

Gabel NW, Thomas V (1976) Inorganic polyphosphate as an integral part of alkaline phosphatase preparations. Bioinorg Chem 5: 189–197

Gillies RJ, Ugurbil K, den Hollander JA, Shulman RG (1981) ^{31}P NMR studies of intracellular pH and phosphate metabolism during cell division cycle of *Saccharomyces cerevisiae*. Proc Natl Acad Sci USA 78: 2125–2129

Griffin JB, Davidian NM, Penniall R (1965) Studies of phosphorus metabolism by isolated nuclei. VII. Identification of polyphosphate as a product. J Biol Chem 240: 4427–4434

Grossmann D, Lang K (1962) Anorganische Poly- and Metaphosphatasen sowie Polyphosphate im tierischen Zellkern. Biochem Z 336: 351–370

Holahan PK, Knizner SA, Gabriel CM, Swenberg CE (1988) Alterations in phosphate metabolism during cellular recovery of radiation damage in yeast. Int J Radiat Biol 54: 545–562

Hostalek Z, Tobek I, Bobyk MA, Kulaev IS (1976) Role of ATP-glucokinase and polyphosphate glucokinase in *Streptomyces aureofaciens*. Folia Microbiol (Praha) 21: 131–138

Hsieh, PC, Shenoy BC, Haase FC, Jentoft JE, Phillips NF (1993a) Involvement of tryptophan(s) at the active site of polyphosphate/ATP glucokinase from *Mycobacterium tuberculosis*. Biochemistry 32: 6243–6249

Hsieh PC, Shenoy BC, Jentoft JE, Phillips NF (1993b) Purification of polyphosphate and ATP glucose phosphotransferase from *Mycobacterium tuberculosis* H37Ra: evidence that poly(P) and ATP glucokinase activities are catalyzed by the same enzyme. Protein Expr Purif 4: 76–84

Hsieh PC, Shenoy BC, Samols D, Phillips NF (1996) Cloning, expression, and characterization of polyphosphate glucokinase from *Mycobacterium tuberculosis*. J Biol Chem 271: 4909–4915

Huang R, Reusch RN (1995) Genetic competence in *Escherichia coli* requires poly-β-hydroxybutyrate/calcium polyphosphate membrane complexes and certain divalent cations. J Bacteriol 177: 486–490

Imsiecke G, Münkner J, Lorenz B, Bachinski N, Müller WEG, Schröder HC (1996) Inorganic polyphosphates in the developing freshwater sponge *Ephydatia muelleri*: effect of stress by polluted waters. Environ Toxicol Chem 15: 1329–1334

Kato J, Yamamoto T, Yamada K, Ohtake H (1993) Cloning, sequence and characterization of the polyphosphate kinase-encoding gene (ppk) of *Klebsiella aerogenes*. Gene 137: 237–242

Keasling JD (1997) Regulation of intracellular toxic metals and other cations by hydrolysis of polyphosphate. Ann N Y Acad Sci 829: 242–249

Keasling JD, Hupf GA (1996) Genetic manipulation of polyphosphate metabolism affects cadmium tolerance in *Escherichia coli*. Appl Environ Microbiol 62: 743–746

Keasling JD, Bertsch L, Kornberg A (1993) Guanosine pentaphosphate phosphohydrolase of *Escherichia coli* is a long-chain exopolyphosphatase. Proc Natl Acad Sci USA 90: 7029–7033

Kjeldstadt B, Heldal M, Nissen H, Bergan AS, Evjen K (1991) Changes in polyphosphate composition and localization in *Propionibacterium acnes* after near-ultraviolet irradiation. Can J Microbiol 37: 562–567

Kornberg A (1994) Inorganic polyphosphate: a molecular fossil come to life. In: Torriani-Gorini AM, Yagil E, Silver S (eds) Phosphate in microorganisms: cellular and molecular biology. American Society for Microbiology, Washington, DC, pp 204–209

Kornberg A (1995) Inorganic polyphosphate: toward making a forgotten polymer unforgettable. J Bacteriol 177: 491–496

Kornberg A, Kornberg SR, Simms ES (1956) Metaphosphate synthesis by an enzyme from *Escherichia coli*. Biochim Biophys Acta 20: 215–227

Kornberg SR (1957) Adenosine triphosphate synthesis from polyphosphate by an enzyme from *Escherichia coli*. Biochim Biophys Acta 26: 294–300

Kowalczyk TH, Phillips NF (1993) Determination of endopolyphosphatase using polyphosphate glucokinase. Anal Biochem 212: 194–205

Kowalczyk TH, Horn PJ, Pan WH, Phillips NF (1996) Initial rate and equilibrium isotope exchange studies on the ATP-dependent activity of polyphosphate glucokinase from *Propionibacterium shermanii*. Biochemistry 35: 6777–6785

Kritskii MS, Chernysheva EK, Kulaev IS (1972) Polyphosphate depolymerase activity in cells of the fungus *Neurospora crassa*. Biokhimiya 37: 983–990 (in Russian)

Kulaev IS (1979) The biochemistry of inorganic polyphosphates. New York

Kulaev IS, Belozersky AN (1957) An investigation using ^{32}P of the physiological role of polyphosphates in the development of *Aspergillus niger*. Biokhimiya 22: 587–597 (in Russian)

Kulaev IS, Bobyk MA (1971) The detection of a new enzyme, 1,3-phosphoglyceratepolyphosphate phosphotransferase, in *Neurospora crassa*. Biokhimiya 36: 426–429 (in Russian)

Kulaev IS, Konoshenko G (1971) The localization of 1,3-diphosphoglycerate-polyphosphate phosphotransferase in cells of *Neurospora crassa*. Dokl Akad Nauk SSSR 200: 477–480 (in Russian)

Kulaev IS, Vagabov VM (1983) Polyphosphate metabolism in microorganism. Adv Microb Physiol 24: 83–171

Kulaev IS, Krasheninnikov IA, Tyrsin YA (1973) Nucleic acids, high-molecular polyphosphates and some polyphosphate metabolic enzymes in a synchronized culture of *Schizosaccharomyces pombe* yeasts. Mikrobiologiya 38: 613–619 (in Russian)

Kulaev IS, Bobyk AM, Tobek I, Goshtialek Z (1976) The possible role of high molecular weight polyphosphate in chlortetracycline biosynthesis by *Streptomyces aureofaciens*. Biokhimiya 41: 343–348 (in Russian)

Kulaev IS, Andreeva NA, Lichko LP, Kulakovskaya TV (1995) Organelle specificity of polyphosphatases in yeast cells. Biochemistry (Moscow) 60: 1061–1065

Kulaev IS, Andreeva NA, Lichlo LP, Kulakovskaya TV (1997) Comparison of exopolyphosphatases of different yeast cell. Microbiol Res 152: 221–226

Kulakovskaya TV, Andreeva NA, Lichko LP, Kulaev IS (1995) Immunoassay of polyphosphatases from different compartments of *Saccharomyces cerevisiae* yeast cells. Biochemistry (Moscow) 60: 1559–1561

Kulakovskaya TV, Andreeva NA, Kulaev IS (1997) Adenosine-5'-tetraphosphate and guanosine-5'-tetraphosphate: new substrates of the cytosolic exopolyphosphatase of the yeast *Saccharomyces cerevisiae*. Biochemistry (Moscow) 62: 1051–1052

Kumble KD, Kornberg A (1995) Inorganic polyphosphate in mammalian cells and tissues. J Biol Chem 270: 5818–5822

Kumble KD, Kornberg A (1996) Endopolyphosphatase for long chain inorganic polyphosphate in yeast and mammals. J Biol Chem 271: 27146–27151

Kumble KD, Ahn K, Kornberg A (1996) Phosphohistidyl active sites in polyphosphate kinase of *Escherichia coli*. Proc Natl Acad Sci USA 93: 14391–14395

Kuroda A, Kornberg A (1997) Polyphosphate kinase as a nucleoside diphosphate kinase in *Escherichia coli* and *Pseudomonas aeruginosa*. Proc Natl Acad Sci USA 94: 439–442

Kuroda A, Murphy H, Cashel M, Kornberg A (1997) Guanosine tetra- and pentaphosphate promote accumulation of inorganic polyphosphate in *Escherichia coli*. J Biol Chem 272: 21240–21243

Kusano S, Ishihama A (1997) Functional interaction of *Escherichia coli* RNA polymerase with inorganic polyphosphate. Genes Cells 2: 433–441

Langen P (1958) Über Polyphosphate in Ostsee-Algen. Acta Biol Med Ger 1: 368–372

Langen P (1961) Über Unterschiede im Gehalt an labilem Phosphat zwischen Rot- und Braunalgen. Pubbl Staz Zool Napoli 32: 130–133

Langen P, Liss E (1958) Über Bildung und Umsatz der Polyphosphate der Hefe. Biochem Z 330: 455–466

LeFurgey A, Ingram P, Blum JJ (1990) Elemental composition of polyphosphate-containing vacuoles and cytoplasm of *Leishmania major*. Mol Biochem Parasitol 40: 77–86

Leitão JM, Lorenz B, Bachinski N, Wilhelm C, Müller WEG, Schröder HC (1995) Osmotic-stress-induced synthesis and degradation of inorganic polyphosphates in the alga *Phaeodactylum tricornutum*. Mar Ecol Prog Ser 121: 279–288

Leyhausen G, Lorenz B, Zhu H, Geurtsen W, Bohnensack R, Müller WEG, Schröder HC (1998) Inorganic polyphosphate in human osteoblast-like cells. J Bone Miner Res 13: 803–812

Lichko LP, Kulakovskaya TV, Dmitriev VV, Kulaev IS (1995) *Saccharomyces cerevisiae* nuclei possess polyphosphatase activity. Biochemistry (Moscow) 60: 1465–1468

Lichko, LP, Kulakovskaya TV, Kulaev S (1996a) Characterization of the polyphosphatase activity of *Saccharomyces cerevisiae* nuclei. Biochemistry (Moscow) 61: 361–366

Lichko LP, Kulakovskaya TV, Kulaev IS (1996b) Characterization of polyphosphatase activity in isolated mitochondria of the yeast *Saccharomyces cerevisiae*. Biochemistry (Moscow) 61: 1176–1181

Lichko LP, Kulakovskaya TV, Kulaev IS (1997) Detection of some properties of membrane-bound and soluble polyphosphatases in mitochondria of the yeast *Saccharomyces cerevisiae*. Biochemistry (Moscow) 62: 1146–1151

Lopez-Revilla R, Gomez-Dominguez R (1985) Incorporation and toxicity of ^{32}P-orthophosphate and occurrence of polyphosphate in *Entamoeba trophozoites*. J Protozool 32: 353–355

Lorenz B, Marmé S, Müller WEG, Unger K, Schröder HC (1994a) Preparation and use of polyphosphate-modified zirconia for purification of nucleic acids and proteins. Anal Biochem 216: 118–126

Lorenz B, Müller WEG, Kulaev IS, Schröder HC (1994b) Purification and characterization of an exopolyphosphatase from *Saccharomyces cerevisiae*. J Biol Chem 269: 22198–22204

Lorenz B, Batel R, Bachinski N, Müller WEG, Schröder HC (1995) Purification and characterization of two exopolyphosphatases from the marine sponge *Tethya lyncurium*. Biochim Biophys Acta 1245: 17–28

Lorenz B, Leuck J, Köhl D, Müller WEG, Schröder HC (1997a) Anti-HIV-1 activity of inorganic polyphosphates. J Acquir Immune Defic Syndr Hum Retrovirol 14: 110–118

Lorenz B, Münkner J, Oliveira MP, Kuusksalu A, Leitão JM, Müller WEG, Schröder HC (1997b) Changes in metabolism of inorganic polyphosphate in rat tissues and human cells during development and apoptosis. Biochim Biophys Acta 1335: 51–60

Lorenz B, Münkner J, Oliveira MP, Leitão JM, Müller WEG, Schröder HC (1997c) A novel method for determination of inorganic polyphosphates using the fluorescent dye fura-2. Anal Biochem 246: 176–184

Mansurova SE, Shama AM, Sokolovskii VY, Kulaev IS (1975) High-molecular polyphosphates in rat liver nuclei. Their behavior in the liver regeneration process. Dokl Akad Nauk SSSR 225: 717–720

Matsuoka A, Tsutsumi M, Watanabe T (1995) Inhibitory effect of hexametaphosphate on the growth of *Staphylococcus aureus*. J Food Hyg Soc Jpn 36: 588–594

Mattenheimer H (1956a) Die Substratspezifität "anorganischer" Poly- and Metaphosphatasen. I. Optimale Wirkungsbedingungen für den enzymatischen Abbau von Poly- und Metaphosphaten. Hoppe Seylers Z Physiol Chem 303: 107–114

Mattenheimer H (1956b) Die Substratspezifität "anorganischer" Poly- und Metaphosphatasen. II. Trennung der Enzyme. Hoppe Seylers Z Physiol Chem 303: 115–124

Mattenheimer H (1956c) Die Substratspezifität "anorganischer" Poly- und Metaphosphatasen. III. Papierchromatographische Untersuchungen beim enzymatischen Abbau von anorganischen Poly- und Metaphosphaten. Hoppe Seylers Z Physiol Chem 303: 125–139

Muhammed A, Rodgers A, Hughes DE (1959) Purification and properties of a polymetaphosphatase from *Corynebacterium xerosis*. J Gen Microbiol 20: 482–495

Murata K, Kato J, Chibata I (1979) Continuous production of NADP by immobilized *Brevibacterium ammoniagenes* cells. Biotechnol Bioeng 21: 887–895

Noegel A, Gotschlich EC (1983) Isolation of a high molecular weight polyphosphate from *Neisseria gonorrhoeae*. J Exp Med 157: 2049–2060

Offenbacher S, Kline ES (1984) Evidence for polyphosphate in phosphorylated nonhistone nuclear proteins. Arch Biochem Biophys 231: 114–123

Ohsumi Y, Kitamoto K, Anraku Y (1988) Changes induced in the permeability barrier of the yeast plasma membrane by cupric ion. J Bacteriol 170: 2676–2682

Oyajobi BO, Russell RG, Caswell AM (1994) Modulation of ecto-nucleoside triphosphate pyrophosphatase activity of human osteoblast-like bone cells by $1\alpha,25$-dihydroxyvitamin D3, 24R,25-dihydroxyvitamin D3, parathyroid hormone, and dexamethasone. J Bone Miner Res 9: 1259–1266

Palumbo SA, Call JE, Cooke PH, Williams AC (1995) Effect of polyphosphates and NaCl on *Aeromonas hydrophila* K144. J Food Safety 15: 77–87

Penniall R, Griffin JB (1984) Studies of phosphorus metabolism by isolated nuclei. XII. Some fundamental properties of the incorporation of $^{32}P_i$ into polyphosphate by rat liver nuclei. Biosci Rep 4: 957–962

Pepin CA, Wood HG (1986) Polyphosphate glucokinase from *Propionibacterium shermanii*. Kinetics and demonstration that the mechanism involves both processive and nonprocessive type reactions. J Biol Chem 261: 4476–4480

Pepin CA, Wood HG (1987) The mechanism of utilization of polyphosphate by polyphosphate glucokinase from *Propionibacterium shermanii*. J Biol Chem 262: 5223–5226

Phillips NF, Horn PJ, Wood HG (1993) The polyphosphate- and ATP-dependent glucokinase from *Propionibacterium shermanii*: both activities are catalyzed by the same protein. Arch Biochem Biophys 300: 309–319

Pick U, Bental M, Chitlaru E, Weiss M (1990) Polyphosphate-hydrolysis – a protective mechanism against alkaline stress? FEBS Lett 274: 15–18

Pilatus U, Mayer A, Hildebrandt A (1989) Nuclear polyphosphate as a possible source of energy during the sporulation of *Physarum polycephalum*. Arch Biochem Biophys 275: 215–223

Pisoni RL, Lindley ER (1992) Incorporation of [^{32}P]orthophosphate into long chains of inorganic polyphosphate within lysosomes of human fibroblasts. J Biol Chem 267: 3626–3631

Rao NN, Kornberg A (1996) Inorganic polyphosphate supports resistance and survival of stationary-phase *Escherichia coli*. J Bacteriol 178: 1394–1400

Rao NN, Roberts MF, Torriani A (1985) Amount and chain length of polyphosphates from *Escherichia coli* depend on cell growth conditions. J Bacteriol 162: 242–247

Rao NN, Liu S, Kornberg A (1998) Inorganic polyphosphate in *Escherichia coli*: the phosphate regulon and the stringent response. J Bacteriol 180: 2186–2193

Regen DM, Davis WW, Morgan HE, Park CR (1964) The regulation of hexokinase and phosphofructokinase activity in heart muscle. J Biol Chem 239: 43–49

Reusch RN (1989) Poly-β-hydroxybutyrate/calcium polyphosphate complexes in eukaryotic membranes. Proc Soc Exp Biol Med 191: 377–381

Reusch RN, Sadoff HL (1988) Putative structure and functions of a poly-β-hydroxybutyrate/calcium polyphosphate channel in bacterial plasma membranes. Proc Natl Acad Sci USA 85: 4176–4180

Reusch RN, Huang R, Bramble KK (1995) Poly-3-hydroxybutyrate/polyphosphate complexes form voltage-activated Ca^{2+} channels in the plasma membranes of *Escherichia coli*. Biophys J 69: 754–766

Reusch RN, Huang R, Kosk-Kosicka D (1997) Novel components and enzymatic activities of the human erythrocyte plasma membrane calcium pump. FEBS Lett 412: 592–596

Robinson NA, Wood HG (1986) Polyphosphate kinase from *Propionibacterium shermanii*. Demonstration that the synthesis and utilization of polyphosphate is by a processive mechanism. J Biol Chem 261: 4481–4485

Robinson NA, Clark JE, Wood HG (1987) Polyphosphate kinase from *Propionibacterium shermanii*. Demonstration that polyphosphates are primers and determination of the size of the synthesized polyphosphate. J Biol Chem 262: 5216–5222

Rodriguez RJ (1993) Polyphosphate present in DNA preparations from filamentous fungal species of *Colletotrichum* inhibits restriction endonucleases and other enzymes. Anal Biochem 209: 291–297

Russell RGG, Muhlbauer RC, Bisaz S, Williams DA, Fleisch H (1970) The influence of pyrophosphate, condensed phosphates, phosphonates and other phosphate compounds on the dissolution of hydroxyapatite in vitro and on bone resorption induced by parathyroid hormone in tissue culture and in thyroparathyroidectomised rats. Calcif Tissue Res 6: 183–196

Schneider DL (1983) ATP-dependent acidification of membrane vesicles isolated from purified rat liver lysosomes. Acidification activity requires phosphate. J Biol Chem 258: 1833–1838

Schuddemat J, de Boo R, van Leeuwen CCM, van den Broek PJA, van Steveninck J (1989) Polyphosphate synthesis in yeast. Biochim Biophys Acta 1010: 191–198

Shabalin YA, Kulaev IS (1989) Solubilization and properties of yeast dolichylpyrophosphate:polyphosphate phosphotransferase. Biokhimiya 54: 68–75 (in Russian)

Shiba T, Tsutsumi K, Yano H, Ihara Y, Kameda A, Tanaka K, Takahashi H, Munekata M, Rao NN, Kornberg A (1997) Inorganic polyphosphate and the induction of rpoS expression. Proc Natl Acad Sci USA 94: 11210–11215

Shirahama K, Yazaki Y, Sakano K, Wada Y, Ohsumi Y (1996) Vacuolar function in the phosphate homeostasis of the yeast *Saccharomyces cerevisiae*. Plant Cell Physiol 37: 1090–1093

Singer FR, Minoofar PN (1995) Bisphosphonates in the treatmet of disorders of mineral metabolism. Adv Endocrinol Metab 6: 259–288

Skorko R (1989) Polyphosphate as a source of phosphoryl group in protein modification in the archaebacterium *Sulfolobus acidocaldarius*. Biochimie 71: 1089–1093

Skorko R, Osipiuk J, Stetter KO (1989) Glycogen-bound polyphosphate kinase from the archaebacterium *Sulfolobus acidocaldarius*. J Bacteriol 171: 5162–5164

Speziali GA, Van Wijk R (1971) Cyclic 3',5'-AMP phosphodiesterase of *Saccharomyces carlsbergensis*. Inhibition by adenosine 5'-triphosphate, inorganic-pyrophosphate and inorganic polyphosphate. Biochim Biophys Acta 235: 466–472

Stros M, Skalka M, Matyasova J, Cejkova M (1984) On the interaction of histones with polyanions. Gen Physiol Biophys 3: 307–316

Sultana GN, Watanabe Y, Tamai Y (1995) Effects of inorganic phosphorus compounds on the hydrolysis of phosphatidylcholine liposomes by phospholipid-deacylating enzymes. Biotechnol Appl Biochem 21 (1): 101–110

Syzmona M, Ostrowski W (1964) Inorganic polyphosphate glucokinase of *Mycobacterium phlei*. Biochim Biophys Acta 85: 283–295

Szymona M, Kowalska H, Pastuszak I (1977) Polyphosphate-glucose phosphotransferase. Purification of *Mycobacterium tuberculosis* H37Ra enzyme to apparent homogeneity. Acta Biochim Pol 24: 133–142

Szymona O, Szymona M (1979) Polyphosphate- and ATP-glucose phosphotransferase activities of *Nocardia minima*. Acta Microbiol Pol 28: 153–160

Telesnina GN, Krakhmaleva IN, Anisova LN, Bartoshevich IE, Sazykin IO (1986) Valinomycin biosynthesis and the dynamics of the content of macroergic phosphorus compounds in *Streptomyces cyaneofuscatus*. Antibiot Med Biotekhnol 31: 3–7 (in Russian)

Tijssen JP, Dubbelman TM, Van Steveninck J (1983) Isolation and characterization of polyphosphates from the yeast cell surface. Biochim Biophys Acta 760: 143–148

Tinsley CR, Gotschlich EC (1995) Cloning and characterization of the meningococcal polyphosphate kinase gene: production of polyphosphate synthesis mutants. Infect Immun 63: 1624–1630

Tinsley CR, Manjula BN, Gotschlich EC (1993) Purification and characterization of polyphosphate kinase from *Neisseria meningitidis*. Infect Immun 61: 3703–3710

Umnov AM, Umnova NS, Kulaev IS (1975) Isolation and properties of polyphosphatase of *Neurospora crassa*. Mol Biol (Moscow) 9: 594–601

Urech K, Dürr M, Boller T, Wiemken A, Schwencke J (1978) Localization of polyphosphate in vacuoles of *Saccharomyces cerevisiae*. Arch Microbiol 116: 275–278

Vagabov VM, Trilisenko LV, Shchipanova IN, Sibeldina LA, Kulaev IS (1988) Changes in inorganic polyphosphate length during the growth of *Saccharomyces cerevisiae*. Microbiology 67: 153–157

Vorisek J, Knotkova A, Kotyk A (1982) Fine cytochemical localization of polyphosphates in the yeast *Saccharomyces cerevisiae*. Zentralbl Mikrobiol 137: 421–432

Watabe S, Terada A, Ikeda T, Kouyama H, Taguchi S, Yago N (1979) Polyphosphate anions increase the activity of bovine spleen cathepsin D. Biochem Biophys Res Commun 89: 1161–1167

Weinstein BI, Li H-C (1976) Stimulation of chromatin template activity by the physiological macromolecule polyphosphate: a possible mechanism for eukaryotic gene derepression. Arch Biochem Biophys 175: 114–120

Wiemken A, Schellenberg M, Urech K (1979) Vacuoles: the sole compartments of digestive enzymes in yeast (*Saccharomyces cerevisiae*)? Arch Microbiol 123: 23–35

Winder FG, Denneny JM (1957) The metabolism of inorganic polyphosphate in mycobacteria. J Gen Microbiol 17: 573–585

Wood HG, Clark JE (1988) Biological aspects of inorganic polyphosphates. Annu Rev Biochem 57: 235–260

Wurst H, Kornberg A (1994) A soluble exopolyphosphatase of *Saccharomyces cerevisiae*. J Biol Chem 269: 10996–11001

Wurst H, Shiba T, Kornberg A (1995) The gene for a major exopolyphosphatase of *Saccharomyces cerevisiae*. J Bacteriol 177: 898–906

Yang YC, Bastos M, Chen KY (1993) Effects of osmotic stress and growth stage on cellular pH and polyphosphate metabolism in *Neurospora crassa* as studied by ^{31}P nuclear magnetic resonance spectroscopy. Biochim Biophys Acta 1179: 141–147

Yoshida A (1955) Metaphosphate. II. Heat of hydrolysis of metaphosphate extracted from yeast cells. J Biochem (Tokyo) 42: 163–168

Yoshino M, Murakami K (1988) A kinetic study of the inhibition of yeast AMP deaminase by polyphosphate. Biochim Biophys Acta 954: 271–276

Cyclic Condensed Metaphosphates in Plants and the Possible Correlations Between Inorganic Polyphosphates and Other Compounds

R. Niemeyer[1]

1
Nomenclature and Structure of Inorganic Condensed Phosphates

High molecular inorganic phosphates have been known for a long time. In the last century Liebermann (1888, 1890) reported on the isolation of these compounds together with nucleic acids and proteins from yeast. Although the existence of high molecular inorganic phosphates has been shown, their presence in the cells of living micro-organisms was not accepted by biochemists until the basis works of Wiame (1947, 1949) and Ebel (1952a–d).

The name changed from metaphosphoric acid to hexametaphosphate and then to inorganic polyphosphates synonymously until the 1950s because nobody distinguished between the different types of inorganic phosphates. According to most recent knowledge it was necessary to bring order into the classification of the different inorganic phosphates. Thilo (1959), who investigated the chemical structures of inorganic phosphate compounds, suggested an new nomenclature for the condensed inorganic phosphates.

Following the works by Niemeyer (1975, 1976, 1977a,b), Niemeyer and Richter (1969, 1972), Inhülsen and Niemeyer (1975, 1978) and Niemeyer and Selle (1987, 1989) concerning the biological activity and the metabolism of the condensed inorganic poly- and metaphosphates, it was necessary to proceed with this classification, too. They convincingly demonstrated the existence of cyclic condensed metaphosphates in many living organisms.

1.1
Linear Condensed Polyphosphates

Polyphosphates in the narrow sense have the common formula $[Me + H]_{n+2} P_n O_{3n+1}$ (e.g. Me is a monovalent or divalent cation) and consist of up to 1000 inorganic phosphoric acid residues condensed to linear chains. In these polyanion chains the phosphorus atoms are linked by oxygen atoms forming an unbranched structure.

[1] Universität Hannover, Fachbereich Biologie, Institut für Botanik, Herrenhäuser Str. 2, 30419 Hannover, Germany.

Progress in Molecular and Subcellular Biology, Vol. 23
H. C. Schröder, W. E. G. Müller (Eds.)
© Springer-Verlag Berlin Heidelberg 1999

1.2
Cyclic Condensed Metaphosphates

These cyclic phosphates have the general formula $[Me + H]_n \, P_n \, O_{3n}$ and the smallest compound is the trimetaphosphate which consists of three phosphate residues. Only synthetic products were known until Niemeyer and Richter (1969) found the trimeta- and tetrametaphosphate as well as the penta- and hexametaphosphate in living organisms. Niemeyer (1975, 1976, 1977a,b) and Niemeyer and Richter (1969, 1972) separated the metaphosphates by two-dimensional thin-layer chromatography on a mixture of cellulose and microcrystalline cellulose. The labeled points of cyclic metaphosphates were identified by comparison with coloured synthetic substances thus proving the existence of naturally occurring cyclic metaphosphates.

1.3
Ultraphosphates

In contrast to the pure condensed inorganic polyphosphates the high molecular weight ultraphosphates are branched and connected via linking points with other linear chains like a network. The ratio of Me_2O to P_2O_5 ist below 1. This type of inorganic phosphate is known as Kurrol's or Graham's salt in the chemical industry. The evidence that they are part of the metabolism in living organisms is improbable and difficult to prove, because they are unusually rapidly hydrolyzed in aqueous solution. So far, this property implies that they may not play a very important role in biological systems, their existence in living systems is very doubtful altogether.

2
Extraction Methods of Condensed Inorganic Phosphates

2.1
Hydrolysis Free Extraction Method as a Prerequisite
for Metabolic Investigation

In order to study the metabolism of the condensed inorganic phosphate, it is essential to separate the high-molar mass of condensed phosphates under native conditions, not by performing fractionation using strong alkaline or acid solvents, thus preventing partial hydrolysis, preferentially of the labile high-molar mass polyphosphates (Niemeyer and Richter 1969, 1972). These authors studied not only the effect of 5 % TCA, but also the conventional extraction method by Kanai and Simonis (1968) using hot HCl, KOH or NaOH (cf. summary in Kulaev 1979). Niemeyer and Richter found that acidic as well as alkaline extraction reduces the chance of receiving the high-molar mass of polyphosphates in the native state considerably. The application of the classical extraction methods to algae cells caused clearly hydro-lytic cleavage of the high-molar mass of polyphosphates. Moreover, trimeta-phosphate can be a hydrolytic product of an acidic extraction. Up until the 1980s extraction methods were carried out in five steps, but in the author's opinion this method cannot be free of hydrolysis (Clark et al. 1986; Wood et al. 1987; Wood and Clark 1988). Using a lysis buffer (Tris/HCl/urea/SDS/EDTA) and using phenol/chloroform as described by Kumble and Kornberg (1995), it was possible to isolate and separate condensed phosphates under native conditions. On the other hand, Kumble and Kornberg (1995) could not find cyclic metaphosphates. Perhaps it is the consequence of the precipi-tation with barium acetate, because this precipitation cannot be complete.

Niemeyer and Richter (1969, 1972) were the first to demonstrated that the cyclic condensed metaphosphates exist and play an important role in the metabolism of all living cells that were tested. Now the time has come to investigate the function of the metaphosphates and the connections between inorganic condensed phosphates and inositol phosphates. The latter seem to be involved in the phosphate balance of living organisms. In order to come to a conclusion concerning high-molar and how-molar mass phosphates, metaphosphates and inositol phosphates, it was necessary to separate this phosphate mixture even more by means of gel electrophoresis preserving under native conditions with a longer finite length than before (Niemeyer 1997b).

2.2
Hydrolysis Free Extraction, Separation and Identification of Metaphosphates Together with Inositol Phosphates and Nucleic Acids

The green thermophilic and acidophilic red alga *Cyanidium caldarium* from the alga collection of the University of Göttingen was cultured under sterile conditions (Niemeyer and Richter 1969). The pulse-labeling experiments, extraction of the condensed inorganic phosphates, column chromatography on MAK, separation of the labeled condensed phosphates form the nucleic acids by ion exchange chromatography and identification of the different inorganic phosphates were carried out (Niemeyer and Richter 1972; Nie-meyer 1976, 1977b). Four synthetic oligopoly- and metaphosphates were used as reference substances for identification. Their positions on the chro-matograms were estimated by means of a specific staining reaction.

With two-dimensional thin-layer chromatography (TLC) the cyclic meta-phosphates were separated from the oligopolyphosphates with chains of up to seven phosphate acid residues (Fig. 1). In each case trimetaphosphate showed the highest labeling of all metaphosphates which indicates its key position in connecting trimetaphosphate with other phosphate using sys-tems. It seems that with the increasing number of phosphate residues the ^{32}P-activity of the oligopolyphosphates goes up, too. The different incuba-

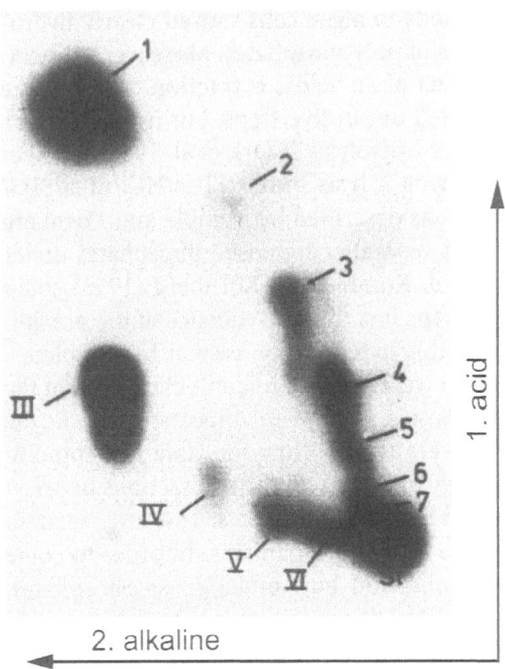

Fig. 1. Autoradiogramm of TLC plate with ^{32}P-labeled inorganic condensed phosphates and inositol phosphates. *1.* Acide solvent; *2.* alkaline solvent; *1* inositol phosphates; *2–7* linear oligophosphates with two to seven phosphate residues; *III–VI* cyclic metaphosphates with three to six phosphate residues; *St* starting point, with the bulk of polyphosphates with eight and more phosphate residues

tion patterns of all the metaphosphates in various experiments signify their metabolic activity in living organisms.

The bulk of the condensed phosphates contained more than seven phosphate residues remaining at the starting point of the thin-layer chromatograhy plates, but which may be separated by gel electrophoresis into several groups depending on their grade of condensation. They were regarded as one fraction on the thin-layer chromatogram and represent a mixture of the fractions I to IV from flat gels, the latter only partially.

Fractionized condensed phosphates always include inositol phosphates and vice versa, since there are only small differences between both compound groups regarding charge, molecular mass and the behavior in an electric field. The mixture of inositol and condensed phosphates was dialysed for 20 h to remove orthophosphate. In two-dimensional thin-layer chromatograms inositol phosphates are separated from metaphosphates with three to six and from oligopolyphosphates with up to seven phosphate residues, but all inositol phosphates migrate to the area of the missing orthophosphate. When inositol phosphates were eluted from the spots of several chromatograms, one could estimate the phosphate values of all inositol phosphates without any influence of other compounds. All the six differently phosphorylated inositol phosphates could be identified by comparison with commercially available reference compounds (Inhülsen 1977; Inhülsen and Niemeyer 1978).

3
Radioactivity Pattern of Condensed Phosphates and Inositol Phosphates After Fractionation on Slab Gels

Low-molar and high-molar mass condensed phosphates as well as the inositol phosphates were separated into six different fractions (F I–F VI) by 250 mm-long slab gels of 40 % polyacrylamide, because Niemeyer (1977b) showed that the distance was too short with 180 mm in the running gel for

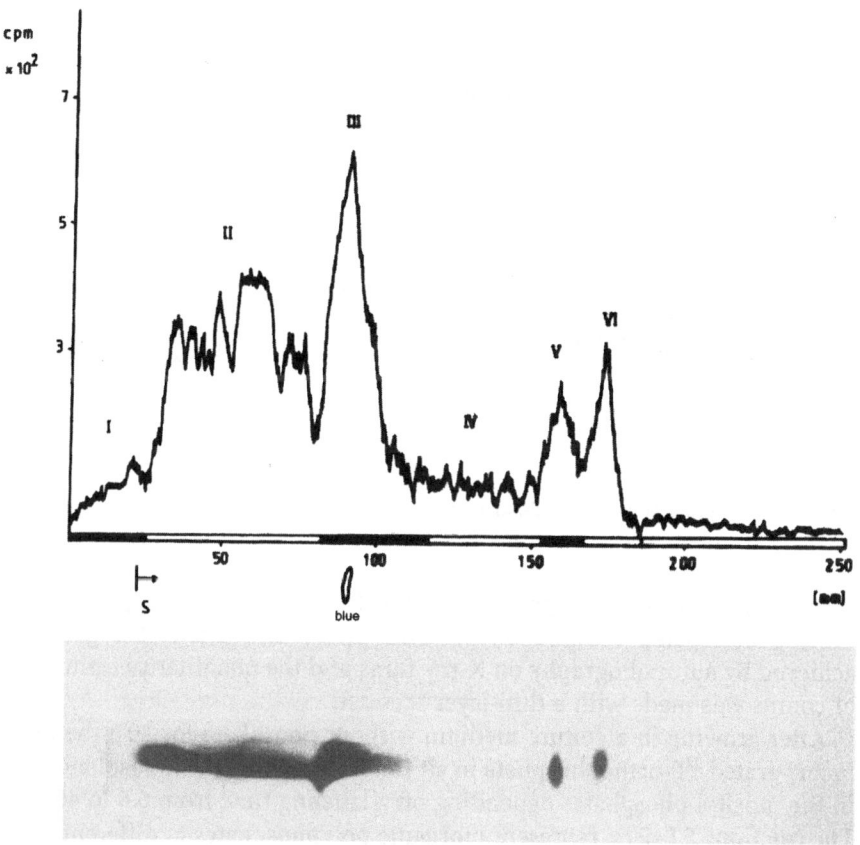

Fig. 2. Scan diagram after 0.5 min ^{32}P-labeling time

Fig. 2–7. Scan diagrams and autoradiogramms from flat gels. Elektrophoresis of the total condensed inorganic phosphates and inositol phosphates after incubation with ^{32}P$_i$-orthophosphate for six different labeling times. S indicates transition from start gel (5 % polyacrylamide) to running gel (40 % polyacrylamide). Six various fractions (F I–F VI) were separated: F I–F IV polyphosphates with different chain lengths; F V metaphosphates; F VI inositol phosphates. In comparison to the scan diagrams, the exposed X-ray films are given *below*

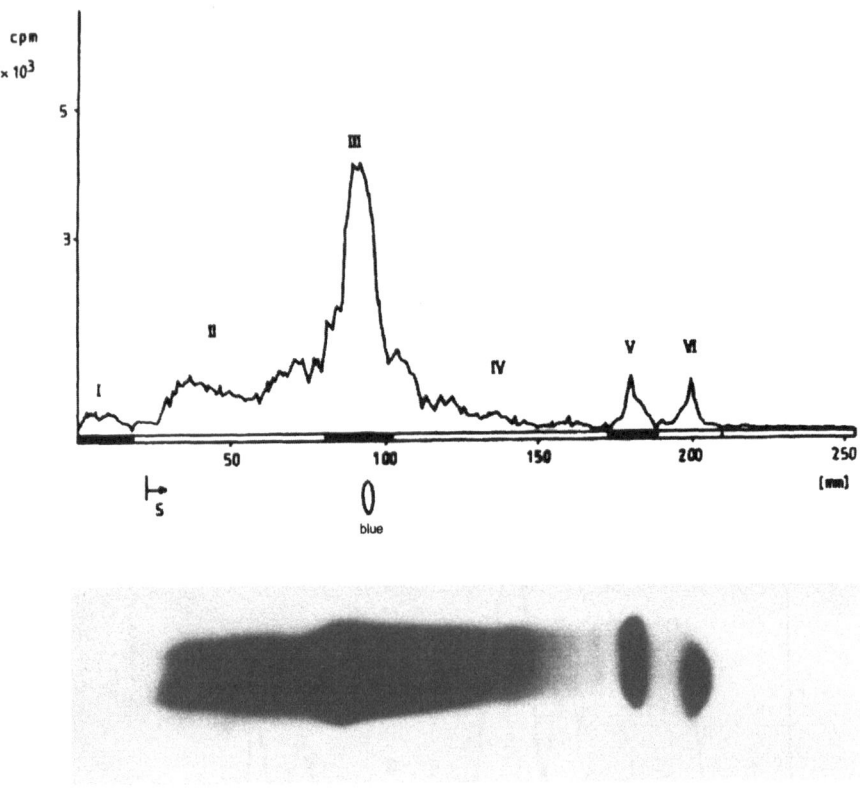

Fig. 3. Scan diagram after 4 min ^{32}P-labeling time

a sufficient separation, especially for the low-molar mass of phosphates. Tracing of labeled compounds on TLC plates and dried slab gels was achieved by autoradiography on X-ray films and the quantitative estimation of counts was made with a thin-layer scanner.

After growing in a culture medium without phosphate for 20 h the algae incorporated ^{32}P-orthophosphate in all condensed inorganic phosphates and in the inositol phosphates depending on a labeling time from 0.5 to 40 mm. The fractions F I–F IV represent inorganic polyphosphates in different complexity, the fraction F V contains the metaphosphates, mainly trimetaphosphate, and the fraction F VI consists of inositol phosphates. The identification of the latter two fractions was achieved by rechromatography of thin-layer plates.

Figures 2–7 depict scan diagrams. The counts of ^{32}P-incorporation were quantitatively measured and documented by scanning diagrams. At the bottom the different fractions can be seen parallel to the gel in the form of black

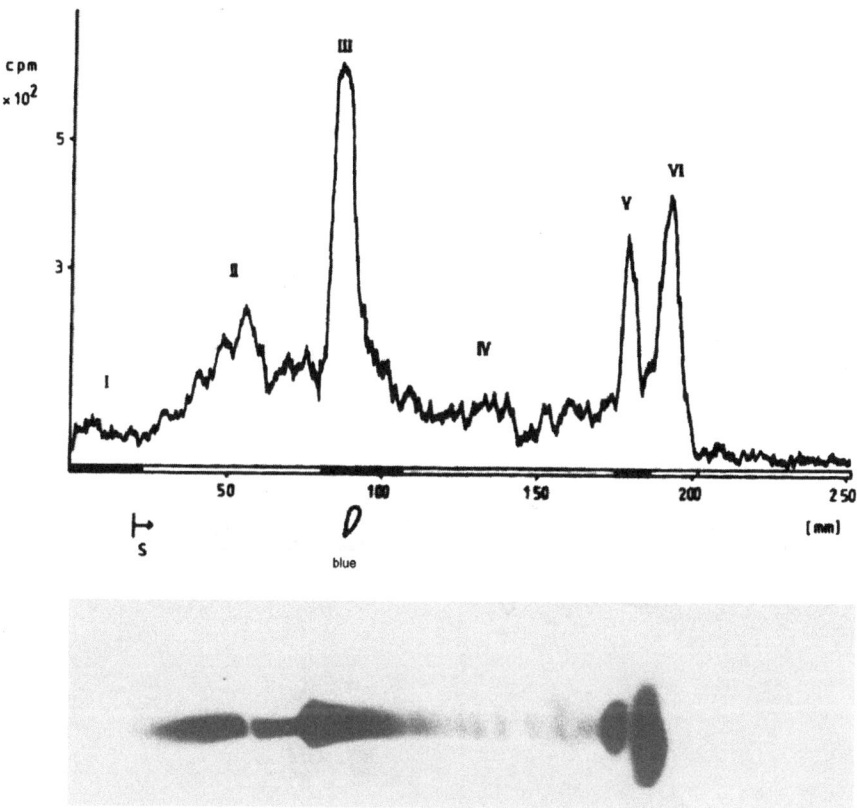

Fig. 4. Scan diagram after 5 min ^{32}P-labeling time

spots in X-ray films. Six fractions can be recognized on all the measured flat gels in the direction of their development form the left to the right.

Figure 2 shows the findings after 0.5 min ^{32}P-labeling time. The amount of ^{32}P-radioactivity from F I (6.1 %), which is close to the transient point from the stacking gel to the running gel, is relatively low. This fraction contains the chains with the highest grade of condensation, which did not enter in the running gel. The greatest ^{32}P-radioactivity of 42.8 % relates to F II between 25 and 80 mm migration running distance, but this fraction consists of several peaks representing a wide spectrum of chains with different grades of condensation. Fraction F III (35.9 %) around 100 mm from point S migrates together with the marker bromophenol blue. The next fraction, F IV (8.1 %), is very likely a group of low-molar mass condensed polyphosphates. Trimetaphosphate, the main component of F V (4.4 %), shows only small activity. The inositol phosphates (F VI) at the front of the running gel contain 4.7 % of the ^{32}P-radioactivity.

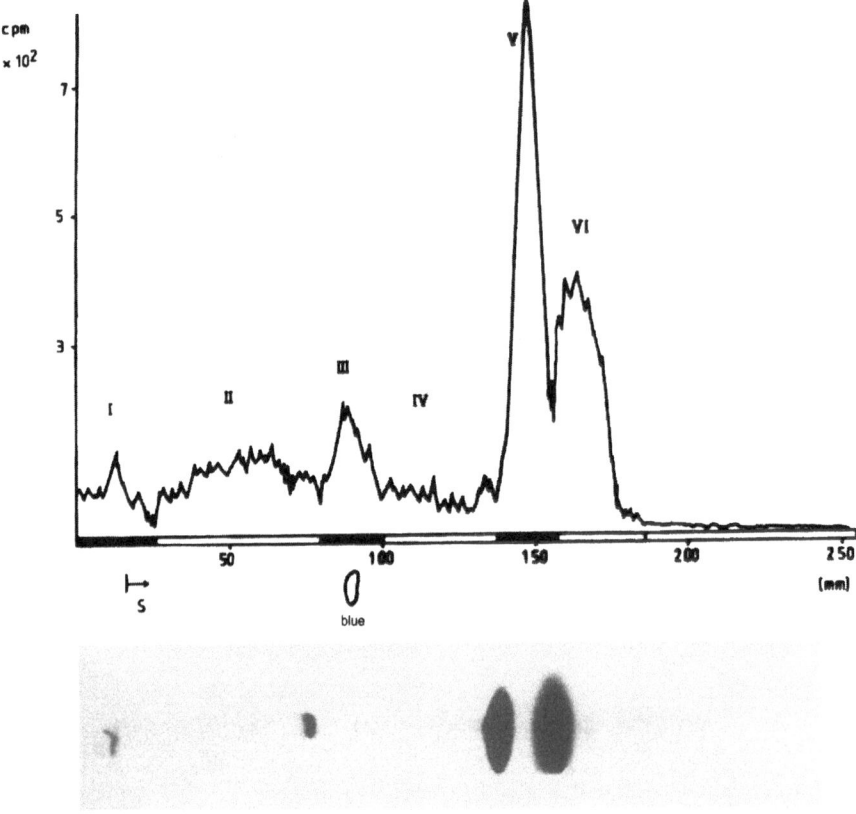

Fig. 5. Scan diagram after 19 min ^{32}P-labeling time

After 1 min ^{32}P-labeling time all the fractions are labeled. Fraction F VI has remarkable increase (26.3 %) in ^{32}P-radioactivity, whereas those of F II and F III decrease. Fraction F III accumulates the largest amount of ^{32}P-radioactivity after incubation of 2 min with ^{32}P-orthophosphate, whereas with 3 min of ^{32}P-labeling that of F III decreases and alternatively at this time the amount of ^{32}P-radioactivity increases in F II and F IV. On the other hand, the algae incorporated the highest amount of ^{32}P of all the tested labeling times in fraction F IV.

After 4 min ^{32}P-labeling time (Fig. 3) the scanner registered a very high amount of ^{32}P-labeling in F III (54.6 %), which gradually shifts to F V (12.7 %) and F VI (10.5 %) after 5 min ^{32}P-labeling time (Fig. 4). After 10 min ^{32}P-labeling time (Fig. 5) the lowest amount of ^{32}P-orthophosphate was incorporated in F II (only 12 %) and in F III (14.5 %) and the highest was found in F V with 31 %. The activity in F VI also increased to 27.1 %. After 20 min ^{32}P-labeling time it was seen that F III reaches the maximum of ^{32}P-labeling (58.1 %) of all the tested times and F V (6 %) and F VI (3.4 %) con-

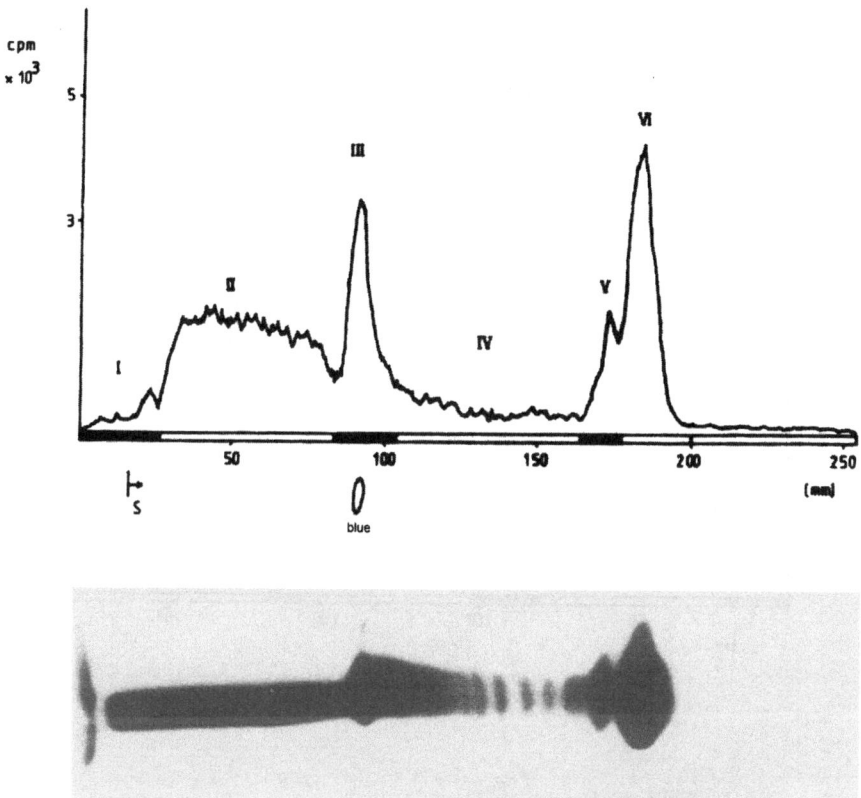

Fig. 6. Scan diagram after 30 min ^{32}P-labeling time

tain only low amounts. In Fig. 6 (30 min) the characteristic phenomenon is demonstrated that now the maximum amount of ^{32}P-orthophosphate is in the bulk of inositol phosphates (F VI, 32.4 %) and that it is relatively low in metaphosphates (F V with 7.3 %). After 40 min ^{32}P-labeling (Fig. 7) F II increases to the largest incorporation (37.7 %), while three fractions (F IV, F V and F VI) have roughly the same ^{32}P-labeling.

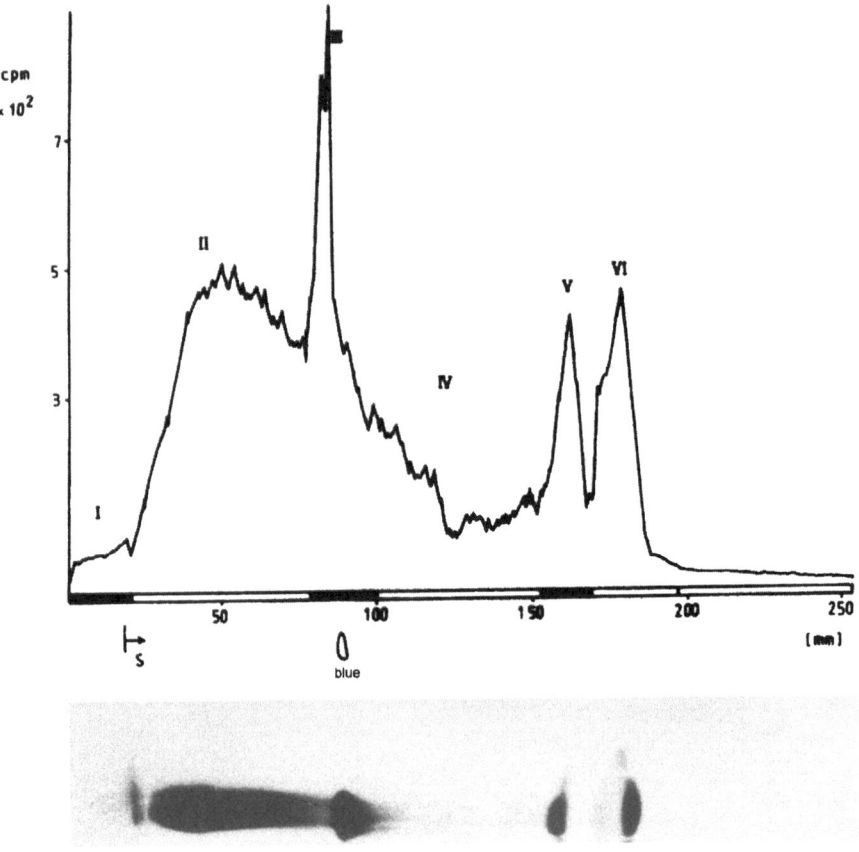

Fig. 7. Scan diagram after 40 min ^{32}P-labeling time

4
Quantitative Distribution of Incorporated ^{32}P-Orthophosphate in Six Different Fractions

In comparison with the percentage distribution of ^{32}P-radioactivity in the six separated fractions (F I–F VI) the figures are presented in Table 1 and also in Fig. 8. The turnover of ^{32}P-activity in the six different phosphate fractions has been observed for ten different labeling times. Maximum incorporation was observed after 4 min and the minimum after the shortest labeling time of 0.5 min. The ^{32}P-radioactivity of the six pool dimensions (F I–F VI) under different conditions was compared and the counts quantitatively estimated in percentage.

The absolute amount of ^{32}P-radioactivity of each gel was determined (= 100 %) and, following the distribution of the activity amongst the various fractions, was registered for different labeling times. Depending on these

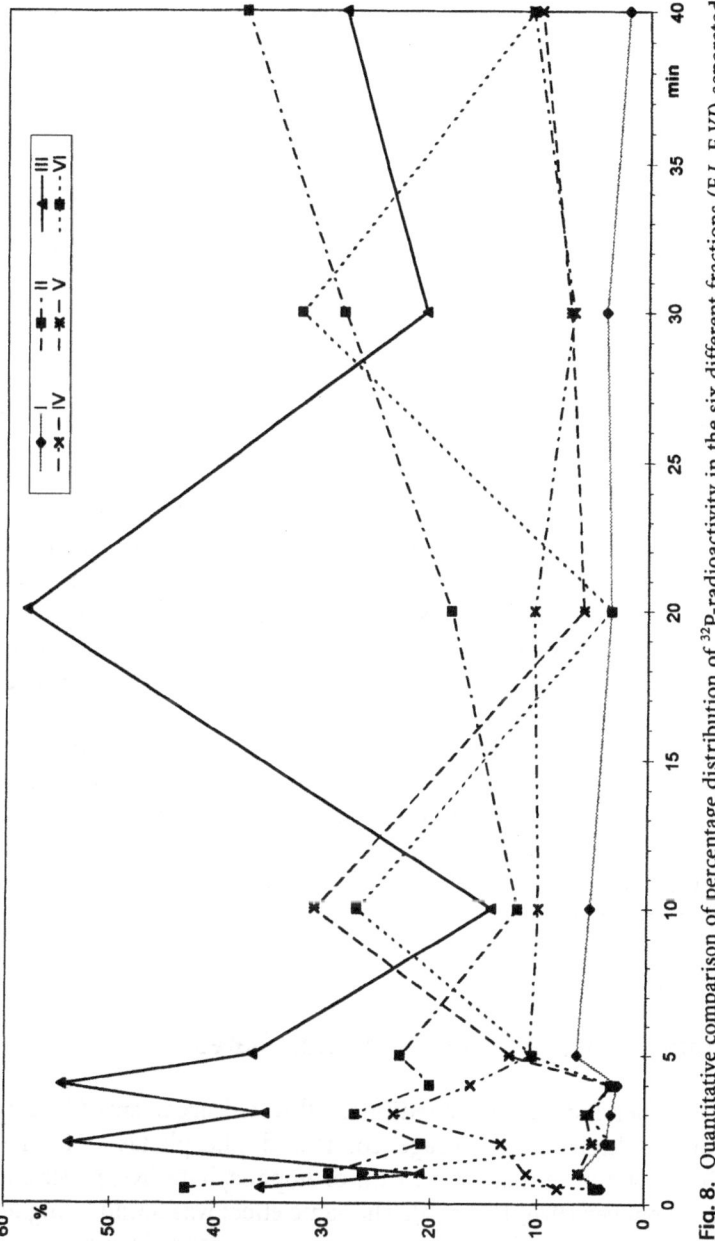

Fig. 8. Quantitative comparison of percentage distribution of ^{32}P-radioactivity in the six different fractions (F I–F VI) separated by gel electrophoresis after ten different labeling times. Data are taken from Table 1

Table 1. Quantitative comparison of percentage distribution of ^{32}P-radioactivity in the different fractions (F I–F VI) separated by gel electrophoresis after ten different labeling times

Fraction	Labeling time (min)									
	0.5	1	2	3	4	5	10	20	30	40
I	4.1	6.1	3.7	3.2	2.6	6.4	5.3	3.5	4.0	2.0
II	42.8	29.4	20.9	27.1	20.1	22.9	12.0	18.4	28.5	37.7
III	35.9	21.0	53.9	35.5	54.6	36.7	14.5	58.1	20.8	28.5
IV	8.1	11.0	13.4	23.5	16.3	10.8	10.1	10.6	7.0	10.9
V	4.4	6.2	4.9	5.5	3.2	12.7	31.0	6.0	7.3	10.1
VI	4.7	26.3	3.2	5.2	3.2	10.5	27.1	3.4	32.4	10.8

times, extensive changes in the distribution of the ^{32}P-radioactivity were observed, especially in F II, F III, F V and F VI. By comparing variable labeling times it can be seen that the maximum of ^{32}P-activity shifts from the relatively high-molar mass of inorganic polyphosphates (F III and F II) more and more to the low-molar mass of metaphosphates (F V) and finally to the inositol phosphates F VI) and other compounds, e.g. nucleic acids (Inhülsen and Niemeyer 1975; Niemeyer and Selle 1987, 1989; Selle and Niemeyer 1987; Kornberg 1995). Obviously, the highest molar inorganic phosphates (F I) have only a low but constant level of 4.09 % of radioactivity on average at all the different labeling times. The largest changes were registered in F II and F III: both fractions show an alternating amount of radioactivity. After short labeling times of up to 5 min and once more after 20 min a very high incubation was observed in F III; however, after 10 and 30 min there was a minimum. The curve of F III is similar to a sinus curve, indicating a high metabolic activity, on average 36 % of the whole amount of ^{32}P-radioactivity over all incubation times.

5
The Metabolic Model of Condensed Phosphates

After growing under P-deficiency conditions all tested organisms filled up their pool of the high-molar storage condensed phosphates (F I–F III) within the first 2–4 min of incubation with orthophosphate (Niemeyer and Richter 1972; Niemeyer 1976, 1977a, b). The same effect was found in higher plants, occurring remarkably slower. A smaller second increase after 10–20 min, depending on the tested organism, corresponds to the labeling of rRNA components and other nucleic acids (Inhülsen and Niemeyer 1975; Niemeyer 1975; Niemeyer and Selle 1987, 1989; Selle and Niemeyer 1987).

It was shown that five different fractions of inorganic condensed phosphates (F I–F V) take part in the phosphate metabolism of plant cells. While fraction F I functions as an absolute last phosphate storage that has to be filled up first, phosphate is transferred to low-molar mass inorganic poly-phosphate subsequently. Therefore fraction FI was named the 'very high-molar constant storage'. This fraction FI and perhaps also FII ('high-molar storage') are constituents of the polyphosphate bodies as observed by other authors in electron micrographs (Harold 1966; Wood and Clark 1988). The turnover of the 'very high-molar constant storage' (FI) is very constant on a low level and the percentage portion of the whole labeling is relatively small with an average of only 4 % over all tested times of ^{32}P-incorporation.

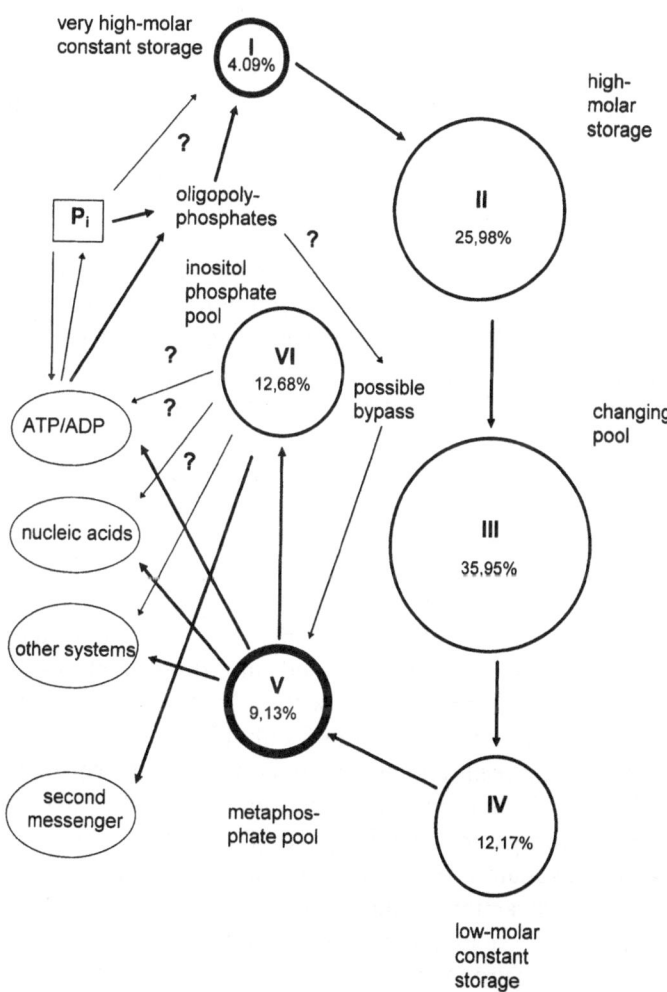

Fig. 9. Hypothetical metabolic pathway of condensed inorganic poly- and metaphosphates in cells from algae. FI–FIV fractions of polyphosphates with different lengths of phosphate residues; FV metaphosphates; FVI inositol phosphates; P_i orthophosphate pool. *Numbers* show percentage distribution of ^{32}P-radioactivity in the six fractions on average

Conspicuously, the flow of incorporated ^{32}P from a maximum to a minimum is correlated between F II and F III, on the one hand, and F V and F VI, on the other. This phenomenon ist interpreted as a continuous shifting of phosphate from high-molar to low-molar chains of inorganic condensed phosphates. It is remarkable that the peaks are better separated when the gel chromatograms were developed for a longer time. Under this condition the low-molar mass phosphates from F IV, F V and F VI left the gel and it is perceptible that the inorganic phosphates of all fractions consist of a mixture of phosphate chains of similar length within each fraction. In contrast to other authors (Rao et al. 1985; Clark et al. 1986; Wood and Clark 1988; Kumble and Kornberg 1995; Lorenz et al. 1997a), it seems impossible to give precise figures for the length and numbers of phosphate residues of the chains in the different fractions. It has to be supposed that the transitions between the fractions occur continuously, corresponding to the metabolic turnover of condensed phosphates.

Fraction F II has a very high labeling at 1 min, decreasing afterwards in contrast to F III which shows a high activity (see 2 min labeling time). It is postulated that F III functions as a buffer for the low-molar mass phosphate (F IV), because it has the greatest fluctuations between 14.5 % and 58.1 % of the incorporated ^{32}P; therefore the fraction F III is described as the 'changing pool' (Fig. 9). F III plays an important role as connecting pool between high-molar and low-molar mass of the inorganic phosphate that shifts via metaphosphates to other pools which use PO_4 groups for the intermediate phosphate metabolism. This is supported by the relatively low and constant activity of F IV ('low-molar constant storage'). Once this pool is filled up within the first 4 min it guarantees the continuous path to the metaphosphates. Metaphosphate may play an important role by transferring inorganic pyrophosphate. Particularly, Petty et al. (1985) found an active trimetaphosphatase in guiena pig and human macrophages.

Fraction F V consists of cyclic metaphosphates only, mainly trimetaphosphate. Niemeyer (1997b) obtained only three fractions and inositol phosphates by gel chromatograms with only 180 mm running distance. In contrast, Langen et al. (1962) found only high-molar mass of inorganic phosphates in yeast cells and no low-molar mass or intermediate phosphates. The authors could not give any reason for this result. They supposed that ATP transfers inorganic orthophosphate (P_i) directly to high-molar mass of inorganic phosphates. This assumption is certainly not correct, since, later on, using an extraction method which prevents hydrolysis, labeled oligopolyphosphates were found in all of the author's experiments every time. On the other hand, Kuroda and Kornberg (1997) observed that the ATP/ADP system connects to condensed phosphates.

Many experiments have demonstrated that the metabolism of condensed phosphate interacts with the metabolism of nuclei acids (Inhülsen and Niemeyer 1975; Wang et al. 1986; Niemeyer and Selle 1987, 1989; Selle and Niemeyer 1987; Kuroda and Kornberg 1997; Kusaono and Ishihama 1997; Niemeyer in prep.). In addition, the interest is more and more the metabolism of the inorganic condensed phosphates when the existence of these compounds was proved not only in bacteria and plants but also in sponges and in mammalian and human cells (e.g. Petty et al. 1985; Chaussepied et al. 1986; Cowling and Birnboim 1994; Kornberg 1995; Kumble and Kornberg 1995; Lorenz et al. 1995).

Reviewing the different distinctive functions of condensed phosphates, Kornberg (1995) described them as an energy supply and ATP substitute, a reservoir for P_i, a chelator of metals, a buffer against alkali, a channel for DNA entry, a cell capsule and a regulator of responses to stress and adjustments for survival in the stationary phase of cell culture.

Another ^{32}P-orthoposphate incorporating pool consists of different inositol phosphates (F VI), which were identified by Inhülsen (1977) and Inhülsen and Niemeyer (1978), though they did not determine the isomeric structure of the isolated compounds. The hexainositol phosphate, the so-called phytic acid, occurs especially in seeds of higher plants. It is generally believed that in higher plants phytic acid has the same function as the condensed phosphates fulfill in lower organisms. On the other hand, meta- and polyphosphates were found in higher plants, too (Niemeyer 1975; Inhülsen and Niemeyer 1975, 1978). Furthermore, the inositol phosphates (F VI) are regarded as a part of the signal transduction pathway, being various second messengers (Berridge 1987; Neuhoff and Friend 1991; Billington 1993; Vincent and Crowder 1996).

Further investigations of the function of different isomeric inositol phosphates rely on the separation and identification of compounds isolated from plants. Recently, Grunwald and Niemeyer (1997) and Niemeyer and Grunwald (1997) identified inositol-2,4,5-trisphosphate, having used a high-performance liquid chromatography (HPLC) method for separation which allows detection of concentrations as low as 80–100 pmol.

This method allows identification not only of inositol phosphates in picomoles but also of different isomeric forms without the use of radioactivity. By comparing the chromatogram with that of standard substances of a well-defined mixture it is possible to investigate the spectrum of inositol phosphates appearing in different plants and to assign them to the different organelles, too. Especially this aspect must be emphasized when examining plant material, since cellular inositol phosphates exist in very small concentrations only within the various cells compartments, where they many fulfill different functions (Niemeyer and Selle 1987, 1989; Selle and Niemeyer 1987).

The occurrence of inorganic condensed phosphates together with inositol phosphates in such a wide range of different organisms suggests they may have an important physiological role that is fundamental to life in prokaryota as well as in eukaryota. However, a universal role for these compounds, especially their physiological connections, has not been demonstrated. Different biological functions have been proposed, based on circumstantial evidence, but now it will be possible to investigate their main role, perhaps as a regulatory system for building the second messengers via the inositol phosphate pool besides the phosphate and energy storage.

References

Berridge MJ (1987) Inositol triphosphate and diacylglycerol: two interacting second messengers. Annu Rev Biochem 56: 159–193

Billington C (1993) The inositol phosphates, chemical synthesis and biological significance. VCH, Weinheim

Chaussepied P, Mornet D, Kassab R (1986) Identification of polyphosphate recognition sites communicating with actin sites on the skeletal myosin subfragment 1 heavy chain. Biochemistry 25: 6426–6432

Clark, JE, Beegen H, Wood HG (1986) Isolation of intact chains of polyphosphate from *Propionibacterium shermanii* grown on glucose or lactate. J Bacteriol 168: 1212–1219

Cowling RT, Birnboim HC (1994) Incorporation of ^{32}P-orthophosphate into inorganic polyphosphates by human granulocytes and other human cell types. J Biol Chem 269: 9480–9485

Ebel JP (1952a) Recherches sur les polyphosphates contenus dans diverses cellules vivantes. I. Mise au point d'une méthode d'extraction. Bull Soc Chim Biol 34: 321–329

Ebel JP (1952b) Recherches sur les polyphosphates contenus dans diverses cellules vivantes. II. Étude chromatographique et potentiometric des polyphosphates de levure. Bull Soc Chim Biol 34: 330–338

Ebel JP (1952c) Recherches sur les polyphosphates contenus dans diverses cellules vivantes. III. Recherche et dosage des polyphosphates dans les cellules de diverses microorganismes et animaux superieures. Bull Soc Chim Biol 34: 491–497

Ebel JP (1952d) Recherches sur les polyphosphates contenus dans diverses cellules vivantes. IV. Localisation cytologique et rôle physiologique des polyphosphates dans la cellule vivante. Bull Soc Chim Biol 34: 498–506

Grunwald I, Niemeyer R (1997) Etablierung einer Methode zur nichtradioaktiven, isomerspezifischen Untersuchung von Inositophosphaten mit Hilfe der Hochdruck-Flüssigkeits-Chromatography. German Soc Cell Biol, 2. DGZ-Nachwuchswissenschaftler-Tagung Zelluläre Signaltransduktion, Jena, S 10

Harold FM (1966) Inorganic polyphosphates in biology: structure, metabolism and functions. Bacteriol Rev 30: 772–794

Inhülsen D (1977) Inosit-Phosphate aus *Lemna minor* L. und ihre Abtrennung von den kondensierten Phosphaten. PhD Thesis, University of Hannover

Inhülsen D, Niemeyer R (1975) Kondensierte Phosphate in *Lemna minor* L. und ihre Beziehungen zu den Nucleinsäuren. Planta 124: 159–167

Inhülsen D, Niemeyer R (1978) Inosit-Phosphate aus *Lemna minor* L. Z Pflanzenphysiol 88: 103–116

Kanai R, Simonis W (1968) Einbau von ^{32}P in verschiedene Phosphatfraktionen besonders Polyphosphate bei einzelligen Grünalgen (*Ankistrodesmus braunii*) im Licht und im Dunkeln. Arch Mikrobiol 62: 56–64

Kornberg A (1994) Phosphate in microorganisms. In: Torrriani-Gorini A, Silver S, Yagil E (eds) Cellular and molecular biology. American Society of Microbiology, Washington, DC, pp 204–209

Kornberg A (1995) Inorganic polyphosphate: toward making a forgotten polymer unforgettable. J Bacteriol 177: 491–496

Kulaev IS (1979) The biochemistry of inorganic polyphosphates. Wiley, New York

Kumble KD, Kornberg A (1995) Inorganic polyphosphate in mammalian cells and tissues. J Biol Chem 270: 5818–5822

Kumble KD, Kornberg A (1996) Endopolyphosphatases for long chain inorganic polyphosphate in yeast and mammals. J Biol Chem 271: 27146–27151

Kuroda A, Kornberg A (1997) Polyphosphate kinase as a nucleoside diphosphate kinase in *Escherichia coli* and *Pseudomonas aeruginosa.* Proc Natl Acad Sci USA 94: 439–442

Kusano S, Ishihama A (1997) Functional interaction of *Escherichia coli* RNA polymerase with inorganic polyphosphate. Genes Cells 7: 433–441

Langen P, Liss E, Lohmann K (1962) Art, Bildung und Umsatz der Polyphosphate der Hefe. Coloq Int Centre Natl Rech Sci (Paris) 106: 603–614

Liebermann L (1888) Über das Nuclein der Hefe und künstliche Darstellung eines Nuclein-Eiweiß und Metaphosphorsäure. Ber Dtsch Chem Ges 21: 598–607

Liebermann L (1890) Nachweis der Metaphosphorsäure im Nuclein der Hefe. Pflugers Arch 47: 155–164

Lorenz B, Batel R, Bachinski N, Müller WEG, Schröder HC (1995) Purification and characterization of two exopolyphosphatases from the marine sponge *Tethya lyncurium.* BBA 1245: 17–28

Lorenz B, Münker J, Oliveira MP, Kuusksalu A, Leitao JM, Müller WEG, Schröder HC (1997a) Changes in metabolism of inorganic polyphosphate in rat tissues and human cells during development and apoptosis. Biochem Biophys Acta 1335: 51–60

Lorenz B, Münkner J, Oliveira MP, Leitao JM, Müller WEG, Schröder HC (1997b) A novel method for determination of inorganic polyphosphates using the fluorescent dye fura-2. Anal Biochem 246: 176–184

Neuhoff V, Friend J (1991) Cell to cell signals in plants and animals. Nato ASI series H: Cell biology 51. Springer, Berlin Heidelberg New York

Niemeyer R (1975) Poly- and Metaphosphaten in Höheren Pflanzen (Lemnaceae). Planta 122: 303–305

Niemeyer R (1976) Cyclic condensed metaphosphates and linear polyphosphates in brown and red algae. Arch Microbiol 108: 243–247

Niemeyer R (1977a) New concept on biosynthesis of condensed phosphates. Protoplasma 91: 221–224

Niemeyer R (1977b) On the biosynthetic of condensed phosphates. In: Woodcock CLF (ed) Progress in *Acetabularia* research. Academic Press, New York, pp 95–104

Niemeyer R (1999) Inositol 1,4,5 trisphosphate enhances DNA synthesis in isolated plant nuclei. In preparation

Niemeyer R, Grunwald I (1997) Non-radioactive and isomer-specific analysis of inositol phosphates by high-performance liquid chromatography. In: Janssen O, Hass R, Friedrich K (eds) 1. Joint meeting, intracellular signal transduction: receptors, mediators and genes. Tagungsband, Langen, p 23

Niemeyer R, Richter G (1969) Schnellmarkierte Poly- and Metaphosphate bei der Blaualge *Anacystis nidulans.* Arch Mikrobiol 69: 54–59

Niemeyer R, Richter G (1972) Rapidly labeled polyphosphates in *Acetabularia.* In: Bonotto S, Goutier R, Kirchmann R, Maisin JR (eds) Biology and radiobiology of anucleate systems II. Plant cells. Academic Press, New York, pp 225–236

Niemeyer R, Selle H (1987) Inorganic condensed phosphates and inositol phosphates in plant nuclei. Biol Chem Hoppe Seyler 368: 1088

Niemeyer R, Selle H (1989) Insp3 enhances DNA synthesis. Biol Chem Hoppe Seyler 370: 940

Petty HR, Hermann W, McConnell HM (1985) Cytochemical study of macrophage lysosomal inorganic trimetaphosphatase and acid phosphatase. J Ultrastruct Res 90: 80–88

Rao NN, Kornberg A (1996) Inorganic polyphosphate supports resistance and survival of stationary-phase *Escherichia coli.* J Bacteriol 178: 1394–1400

Rao NN, Roberts MF, Torriani A (1985) Amount and chain length of polyphosphates in *Escherichia coli* depend on cell growth conditions. J Bacteriol 162: 242–247

Selle H, Niemeyer R (1987) Inositol phosphates in nuclei of barley seedlings. Biol Chem Hoppe Seyler 368: 568

Thilo E (1955) Die kondensierten Phosphate. Angew Chem 67: 141–149

Thilo E (1959) Die kondensierten Phosphate. Naturwissenschaften 46: 367–374

Vicent JB, Crowder MW (1996) Phosphatases in cell metabolism and signal transduction: structure, function and mechanism of action. Springer, Berlin Heidelberg New York; Landes, Austin

Wang Y, Yang ZW, Wang QW, Xu YZ, Liu XY, Xu JF, Chen CH (1986) The role of metaphosphate in the activation of the nucleotide by TPS and DCC in the oligonucleotide synthesis. Nucleic Acids Res 14: 2699–2706

Wiame JM (1947) Metaphosphate of yeast. Fed Proc 6: 302–309

Wiame JM (1949) The occurrence and physiological behaviour of two metaphosphate fractions in yeast. J Biol Chem 178: 919–925

Wood HG, Clark JE (1988) Biological aspects of inorganic polyphosphates. Annu Rev Biochem 57: 235–260

Wood HG, Robinson NA, Pepin CA, Clark JE (1987) Polyphosphate kinase and polyphosphate glucokinase of *Propionibacterium shermanii*. In: Torriani-Gorini A, Rothman FG, Silver S, Wright A, Yagil E (eds) Phosphate metabolism and cellular regulation in microorganisms. American Society of Microbiology, Washington, DC, pp 225–232

Polyphosphate Glucokinase

N. F. B. Phillips[1], Pei Chung Hsieh, and T. H. Kowalczyk

1
Introduction

Reports on inorganic polyphosphates (polyP) in living organisms date back to 1888 (Liebermann 1888). PolyP is recognized as one of the earliest biopolymers and most likely a prominent precursor in prebiotic evolution (Yamagata et al. 1991). This polymer has been found in all representations of living cells, including humans (Kulaev 1979; Kulaev and Vagabov 1983), and Archaebacteria (Scherer et al. 1983; Rudnik et al. 1990). The ubiquitous occurrence of polyPs suggests that they may have a physiological role that is fundamental to life. Although numerous roles have been attributed to polyP in cellular functions from as early as 1966 (Harold 1966) to as recently as 1995 (Kornberg 1995), the status of these roles, with a few additions, has remained essentially unchanged. PolyP has been regarded at best as a "metabolic fossil". Futhermore, a universal role for polyP has yet to be demonstrated.

Of the numerous functions attributed to polyP, its role as a source of high energy phosphate was the most attractive, since polyP was found to be utilized in certain metabolic reactions without the intervention of ATP in contemporary organisms and owing to the fact that the energy of hydrolysis of the phosphoanhydride bonds of polyP is thermodynamically comparable to that of ATP. Lipmann (1965) proposed that the generation of the high energy phosphate group potential might have originated with inorganic pyrophosphate and polyphosphate and ancient organisms may principally have utilized polyP in the metabolic roles that are contemporarily fulfilled by ATP. He theorized that this ability to utilize polyP was probably replaced by ATP throughout the course of evolution. However, the discovery of enzymes that utilize and synthesize polyP in contemporary living organisms suggests that polyP, far from being a "metabolic fossil", may still be crucial in the regulation of phosphorous metabolism during growth and development as pointed out by Kornberg (1995).

---

[1] Case Western University, Departments of Biochemistry and Medicine, School of Medicine, Room W127, 10900 Euclid, Cleveland, Ohio 44106-4983, USA.
This article is dedicated to Drs. Marian and Olga Szymona for their pioneering contributions to the field of polyphosphate glucokinases.

Progress in Molecular and Subcellular Biology, Vol. 23
H. C. Schröder, W. E. G. Müller (Eds.)
© Springer-Verlag Berlin Heidelberg 1999

One such enzyme that utilizes polyP is the polyP-dependent glucokinase found in certain organisms.

2
A Historical Perspective

PolyP glucokinase (polyphosphate:D-glucose 6-phosphotransferase) cata-lyzes the phosphorylation of glucose using polyP without the involvement of ATP [Eq. (1)]:

$$\text{poly(P)n} + \text{glucose} \rightarrow \text{glucose-6-phosphate} + \text{poly(P)n}_{-1}. \tag{1}$$

The enzyme was first observed in *Mycobacterium phlei* by Szymona in 1957. Since then, M. Szymona and his wife O. Szymona, along with other cowork-ers, detected glucokinase activities in numerous bacteria (Szymona et al. 1962), including a number of other mycobacterial species (Szymona and Szymona 1978), *Nocardia minima* (Szymona and Szymona 1979) and *Cory-nebacterium diphtheriae* (Szymona and Szymona 1961). Subsequently, Szy-mona, Uryson, Kulaev and coworkers screened for polyP glucokinase activi-ties in a variety of different organisms (see review by Kulaev 1979) and detected this enzyme only in the phylogenetically ancient bacteria belonging to the *Actinomycetales* order, according to the classification of Krasil'nikov (1949). The enzyme activity was not found in eubacteria, fungi, green or blue–green algae. Besides its presence in the *Actinomycetales*, more recent studies have shown that polyP glucokinase is present in *Myxococcus coral-loides* (Gonzáles et al. 1990), in the bacterial parasite *Bdellovibrio bacteriovo-rus* (Bobyk et al. 1980) and in the oligotrophic bacterium, *Renobacter vacuo-latum* (Nikitin et al. 1983, referred to by Kulaev and Vagabov 1983). A com-mon observation with the discovery of polyP glucokinase was that the enzyme preparations were also able to utilize ATP to phosphorylate glucose.

Tripolyphosphate was reported to be the end product formed from long-chain polyP in glucose phosphorylation (Szymona and Widomski 1974). Recently, the end products of the reaction were characterized in more detail in our laboratory and were found to consist of a mixture of $polyP_4$ and $polyP_3$ (Kowalczyk and Phillips 1993).

2.1
Involvement of PolyP Glucokinase in Metabolism

There are no reports of an organism that contains only polyP glucokinase and no ATP glucokinase. If such an organism were found, it would provide strong evidence that polyP is used physiologically for phosphorylation of glucose. Wood and Goss (1985) have shown that polyP glucokinase is a con-stitutive enzyme of *Propionibacterium shermanii*. The enzyme was present in cells grown on glucose, glycerol or lactate. Other evidence that hexoses are phosphorylated using polyP has been obtained with *Mycobacterium phlei* by Szymona and coworkers, who showed that polyP fructokinase (Szymona and Szumilo 1966), polyP mannokinase and polyP gluconatokinase (Szymona et al. 1969) were only found when the cells were grown in media containing their counterpart sugars, but were absent when the cells were grown on glucose. It is very unlikely that adaptation would occur if the sugars were not being utilized for phosphorylation with polyP. The only indirect in vivo evidence for the utilization of polyP by polyP glucokinase is the demonstration by Wood and Clark (1988, and references therein), who showed that long chain polyPs accumulate in the cells of *P. shermanii*, when grown on lactate but not on glucose. These authors concluded that the polyP accumulates in the lactate-grown cells because there is no substrate for phosphoryl transfer from polyP, whereas in the glucose-grown cells, polyP is utilized by polyP glucokinase. Overall, these observations suggest that polyP glucokinase is involved in the metabolism of glucose in certain bacteria.

3
Multiple Forms of PolyP Glucokinase and Molecular Weight Characterization

There are numerous reports on the presence of various isoenzymes of polyP glucokinase from different microorganisms, and on differences in the molecular weights of the enzyme from the same organism. For instance, Szymona et al. (1977) found that the native M_r of the enzyme from *Mycobacterium tuberculosis* was 118 kDa, while Pastuszak and Szymona (1980) found a larger form of the enzyme. The enzyme from *M. phlei* was found to be a 113-kDa protein (Szymona and Ostrowski 1964), and, more recently, it was reported by Girbal et al. (1989) to be 275–280 kDa. Likewise, the enzyme from *P. shermanii* was reported to have a native M_r of 31 kDa (Clark 1990). The purified enzyme from *P. shermanii* (Phillips et al. 1993a) and *M. tuberculosis* (Hsieh et al. 1993a) also showed multiple proteins by HPLC gel-filtration, native PAGE and isoelectric focusing (IEF)-PAGE, although a single band was observed by SDS-PAGE. To account for the multiple forms of the glucokinase, it is suggested that the enzyme may contain residual amounts of strongly bound polyP of various chain lengths. This could confer different ionic properties to the enzyme species resulting in multiple forms observed during the analytical procedures. A feature that is common to all the polyP glucokinases, which is consistent with the above proposal, is the observation that, although the native molecular weights of the enzyme from different sources (and occasionally from the same source) vary, the subunit size is about 30–35 kDa in each case as determined by SDS-PAGE. As detailed below, using various techniques, the enzymes from *P. shermanii*, *M. tuberculosis* and *Propionibacterium arabinosum* were all found to be homodimers of subunit molecular weights of ~30 kDa.

4
Bifunctionality

A common feature of the polyP glucokinases from different sources was that extracts containing the polyP glucokinase activity also contained an ATP-dependent activity. Kulaev (1979) made an interesting observation that in the more ancient representatives of the *Actinomycetales* group, such as *Micrococci, Tetracocci, Mycococci* and the propionic bacteria, the polyP glucokinase activity exceeded the ATP-dependent activity several-fold; while in the phylogenetically younger representatives of this order, the true *Actinomycetes*, ATP glucokinase activity was substantially greater. From this observation, Kulaev suggested that the phosphorylation of glucose with polyP as the phosphoryl donor is more ancient than that with ATP. This is consistent with the view of Lipmann (1971) that inorganic pyrophosphate and polyP may have been the primitive group carriers for the generation of phosphate group potential.

The Szymonas and coworkers, in their later work on the polyP glucokinase from *N. minima* (Szymona and Szymona 1979), *M. tuberculosis* (Szymona et al. 1977; Pastuszak and Szymona 1980) and a number of other mycobacteria (Szymona and Szymona 1978), showed that electrophoretically homogenous preparations of polyP glucokinase also displayed some activity with ATP as the phosphoryl donor. In addition, Szymona et al. (1977) found that the ratio of polyP to ATP glucokinase activity remained constant at 2 to 1 throughout the 600-fold purification of the enzyme from *M. tuberculosis*. With the glucokinase from *P. shermanii*, Pepin and Wood (1986) found that the polyP- and ATP-dependent activities co-purified with a ratio of 4:1 during the 960-fold purification. These authors (Szymona et al. 1977; Pepin and Wood 1986) suggested that both activities may be catalyzed by a single enzyme.

In contrast, with the enzyme preparation from *Corynebacterium xerosis*, there was a large change in the ratio of polyP to ATP activities during purification, from 1.75 to 15.2 (Dirheimer and Ebel 1968); however, the purified polyP glucokinase still displayed an activity with ATP. Likewise, in the case of *M. corralloides* D, González et al. (1990) showed that the polyP- and ATP-dependent glucokinase were partially separated by conventional gel filtration chromatography.

Nevertheless, in the numerous accounts of the proposed separations of the two activities, there was always some degree of overlap in the two activities. In the case of the apparently homogenous preparations that displayed a separate polyP- or ATP-dependent activity, convincing evidence for a single protein in these preparations (as demonstrated by SDS-PAGE, reverse phase HPLC or by N-terminal sequence analysis) has not been provided.

Thus, the question of whether different enzymes catalyze the polyP- and ATP-dependent glucokinase activities, or if a single protein is responsible for the two activities, remained essentially unanswered.

5
PolyP and ATP Glucokinase Activities Are Catalyzed by a Single Enzyme

To answer the question of bifunctionality, Phillips and coworkers conducted extensive purification of the enzymes from *P. shermanii* (Phillips et al. 1993a), *M. tuberculosis* (Hsieh et al. 1993a) and *P. arabinosum* (Horn and Phillips, unpubl.) and the detailed characterization revealed unequivocally that a single enzyme from these sources catalyzes both the polyP- and ATP-dependent glucokinase activities. Evidence for homogeneity for the purified protein from the three sources containing both the polyP and ATP glucokinase activities was as follows:

1. A single band was observed by SDS-PAGE with a mol. wt. of ~30 kDa.
2. The protein fractionated as a single peak on a C_4 reverse-phase column by HPLC.
3. Both activities co-migrated on acrylamide gels after native gel electrophoresis or isoelectric focusing as observed by glucokinase activity staining.
4. The purified enzyme yielded a single peak by HPLC gel filtration on TSK-G2000 SW. The peak contained both activities and fractionated as a 63 and 66-kDa protein for the *P. shermanii* and *M. tuberculosis* enzymes respectively. The above result suggests that the enzyme is a homodimer.
5. The purified enzymes yielded a single N-terminal sequence by automated gas-phase sequencing.
6. The most convincing evidence was provided by Hsieh et al. (1996a) who cloned the polyP glucokinase gene from *M. tuberculosis* $H_{37}Rv$ and showed that the recombinant protein, expressed and purified from *E. coli*, contained both activities.

6
PolyP and ATP Have Separate Binding Sites

The bifunctionality of polyP glucokinase, i.e. the ability to utilize both inorganic (polyP) and organic (ATP) phosphoryl donors in glucose phosphorylation, suggested that, due to the fundamental differences in the structures of the two phosphate donors, the residues involved in the binding of them may also be different. The only available evidence for separate binding sites for the substrates was provided by Phillips et al. (1993a) and Hsieh et al. (1993a), as follows:

1. With the *P. shermanii* enzyme, Phillips et al. (1993a) demonstrated differential inactivation kinetics using an affinity label of ATP, 2',3'-dialdehyde of ATP (oxidized ATP, oATP). The oATP, whose reactive group is the dial-

dehyde of the ribose ring, inhibited both the ATP- and polyP-dependent reactions. However, while ATP afforded protection against inactivation of the ATP-dependent activity, polyP did not protect against loss of the polyP activity.

2. oATP was found to be competitive with ATP in the ATP-dependent reaction, while it was noncompetitive with $polyP_{35}$ in the polyP-dependent reaction. The latter result suggests that oATP and polyP can bind to the enzyme simultaneously.

3. With the enzyme from *M. tuberculosis*, Hsieh et al. (1993a) found that the dichlorotriazinyl dye, Procion Blue Mx-3G, which has been used as affinity probe for nucleotide-binding enzymes (Clonis and Lowe 1980), including ATP glucokinase (Goward et al. 1987), was competitive vs ATP and noncompetitive with polyP in their respective glucokinase reactions.

7
The Glucose-Phosphorylating Center Is Common for Both PolyP and ATP

Although the findings discussed in the above section suggest that polyP and ATP have separate binding sites, other observations indicate that the catalytic center may be common for both reactions, or that the binding sites for the phosphoryl donors overlap partially.

7.1
Kinetic Evidence

7.1.1
Substrate Competition

It was shown for the enzyme from *P. shermanii* (Phillips 1993a) that the rates of the ATP- and polyP-dependent reactions were not additive when the formation of the common product, glucose 6-phosphate, was measured. In fact, the faster polyP-dependent reaction was inhibited by 60 % in the presence of ATP. Although this result is consistent with competition of both phosphoryl donors for a single site, it can also be explained in terms of separate sites with cross-inhibition. The latter possibility was ruled out for the enzyme from *M. tuberculosis* by constructing the competition plot as described by Chevillard et al. (1993). Horizontal straight lines (Fig. 1) indicate that the two reactions occur at the same site. Separate reactions with cross-inhibition would give curves with either maxima or minima depending on ratios of the Michaelis constants of the two substrates and their inhibition constants in the other reactions. Possible ambiguity in the interpretation of this plot was eliminated by performing measurements at two different sets of substrate concentrations.

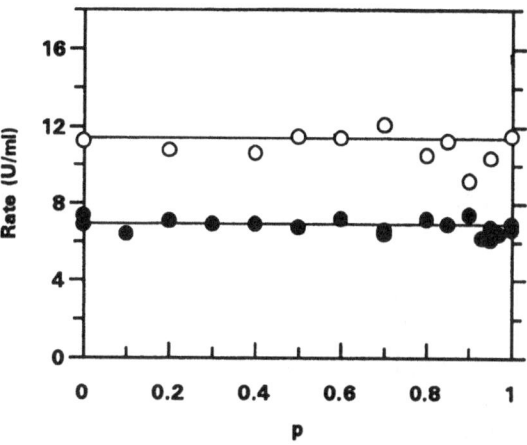

Fig. 1. Competition plot. Total rate of glucose phosphorylation by polyP glucokinase from *M. tuberculosis* is plotted against *p*, where *p* varies from 0 to 1 and specifies concentrations $(1-p)$ $[ATP]_0$ and $p[polyP]_0$ of the two substrates in terms of reference concentrations $[ATP]_0$ and $[polyP]_0$, chosen to give the same rates at $p = 0$ and $p = 1$. $[ATP]_0 = 4\,mM$ (*lower line*) or $8\,mM$ (*upper line*) and $[polyP_{31}]_0 = 4.3\,\mu M$ (*lower line*) of $7\,\mu M$ (*upper line*). Glucose was saturating at $33\,mM$. Rate is expressed in terms of glucose-6-P production

7.1.2
Affinity Labeling with Nucleotide Analogs

To test the single phosphorylating center hypothesis, Phillips et al. (1993a) utilized another along of ATP, 5'-p-fluorosulfonylbenzoyl adenosine (FSBA). The rationale for using this analog was that the reactive sulfonyl fluoride group mimics the γ-phosphoryl group of ATP and was expected to bind at the terminal-phosphate binding site of both substrates. The authors found that FSBA was competitive with both ATP and polyP$_{35}$ in the ATP- and polyP-dependent reactions respectively.

Additional evidence for a common catalytic center was provided by Horn et al. (1991) and Phillips et al. (1993b) with the enzyme from *P. shermanii*. These authors, using a photo-affinity label of ATP, 8-azido ATP, found that after photolysis it inactivated both the ATP- and polyP-dependent glucokinase activity. The limiting inactivation rate constants for both reactions were essentially identical ($0.015\,s^{-1}$ (Pan 1994). In addition, the products of the two reactions, polyP$_4$ and ADP, afforded almost complete protection against inactivation of their respective glucokinase activities by the photo-activated 8-azido-ATP. Since 8-azido-ATP was found to be a reasonably good substrate (Horn 1991), the inactivation is thought to be specific for the active site.

7.2
Structural Evidence

Structural evidence for a common catalytic center has been obtained with the enzyme from *M. tuberculosis* (Hsieh et al. 1993b) and *P. shermanii* (Pan et al., submitted). Using N-bromosuccinimide (NBS) as a probe for trypto-phan residues, Hsieh et al. (1993b) obtained evidence for the involvement of tryptophan(s) in the catalysis of the *M. tuberculosis* enzyme and suggested that they may be located at a common catalytic site for polyP and ATP. The following experimental findings were cited as evidence for their conclusion:

1. NBS-oxidation of the enzyme resulted in quenching of the intrinsic tryptophan fluorescence with concomitant loss of both the polyP- and ATP-dependent glucokinase activities, which were lost in parallel to >95%.
2. PolyP and ATP afforded similar levels of protection (~65%) against the loss of the polyP and ATP-dependent activities respectively.
3. The product, polyP$_4$, protected both the polyP- and ATP-dependent activities.
4. Likewise, AMP conferred 58% protection of both of the activities.
5. The K_m values for polyP and ATP were not altered by NBS-modification; however, the catalytic efficiencies of both reactions were significantly decreased.
6. The same tryptophan-containing peptide (detected at 280 nm) was found to be protected by polyP or ATP against oxidation by NBS as determined by reverse-phase HPLC peptide mapping and N-terminal sequencing. The peptide sequence corresponded to residues 190–201 based on the deduced protein sequence (Hsieh et al. 1996a).

Pan et al. (submitted) also used NBS to probe for the involvement of tryptophans in the *P. shermanii* enzyme. Most of the results were similar to that observed for the *M. tuberculosis* enzyme. The major difference in their result was the finding of two tryptophan-containing peptides that were oxidized by NBS. Nevertheless, the same two peptides were protected by polyP$_4$ or ADP against oxidation by NBS, as determined by reverse-phase HPLC peptide mapping and N-terminal sequence determination.

More recently, Badola et al. (submitted) performed site-directed mutagenesis studies on Aspartyl-24 of the recombinant enzyme from *M. tuberculosis*. This aspartyl residue is highly conserved in the hexokinases and glucokinases and is located in the "phosphate-1" domain (see Sect. 12). In the brain hexokinase, this residue was shown to interact with MgATP by coordinating with the Mg^{2+} ion (Zeng et al. 1996). Badola et al. (submitted) showed that substitution of this amino acid with residues that are neutral or opposite in charge resulted in a dramatic decrease in the catalytic efficiencies of both the polyP and ATP reactions. However, the binding affinities of polyP and ATP were not affected. The "phosphate-1" domain may thus constitute the common catalytic center. As proposed by Phillips et al. (1993a), the adenosine moiety of ATP and the bulk of the phosphates of polyP may have separate binding sites, while the terminal phosphoryl groups of polyP and ATP share a common site in close proximity to the single glucose binding site where phosphoryl transfer occurs.

8
Mechanism of the PolyP Glucokinase Reaction

Until recently, most of the mechanistic analysis of the polyP glucokinase reaction dealt with the mode of utilization of polyP. The only kinetic analysis was restricted to obtaining the Michaelis constants for different chain lengths of polyP (Pepin and Wood 1986).

8.1
Mechanism of the Mycobacterial Enzyme

Detailed kinetic analysis of the polyP- and ATP-dependent glucokinase reactions of the *M. tuberculosis* enzyme has been performed by Hsieh et al. (1996b). The steady-state kinetic mechanism was solved using initial velocity, product-inhibition and dead-end inhibition analysis.

8.1.1
Kinetic Mechanism of the PolyP-Dependent Reaction

Hsieh et al. (1996b) found that initial velocity experiments in the absence of products were consistent with either a Rapid-Equilibrium Random Bi Bi or a Steady-State Ordered Bi Bi mechanism. Product inhibition experiments to distinguish between the two mechanism ruled out random mechanisms and were consistent with an Ordered Bi Bi mechanism with polyP being the first substrate to add and G6P the last product to dissociate from the enzyme. The order of substrate addition was confirmed by Dead-End inhibition studies with substrate competitive inhibitors of glucose and polyP, namely xylose and AMP, respectively.

8.1.2
Kinetic Mechanism of the ATP-Dependent Reaction

Similar experiments to those described above were also employed by Hsieh et al. (1996b) to solve the ATP-dependent reaction. The analyses were all consistent with an Ordered Bi Bi mechanism, with ATP binding first followed by glucose, and G6P being the last product to dissociate. More recent findings (Kowalczyk and Phillips, unpubl.), however, suggest that the reaction may not be strictly ordered, and could involve a degree of randomness in substrate addition as was found for the *P. shermanii* enzyme (see below). In addition, substrate inhibition by ATP was observed by ATP at pH 7.5 but not at pH 8.6.

8.2
Kinetic Mechanism of the *P. shermanii* Enyzme

The kinetic mechanism for the ATP-dependent glucokinase reaction was solved by Kowalczyk et al. (1996), using steady-state kinetic analysis and isotope exchange at equilibrium. These authors found that the results were consistent with a partially random mechanism (although a kinetically compulsory order of substrate binding was not excluded), where glucose is preferentially bound to free enzyme before ATP, and ADP is preferentially released as the first product followed by G6P. Their results also indicated the formation of abortive complexes.

8.3
Mechanism of PolyP Utilization: Processivity vs Nonprocessivity

The polymeric nature of polyP motivated Pepin and Wood (1986, 1987) to determine whether the utilization of this substrate occurs by a processive or nonprocessive mechanism. They defined a strictly processive mechanism as one in which, following binding, polyP is not released from the enzyme until all the phosphates are transferred to glucose. The nonprocessive phosphorylation, on the other hand, involves dissociation of the polyP from the enzyme after each catalytic event. The authors also defined a quasiprocessive (or not strictly processive) mechanism in which more than one phosphorylation event occurs prior to the dissociation of polyP from the enzyme. Based on the observed gel patterns, Pepin and Wood (1987) concluded that the polyP glucokinase from *P. shermanii* utilized polyP via a quasiprocessive or nonprocessive mechanism. Chain lengths shorter than ~100 were utilized nonprocessively only after the longer chains were consumed.

   The mechanism of utilization of polyP by the *M. tuberculosis* enzyme was recently investigated by Hsieh et al. (1996b) using the product distribution method employed by Pepin and Wood (1986, 1987). For comparison, some of the experiments were also repeated with the enzyme from *P. shermanii* and *P. arabinosum* (unpubl.). They showed that the enzyme from *M. tuberculosis* utilizes a wide range of polyP sizes by a nonprocessive mechanism (Fig. 2). The enzyme from *P. shermanii*, on the other hand, shows a transition from a strictly processive mode with very long polyP (Fig. 3) to a strictly nonprocessive mode with short polyP below ~100 residues. Intermediate sizes of polyP (~100–200 residues) are utilized by a quasiprocessive mechanism, which is evidenced by noticeable broadening of the range of polyP sizes with the reaction time (Hsieh et al. 1996b). The continuous transition from one mode to another indicates that the basis of the observed processivity is kinetic, not structural. This conclusion is supported by the observation (Kowalczyk et al., unpubl.) that the processive mode coincides with the values of the apparent association rate constant, exceeding signifi-

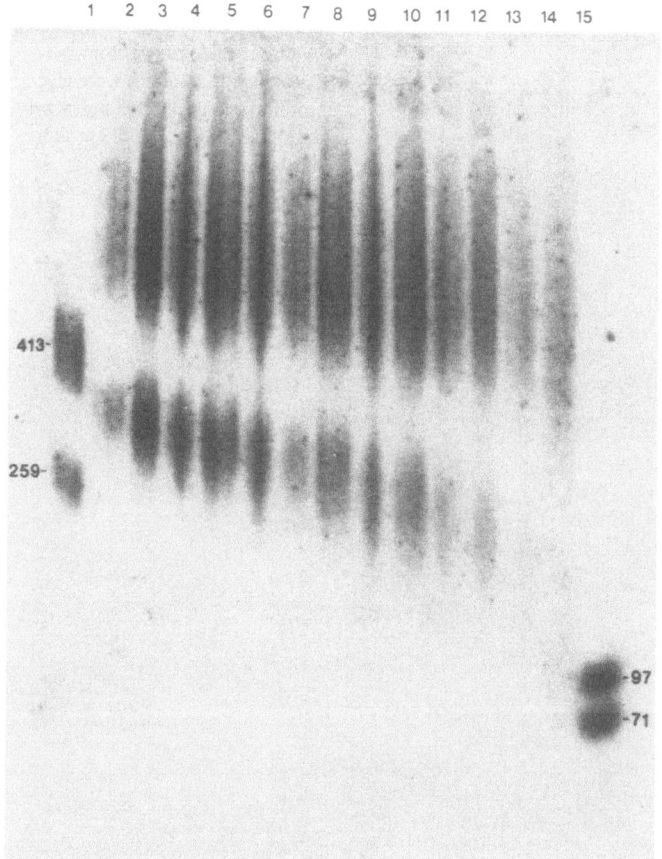

Fig. 2. Nonprocessive utilization of polyP of average chain length 355 and 658 by polyP glucokinase from *M. tuberculosis*. At different time intervals of the reaction (lanes 2–14), aliquots were removed, polyP was isolated by the method of Pepin and Wood (1986) and Clark and Wood (1987), and samples were electrophoresed into a 2% polyacrylamide gel containing 0.75% agarose

cantly the expected diffusion limit: $k_{cat}/K_m = 4 \times 10^{10}$ M^{-1} s^{-1} and 7×10^{10} M^{-1} s^{-1} for polyP$_{300}$ and polyP$_{400}$, respectively. Kowalczyk et al. (submitted) proposed that the kinetic mechanism responsible for this phenomenon is similar to that described for specific protein–nucleic acid interactions (Berg et al. 1981). The enzyme from *P. arabinosum* utilizes polyP by the mechanism identical to that of *M. tuberculosis* (unpubl. observ.)

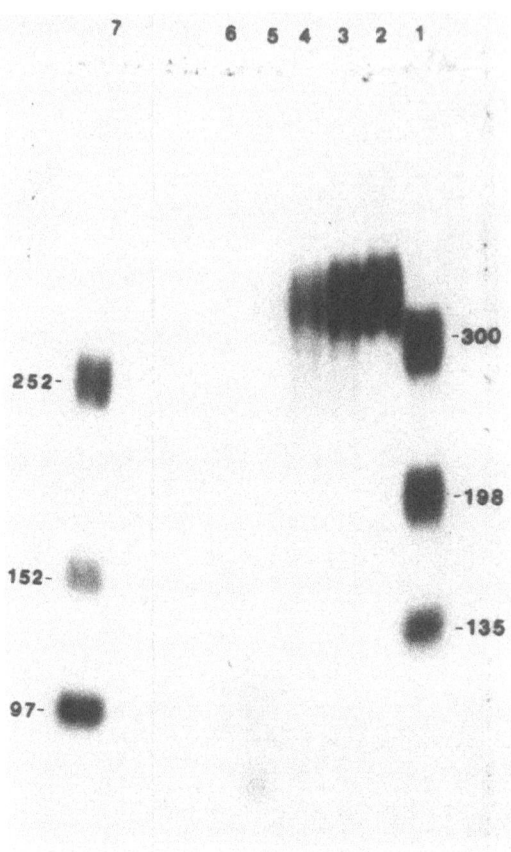

Fig. 3. Processive untilization of polyP of average chain length 355 by polyP glucokinase from *P. shermanii*. Samples were electrophoresed into a 6 % polyacrylamide gel (lanes 1 and 7 = standards)

8.4
Preferential Utilization of Long Chains Over Short Chains Is Due to Better Binding Affinities Rather than Faster Catalysis or Product Release

The preferential utilization of long-chain polyP by the partially purified *P. shermanii* glucokinase was investigated by Pepin and Wood (1986). These authors observed that there was a remarkable decrease in K_m with increasing chain length of polyP. The K_m for polyP$_{700}$ (2×10^{-3} µM) was 2000-fold lower than for polyP$_{30}$. More recently, with the purified enzymes from *M. tuberculosis*, *P. shermanii* and *P. arabinosum* Kowalczyk et al. (submitted) and Pan and Phillips (unpubl.) determined the kinetic parameters for a broad range of polyP of discrete sizes. They found that the apparent association rate constant, k_{cat}/K_m, for polyP increases up to 3 orders of magnitude when the size of polyP increases from 30 to 400 (Fig. 4; Table 1). Since the catalytic con-

Fig. 4. Dependence of k_{cat}/K_m on chain length of polyP for different polyP glucokinases. *Open circles P. shermanii; closed circles M. tuberculosis; squares P. arabinosum*

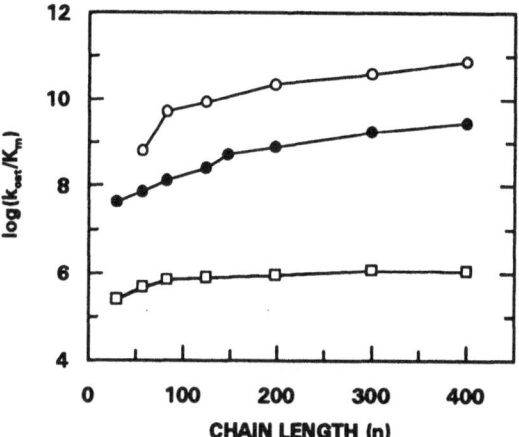

Table 1. Comparison of kinetic parameters for different phosphoryl donors and polyP glucokinases from different species

Species	Substrate	K_m (M)	k_{cat} (s^{-1})	k_{cat}/K_m (M^{-1} s^{-1})
P. shermanii	PolyP$_{35}$	1.2×10^{-6}	57	4.8×10^{7}
	PolyP$_{400}$	5.0×10^{-9}	371	7.4×10^{10}
	ATP	1.5×10^{-3}	25	1.7×10^{4}
M. tuberculosis	PolyP$_{35}$	4.6×10^{-6}	196	4.3×10^{7}
	PolyP$_{400}$	8.0×10^{-8}	163	2.0×10^{9}
	ATP	8.8×10^{-4}	116	1.2×10^{5}
P. arabinosum	PolyP$_{35}$	8.8×10^{-6}	55	6.3×10^{6}
	PolyP$_{400}$	3.7×10^{-5}	41	1.1×10^{6}
	ATP	4.0×10^{-4}	63	1.6×10^{5}

stant, k_{cat}, remains essentially constant, the increase in k_{cat}/K_m results in decreased K_m for longer polyP chains.

The higher k_{cat}/K_m with increasing polyP size cannot be explained by assuming multiple phosphoryl binding sites on the enzyme molecule secondary to the active site. It is more likely that the mechanism for the higher apparent association rate constant for longer polyP compared with short polyP chains is based on faster association of longer polyP chains with the enzyme (high on-rate), and not on tighter polyP binding (small off-rate). Kowalczyk et al. (submitted) proposed that the enzyme could bind to any phosphate residue within the polyP chain. Consequently, the increased target area of longer polyP would account for the increased association rate constant and the apparent stronger binding (lower K_m) of longer substrates. They proposed that the initial binding of polyP to the enzyme is nonspecific. This is followed by a facilitated transfer of the enzyme along the polyP chain by a sliding mechanism to one of its termini where specific binding occurs, followed by catalysis. In support of the facilitated transfer mechanism, they offered the following arguments:

1. In the absence of this mechanism, the nonspecific sites on polyP would compete with the specific sites (i.e., the ends of the chain) for the enzyme and the observed specific binding rate would be lower for longer polyP (Lohmann 1986).
2. The enzyme is more active in the absence of NaCl than in its presence. At low salt concentrations the nonspecific binding constant is expected to be large due to preferred electrostatic interactions between polyP and the enzyme. If the nonspecific binding sites acted solely as a trap, the observed association rate constant would be lower at lower ionic strength due to strong electrostatic interactions between polyP and the enzyme.
3. At 0.6 M NaCl the k_{cat}/K_m is almost identical for polyP, having an average length of 15–132 residues. This agrees with the prediction of the sliding model developed for specific protein–nucleic acid interactions (Berg et al. 1981).

Strong dependence of k_{cat}/K_m and the relative independence of k_{cat} on polyP chain length was also reported for exo-polyphosphatases (Lorenz et al. 1994, Wurst and Kornberg 1994). It is possible that the facilitated transfer mechanism is operating in those cases, too.

8.5
Influence of Glycerol on Kinetic Parameters

Kowalczyk et al. (submitted) analyzed effects of glycerol on the kinetics of polyP glucokinase from *M. tuberculosis*. They found that with ATP as the substrate, the reaction is viscosity independent, while with polyP both k_{cat}/K_m and k_{cat} decrease with increasing viscosity. These results suggest that the binding of polyP is diffusion-controlled and that release of polyP as a product, rather than catalysis, is rate limiting for k_{cat}.

9
Which Reaction Is Favorable in *M. tuberculosis*?

Several observations made by Hsieh et al. (1997) suggest that polyP is the preferred substrate over ATP for the *M. tuberculosis* polyP glucokinase, although the actual preference for polyP in the bacterial cell cannot be evaluated without knowing the cellular polyP concentration. Their observations are as follows:

1. PolyP glucokinase was shown to have broad nucleotide specificity, i.e., besides ATP, it could also efficiently utilize GTP, UTP, TTP, XTP and CTP as the phosphoryl donor (Hsieh et al. 1993a).

2. The overall catalytic efficiency of polyP utilization ($k_{cat}/K_{ia}K_b$) is 1000-fold better than that of ATP.

3. The k_{cat}/K_{polyP} (1.4×10^7 M^{-1} s^{-1}) was 100-fold higher than the k_{cat}/K_{ATP} (1.2×10^5 M^{-1} s^{-1}). For an Ordered Bi kinetic mechanism, k_{cat}/K_m is the true second-order rate constant for the binding of the first substrate (Gawlita et al. 1995), in this case polyP or ATP. Consequently, the true dissociation rate constant for the binary polyP (or ATP)–enzyme complex, which is a measure of binding affinity, can be calculated by multiplying K_{ia} (the dissociation constant) by k_{cat}/K_m. When Hsieh et al. (1996b) compared the dissociation rate constants, they found that the binding of polyP is only ~17-fold tighter than ATP at pH 7.5. At pH 8.6 (the pH optimum for the mycobacterial enzyme) the binding was less than two-fold tighter than that of ATP. Thus, the greater overall catalytic efficiency of polyP utilization over ATP is largely due to faster association of this substrate and not due to its tighter binding.

4. The polyP glucokinase is well adapted to utilize polyP, since the second-order rate constant (k_{cat}/K_m) for the reaction of free enzyme with the mixture of polyP ($k_{cat}/K_m = \sim 10^7$ M^{-1} s^{-1}) is close to the diffusion-controlled limit (Fersht 1985). This means that the fraction of productive collisions is near unity (Simopoulos and Jencks 1964). On the other hand, the rate constant for ATP binding ($k_{cat}/K_m = \sim 10^5$ M^{-1} s^{-1}) is far below the theoretical diffusion-controlled limit, which implies that only a small fraction of collisions are productive.

5. ATP, unlike polyP, was found to cause substrate inhibition at physiological pH and concentration.

10
Cloning and Expression of the PolyP Glucokinase Gene from *M. tuberculosis*

Sequence alignment analysis of glucokinases and hexokinases from a number of eukaryotic cells and from a few prokaryotic glucokinases revealed a number of conserved domains, some of which were proposed to be involved in substrate binding. It was of interest, therefore, to know the relationship of the polyP glucokinase to the ATP-dependent hexokinases and glucokinases. Hsieh et al. (1996a) cloned and sequenced the polyP glucokinase gene (*ppgk*) from *M. tuberculosis* $H_{37}R_v$. They further expressed and purified the recombinant enzyme (re-PPGK) from *Escherichia coli* and showed that the cloned *ppgk* encodes a single polypeptide chain containing both the polyP- and ATP-dependent glucokinase activities.

To clone the *ppgk* gene, two degenerate oligonucleotide primers, based on two internal peptide sequences of the glucokinase from the avirulent $H_{37}R_a$ strain, were used in a PCR reaction with the genomic DNA from the same strain as the template. The 365-bp PCR product was labeled with [^{32}P]CTP and was used to screen a λ gt-11 genomic library of *M. tuberculosis* H₃₇Rᵥ ("Erdman" virulent strain). The λ-phage DNAs that contained the *ppgk* inserts were selected for subcloning into pBluescript vector and identified by restriction enzyme digestion and Southern blotting. A 2.3-kb EcoR1 insert of the pBS-*ppgk* clone was found to contain the open reading frame of the *ppgk* gene.

11
Characterization of the PolyP Glucokinase Gene

The DNA sequence revealed a putative Shine-Delgarno sequence (GAGGAG) located 5 bp upstream of the translation-initiation site. However, an *E. coli*-like consensus promoter region was not observed, which is consistent with the findings from other *M. tuberculosis* genes (Stover et al. 1991; Cirillo et al. 1994; Kremer et al. 1995). The *ppgk* gene translated to a 265 amino acid protein with a calculated mass of 27 400 daltons. Sequence analysis search identified an unknown gene sequence (gi699175) from a *Mycobacterium leprae* B1764 cosmid library that was found to have 71 % sequence identity with that of the *ppkg* gene. The gene sequence of gi699175 from *M. leprae* is in all likelihood the polyP glucokinase of this organism with an open reading frame coding for a protein of 35 kDa as was found for polyP glucokinase subunit from other bacteria.

12
Identification of Substrate Binding Domains
of PolyP Glucokinase and Other ATP Hexokinases

Bork et al. (1992, 1993) found that the three-dimensional structures of hexokinases, actin and heat shock protein (Hsp70) families contained common motifs interacting within the vicinity of the ATP molecule. As depicted in Fig. 5, these are "phosphate-1" and "phosphate-2" motifs contacting the

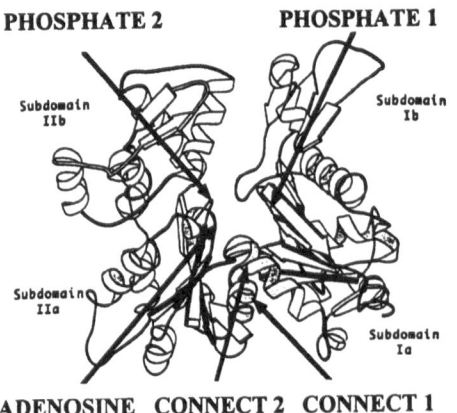

PHOSPHATE 2   **PHOSPHATE 1**

Subdomain IIb

Subdomain Ib

Subdomain IIa

Subdomain Ia

ADENOSINE   CONNECT 2   CONNECT 1

Fig. 5. Common structural motifs of actin and related protein families. (Modification after Bork et al. 1992)

β- and γ-phosphates of ATP, the "connect-1" and "connect-2" motifs which are the two hinge-regions at the interface between the subdomains, and an "adenosine" motif which contacts the adenine ring of ATP. Residues in these common motifs were highly conserved and proposed to be functionally important in the eukaryotic hexokinase family. In order to identify these domains in the polyP glucokinases from *M. tuberculosis*, Hsieh (1996) and Hsieh et al. (1996a) performed sequence alignment analysis using programs such as MACAW (Altschul et al. 1990; available on the web at http://www.ncbi.nlm.nih.gov) and DALI (Holm and Sander 1993; available through the European Molecular Biology Laboratory web site at http://www.sander.embl-heidelberg.de). These authors compared the sequences of polyP glucokinase with the sequences from hexokinase B of yeast, the glucokinase from yeast (Albig and Entian 1988), *Streptomyces coelicolor* (Angell et al. 1994), *Staphyloccus xylosus* (unpubl. data, GenBank accession number gi666116), *E. coli* (unpubl., gi736416), *Brucella abortus* (unpubl., gi719301) and *Zymomonas mobilis* (Barnell et al. 1990; gi123900). Through these analyses, Hsieh (1996) found that the five structurally and functionally conserved domains in the hexokinase family, ("phosphate-1", "phosphate-2", "connect-1", "connect-2" and "adenosine") and a "glucose" site were also identified in the polyP glucokinase and other prokaryotic glucokinases. Based on these sequences conversations, Hsieh et al. (1996a) tentatively identified the substrate-binding domains in the polyP glucokinase from *M. tuberculosis*.

13
Proposed Substrate-Binding Domains of PolyP Glucokinase

13.1
The ATP-Binding Site

The ATP-binding site of polyP glucokinase and other prokaryotic glucokinases was tentatively assigned by Hsieh et al. (1996a) to four regions. As shown in Fig. 6, the proposed motifs are: (1) "phosphate-1", Z-Z-D^{24}-Z-G^{26}-G-S-X-X-K (where Z is a hydrophobic amino acid and X is any amino acid); (2) "phosphate-2", T-X-G^{149}-T^{150}-G-I-G-X-A-X-I; (3) "connect-1" hinge region, Z-X-N^{121}-D^{122}-A-X-A-A; and (4) "connect-2" hinge region, G-I-X-G^{254}-A-A-X.

13.2
The Adenosine Site

Despite the lack of sequence similarities between the eukaryotic hexokinases and prokaryotic glucokinases in the putative adenosine site, Hsieh (1996) proposed an "adenosine site" in the polyP glucokinase based on the following rationale:

Fig. 6. Deduced amino acid sequence of *M. tuberculosis* poly P/ATP glucokinase. Labeled *highlighted areas* denote putative structural and functional domains

(1) this motif was found to contain a cluster of hydrophobic residues as well as a number of glycines which could potentially be involved in adenine binding;

(2) it is located on the N-terminal side of "connect-2" as was found for the eukaryotic hexokinase;

(3) the predicted secondary structure comprising this region is similar to that of the hexokinase (loop-β sheet loop). All these features are common for the proposed adenine site of yeast hexokinase.

13.3
The Glucose-Binding Site

A proposed glucose-binding site peptide was isolated and sequenced from rat brain type I hexokinase (Schirch and Wilson 1987) using the reactive glucose analog. N-(bromoacetyl)D-glucosamine. The peptide sequence was shown to be highly conserved in yeast and mammalian glucokinases and hexokinases (Bork et al. 1993). Analysis of the crystal structure of yeast hexokinase complexed with glucose identified a number of residues that were involved in the binding of glucose (Bennet and Steitz 1980). However, there was very low sequence identity in the "glucose site" with respect to the rest of the prokaryotic glucokinase. Nevertheless, Hsieh (1996) proposed that the sequence depicted in Fig. 6 as the "glucose-site" is the putative glucose-binding site of polyP glucokinase based on the following: (1) the predicted secondary structure of this site (loop-β sheet-loop) is similar to that of the glucose-binding site of yeast hexokinase; (2) this region is highly conserved among the prokaryotic glucokinases; (3) this region is also located between the "phosphate-1" and "connect-1", which is in a similar location to that of the glucose site in the yeast enzyme.

13.4
The PolyP-Binding Site

The identity of sequences involved in polyP binding is less clear. As described in Section 7.2, Hsieh et al. (1993b) identified a tryptophan-containing peptide in the *M. tuberculosis* polyP glucokinase that may be located at a common catalytic center. The deduced sequence in the $H_{37}R_v$ strain (denoted as "phosphate-3" in Fig. 6) may represent part of the polyP-binding site. The tryptophans in this regions were found to be selectively oxidized by N-bromosuccinimide with concomitant loss of enzymatic activity. Because tetrapolyphosphate and long-chain polyP protected against this oxidation and the loss of activity, Hsieh et al. (1996a) proposed that residues 177–216 ("phosphate-3") might be a binding site for polyP. Several charged groups around this region, Lys^{188}, Glu^{189}, Lys^{190}, Asp^{192}, Lys^{187} and Lys^{200}, may fulfill the requirement for MgpolyP binding. Preliminary site-directed mutagenesis of Trp^{193} and Trp^{198} to glycines resulted in complete loss of catalytic activity (Badola and Phillips, unpubl. data).

14
Evolutionary Significance of PolyP Glucokinase

14.1
Transition from PolyP to ATP as the Phosphoryl Donor
in Glucose Phosphorylation

As described earlier, in the phylogenetically older organisms of the *Actinomycetales* (e.g., *Micrococci* and *Propionibacteria*) the polyP glucokinase activity exceeded the ATP glucokinase activity several-fold, whereas in the younger representatives of the *Actinomycetales* (e.g., *Mycobacteria* and *Nocardia* and the true *Actionomycetes*) the ATP-dependent glucokinase activity was greater. The most accurate and state-of-the-art method to establish phylogenetic relationship, namely 16S rRNA sequences, has been used to verify the evolutionary status of different organisms in the *Actinomycetales* (Stackebrandt and Woese 1981). Using this method, *Propionibacteria* was also determined to be phylogenetically older than *Mycobacteria*.

In order to understand why polyP was a better substrate than ATP for the glucokinase from the older organisms compared to the phylogenetically newer *Actinomycetales* bacteria, Phillips et al. (unpubl.) conducted extensive kinetic analyses of these enzymes. They determined various kinetic parameters of the glucokinase from the three different organisms (*P. shermanii*, *M. tuberculosis* and *P. arabinosum*), representing different stages in the phylogenetic order. When the substrate specificity constant, k_{cat}/K_m ratios, for the utilization of polyP and ATP were compared, they found that the ratios decreased progressively with the enzymes from older to younger organisms

(Fig. 4). As shown in Table 1, the k_{cat}/K_m ratios of polyP$_{35}$ to ATP were 2800, 200 and 30 for the glucokinases from *P. shermanii*, *M. tuberculosis* and *P. arabionosum*, respectively. When a longer polyP chain length of 400 is used in comparing the k_{cat}/K_m ratios of polyP/ATP, a more dramatic difference is seen. The k_{cat}/K_m ratio of polyP$_{400}$ to ATP is 6 orders of magnitude for the enzyme from the older *P. shermanii*, while it is 4 orders of magnitude for the more recent *M. tuberculosis* enzyme and only an order of magnitude different for the *P. arabionsum* enzyme. These results support the hypothesis that the glucokinase in the earliest organisms may have predominantly been dependent on polyP rather than ATP. The results shown in Table 1 also demonstrate that the longer polyP$_{400}$ is a better substrate (k_{cat}/K_m) than the shorter polyP$_{35}$ for all three species. The reason for the more efficient utilization of the longer chain length polyP was due to the faster association rather than a more rapid catalysis, as detailed in Section 8.4. Although the longer chain polyPs are better substrates, in terms of k_{cat}/K_m for all three species, the longer chains are more efficiently utilized by the glucokinases from the older organisms. As shown in Table 1, the k_{cat}/K_m for polyP$_{400}$ is ~10^{11} M^{-1} s^{-1} for the enzyme from *P. shermanii* which is a 3-orders for magnitude increase from the k_{cat}/K_m of polyP$_{35}$ (~10^7 M^{-1} s^{-1}). This value far exceeds the diffusion controlled limit of substrate utilization. In contrast, the k_{cat}/K_m of polyP$_{400}$ for the newer *Mycobacterial* enzyme is 10^9 M^{-1} s^{-1}, while that of the *P. arabinosum* enzyme is 10^6 M^{-1} s^{-1}. Hence, there is a progressive decrease in the efficiency of polyP utilization by the glucokinases, from the older to newer organisms. All these observations suggest that there might have been a gradual transition from polyP to ATP as the phosphoryl donor in glucose phosphorylation during the course of evolution.

14.2
Bifunctional PolyP Glucokinase May be an "Intermediate" in the Evolution of Hexokinase and Glucokinases

As described above, the polyP glucokinases from the older organisms had a higher preference for polyP over ATP compared to the enzyme from the more recent organism. The question then arose of how the enzymes evolved to a strictly ATP-utilizing glucokinase in the more evolved bacteria and eukaryotes. To obtain some insights into this hypothetical transition from polyP to ATP as the phosphoryl donor in glucose phosphorylation, Phillips et al. (unpubl.) compared the properties of ATP utilization in the bifunctional polyP/ATP glucokinases with those of the hexokinases and glucokinases that were strictly dependent on either ATP or other nucleoside triphosphates but unable to utilize polyP. A fascinating observation was that the bifunctional polyP glucokinases from *P. shermanii* (Kowalczyk et al. 1996), *M. tuberculosis* (Hsieh et al. 1993a) and *P. arabinosum* (Horn 1991) displayed broad nucleotide specificity, i.e., besides ATP, they could efficiently utilize GTP, UTP, TTP, XTP, CTP and dATP as the phosphoryl donor. The ATP-dependent glucokinase from *Bacillus stearothormophilus* which could not utilize polyP was also found to efficiently utilize GTP, dATD, UTP and CTP. The glucokinase from the more evolved *Zymomonas mobilis* and *Aerobacter aerogenes* could only use GTP albeit poorly, besides its predominant phosphoryl donor, ATP. In the more highly evolved organisms, like the yeast glucokinase, the mammalian hexokinase D and hexokinase A, ATP was the sole phosphoryl donor. None of the other nucleoside triphosphates (NTP) nor polyP could serve as substrates.

One possible scenario to account for the differences in the nucleotide substrate specificities of prokaryotic and eukaryotic glucokinases/hexokinases is that the ATP-binding site of the eukaryotic enzymes originated from a nonspecific hydrophobic region capable of accommodating a variety of structurally related nucleotide substrates (e.g., GTP, TTP, UTP and CTP) and subsequently evolved into a more specific ATP-binding site. It is possible that an ancient glucokinase may have consisted of only the polyP-binding site, while a nonspecific NTP site evolved later to generate the bifunctional polyP/NTP glucokinase. It is conceivable that the bacterial glucokinases that utilize NTP but not polyP (e.g., *B. stearothermophilus*) may represent an intermediate stage between the bifunctional polyP glucokinase and the strictly ATP-dependent eukaryotic hexokinases/glucokinases. As outlined in Eq. (2) below, the bifunctional polyP/NTP glucokinase may be the intermediate between the hypothetical polyP glucokinase (strictly polyP-dependent) and the ATP glucokinases:

$$\text{polyP-GK(?)} \rightarrow \text{polyP/NTP-GK} \rightarrow \text{NTP-GK} \rightarrow \text{ATP-GK.} \tag{2}$$

15
Concluding Remarks

Based on the data on chemical evolution, polyphosphates are known to have existed in pre-biological times and it has been suggested that they may have been the phosphoryl donor for the synthesis of nucleic acids and ATP (West and Ponnamperuma 1970). The biological evidence for their appearance in primordial times comes from the observation that these polymers were utilized by some primitive microorganisms, including those from the Archaebacteria (Woese 1987). In 1965, Lipmann first proposed that ancient organisms may have utilized pyrophosphate and polyP instead of ATP in their metabolic reactions. He also suggested that polyP was probably replaced by ATP during the course of evolution. As detailed in this review, the discovery of the bifunctional polyP/ATP glucokinases from a number of bacteria has given credence to Lipmann's hypothesis. The observation that the ratios of polyP to ATP utilization decreased from the more primitive to more recent members of the bacteria in the *Actinomycetales* order has further strengthened our notion that there might have been a gradual transition from polyP to ATP as the phosphoryl donor in glucose phosphorylation. In order to gain greater insights into the mechanistic and structural basis by which these enzymes have evolved to use polyP less efficiently, while becoming more adapt in utilizing ATP, a greater number of polyP/ATP glucokinases with differing ratios in polyP/ATP activities have to be cloned and their kinetic parameters and substrate-binding sites determined. Site-directed mutagenesis could be used to knock out the adenine-binding site, to generate the hypothetical polyP glucokinase with a strict dependence on polyP but not ATP. This mutant enzyme could represent the ancestral form of the bifunctional polyP glucokinase and ATP hexokinase. The exclusive polyP glucokinase construct can be used to test for the physiological relevance of polyP utilization. This could be achieved by complementation cloning of the ATP glucokinase-deficient mutant into the glucose-deficient *E. coli* ZSC112 strain, and determination of whether glucose utilization is restored in the *E. coli* strain. By these techniques, it may be possible to conclusively demonstrate that polyP is used physiologically in glucose phosphorylation.

Acknowledgements. We thank Pat Amato for secretarial assistance. N.F.B.P. acknowledges the support of the National Institute of Health through Grant GM 29569.

References

Albig, W, Entian KD (1988) Structure of yeast glucokinase, a strongly diverged specific aldo-hexose-phosphorylating isoenzyme. Gene 73: 141–152

Altschul SF, Gish W, Miller W, Myers EW, Lipman DJ (1990) Basic local alignment search tool. J Mol Biol 215: 403–410

Angell S, Lewis CG, Buttner MJ, Bibb MJ (1994) Glucose repression in *Streptomyces coelicolor* A3(2): a likely regulatory role for glucose kinase. Mol Gen Genet 244: 135–143

Barnell WO, Yi KC, Conway T (1990) Sequence and genetic organization of a *Zymomonas mobilis* gene cluster that encodes several enzymes of glucose metabolism. J Bacteriol 172: 7227–7240

Bennet WS, Steitz TA (1980) Structure of a complex between hexokinase A and glucose. J Mol Biol 140: 211–230

Berg OG, Winter RB, von Hippel PH (1981) Diffusion-driven mechanisms of protein translocation on nucleic acids. I. Models and theory. Biochemistry 20: 6929–6948

Bobyk MA, Afinogenova AV, Dubinskaya MV, Lambina VA, Kulaev IS (1980) Detection of polyphosphates and enzymes of polyphosphate metabolism in *Bdellovibrio bacteriovorus*. Zentralbl Bakteriol Parasitenkd Infektionskrankh Hyg 135: 461–466

Bork P, Sander C, Valencia A (1992) An ATPase domain common to prokaryotic cell cycle proteins, sugar kinases, actin, and hsp 70 shock proteins. Proc Natl Acad Sci USA 89: 7290–7294

Bork P, Sander C, Valencia A (1993) Convergent evolution of similar enzymatic function on different protein folds: the hexokinase, ribokinase, and galactokinase families of sugar kinases. Protein Sci 2: 31–40

Chevillard C, Cárdenas ML, Cornish-Bowden A (1993) The competition plot: a simple test of whether two reactions occur at the same active site. Biochem J 289: 599: 604

Cirillo JD, Weisbrod TR, Pascopella L, Bloom BR, Jacobs WR Jr (1994) Isolation and characterization of the aspartokinase and aspartate semialdehyde dehydrogenase operon from mycobacteria. Mol Microbiol 11: 629–639

Clark JE (1990) Purification of polyphosphate glucokinase from *Propionibacterium shermanii*. In: Dawes EA (ed) Novel biodegradable microbial polymers. Kluwer, Dordrecht, pp 213–221

Clark JE, Wood HG (1987) Preparation of standards and determination of sizes of long-chain polyphosphates by gel electrophoresis. Anal Biochem 161: 280–290

Clonis YD, Lowe CR (1980) Triazine dyes, a new class of affinity labels for nucleotide-dependent enzymes. Biochem J 191: 247–251

Dirheimer G, Ebel JP (1968) Purification et propértiés d'une polyphosphate:glucose et glucosamine 6-phosphotransférase à partir de *Corynebacterium xerosis*. Bull Soc Chim Biol 50: 1933–1947

Fersht A (1985) Enzyme structure and mechanism. Freeman, New York

Gawlita E, Caldwell WS, O'Leary MH, Paneth P, Anderson VE (1995) Kinetic isotope effects on substrate association: reactions of phosphoenolpyruvate with phosphoenolpyruvate carboxylase and pyruvate kinase. Biochemistry 34: 2577–2583

Girbal E, Binot RA, Monsan PF (1989) Production, purification and kinetic studies of free and immobilized polyphosphate:glucose-6-phosphotransferase from *Mycobacterium phlei*. Enzyme Micro Technol 11: 519–527

Gonzáles F, Fernández-Vivas A, Arias JM, Montoya E (1990) Polyphosphate glucokinase and ATP glucokinase activities in *Myxococcus coralloides* D. Arch Microbiol 154: 438–442

Goward CR, Scawen MD, Atkinsin T (1987) The inhibition of glucokinase and glycerokinase from *Bacillus stearothermophilus* by the triazine dye procion blue MX-3G. Biochem J 246: 83–86

Harold FM (1966) Inorganic polyphosphates in biology: structure, metabolism and functions. Bacteriol Rev 30: 772–794

Holm L, Sander C (1993) Protein structure comparison by alignment of distance matrices. J Mol Biol 233: 123–138

Horn PJ (1991) Characterization of polyphosphate glucokinase. MS Thesis, Case Western Reserve University, Cleveland, OH

Horn PJ, Phillips NFB, Wood HG (1991) Photoinactivation of a polyphosphate/ATP dependent glucokinase by 8-azido-adenosine-5'-triphosphate. FASEB J 5: A420

Hsieh PC (1996) Molecular cloning and characterization of polyphosphate-glucokinase from *Mycobacterium tuberculosis*. PhD Thesis, Case Western Reserve University, Cleveland, OH

Hsieh PC, Shenoy BC, Jentoft JE, Phillips NFB (1993a) Purification of polyphosphate and ATP glucose phosphotransferase from *Mycobacterium tuberculosis* H$_{37}$Ra: evidence that poly(P) and ATP glucokinase activities are catalyzed by the same enzyme. Protein Expr Purif 4: 76–84

Hsieh PC, Shenoy BC, Haase FC, Jentoft JE, Phillips NFB (1993b) Involvement of tryptophan(s) at the active site of polyphosphate/ATP glucokinase from *Mycobacterium tuberculosis*. Biochemistry 32: 6243–6249

Hsieh PC, Shenoy BC, Samols D, Phillips NFB (1996a) Cloning, expression and characterization of polyphosphate glucokinase from *Mycobacterium tuberculosis*. J Biol Chem 271: 4909–4915

Hsieh PC, Kowalczyk TH, Phillips NFB (1996b) Kinetic mechanisms of polyphosphate glucokinase from *Mycobacterium tuberculosis*. Biochemistry 35: 9772–9781

Kornberg A (1995) Inorganic polyphosphate: toward making a forgotten polymer unforgettable. J Bacteriol 177: 491–496

Kowalczyk TH, Phillips NFB (1993) Determination of endopolyphosphatase using polyphosphate glucokinase. Anal Biochem 212: 194–205

Kowalczyk TH, Horn PJ, Pan WH, Phillips NFB (1996) Initial rate and equilibrium isotope exchange studies on the ATP-dependent activity of polyphosphate glucokinase from *Propionibacterium shermanii*. Biochemistry 35: 6777–6785

Kowalska H, Pastuszak I, Szymona M (1979) Multiple forms of polyphosphate-glucose phosphotransferase from *Mycobacterium tuberculosis* H_{37}Ra. Folia Soc Sci Lublinensis 21–112

Krasil'nikov NA (1949) A guide to bacteria and *Actinomycetes*. Izd Acad Nauk SSSR, Moscow

Kremer L, Baulard A, Estaquier J, Content J, Capron A, Locht C (1995) Analysis of the *Mycobacterium tuberculosis* 85A antigen promoter region. J Bacteriol 177: 642–653

Kulaev IS (1979) The biochemistry of inorganic polyphosphates. Wiley, New York

Kulaev IS, Vagabov VM (1983) Polyphosphate metabolism in microorganisms. Adv Microb Physiol 24: 83–171

Liebermann L (1888) Über das Nuclein der Hefe und künstliche Darstellung eines Nuclein-Eiweiss und Metaphosphorsäure. Ber Dsch Chem Ges 21–598

Lipmann F (1965) In the origins of prebiological system. Academic Press, New York, pp 259–280

Lipmann F (1971) Wanderings of a biochemist. Wiley, New York

Lohman TM (1986) Kinetics of protein–nucleic acid interactions: use of salt effects to probe mechanisms of interaction. CRC Crit Rev Biochem 19: 191–245

Lorenz B, Muller WEG, Kulaev IS, Schroder HC (1994) Purification and characterization of an exopolyphosphatase from *Saccharomyces cerevisiae*. J Biol Chem 269: 22198–22204

Pan WH (1994) The characterization of polyphosphate/ATP glucokinase from *P. shermanii*. MS Thesis, Case Western Reserve University, Cleveland, OH

Pastuszak I, Szymona M (1980) Occurrence of a large molecular size from of polyphosphate-glucose phosphotransferase in extracts of *Mycobacterium tuberculosis* H_{37}Ra. Acta Microbiol Pol 29: 49–56

Pepin CA, Wood HG (1986) Polyphosphate glucokinase from *Propionibacterium shermanii*. Kinetics and demonstration that the mechanism involves both processive and nonprocessive type reactions. J Biol Chem 261: 4476–4480

Pepin CA, Wood HG (1987) The mechanism of utilization of polyphosphate by polyphosphate glucokinase from *Propionibacterium shermanii*. J Biol Chem 262: 5223–5226

Pepin CA, Wood HG, Robinson NA (1986) Determination of the size of polyphosphates with polyphosphate glucokinase. Biochem Int 12: 111–123

Phillips NFB, Horn PJ, Wood HG (1993a) The polyphosphate- and ATP-dependent glucokinase from *Propionibacterium shermanii*: both activities are catalyzed by the same protein. Arch Biochem Biophys 300: 309–319

Phillips NFB, Shenoy BC, Horn PJ, Pan WH (1993b) Photo-affinity labeling of the ATP binding region of poly(P)/ATP glucokinase. FASEB J 7: 1278

Rudnik H, Hendrich S, Pilatus U, Blotevogel KH (1990) Phosphate accumulation and the occurrence of polyphosphates and cyclic 2,3-diphosphoglycerate in *Methanosarcina frisia*. Arch Microbiol 154: 584–588

Scherer PA, Bochem HP (1983) Ultrastructural investigation of 12 *Methanosarcinae* and related species grown on methanol for occurrence of polyphosphatelike inclusions. Can J Microbiol 29: 1190–1199

Schirch DM, Wilson JE (1987) Rat brain hexokinase: amino acid sequence at the substrate hexose binding site is homologous to that of yeast hexokinase. Arch Biochem Biophys 257: 1–12

Simopoulos TT, Jencks WP (1994) Akaline phosphatase is an almost perfect enzyme. Biochemistry 33: 10375–10380

Stackebrandt, E, Woese CR (1981) Towards a phylogeny of the *Actinomycetes* and related organisms. Curr Microbiol 5: 197–202

Stover CK, de la Cruz VF, Fuerst TR, Burlein JE, Benson LA, Bennett LT, Bansal GP, Young JF, Lee MH, Hatfull GE et al. (1991) New use of BCG for recombinant vaccines. Nature 351: 456–460

Szymona M (1957) Utilization of inorganic polyphosphates for phosphorylation of glucose in *Mycobacterium phlei*. Bull Acad Pol Sci Ser Sci Biol 5: 379: 381

Szymona M, Ostrowski W (1964) Inorganic polyphosphate glucokinase of *Mycobacterium phlei*. Biochim Biophys Acta 85: 283–295

Szymona M, Szymona O (1961) Participation of volutin in the hexokinase reaction of *Coryne-bacterium diphtheriae*. Bull Acad Polen Sci Ser Sci Biol 9: 371

Szymona M, Widomski J (1974) A kinetic study on inorganic polyphosphate glucokinase from *Mycobacterium tuberculosis* H_{37}Ra. Physiol Chem Phys 6: 393–404

Szymona M, Szymona O, Kulesza S (1962) On the occurrence of inorganic polyphosphate hexo-kinase in some microorganisms. Acta Microbiol Pol 11: 287–300

Szymona M, Kowalska H, Pastuszak I (1977) Polyphosphate-glucose phosphotransferase. Puri-fication of *Mycobacterium tuberculosis* H_{37}Ra enzyme to apparent homogeneity. Acta Bio-chim Pol 24: 133–142

Szymona O, Szumilo T (1966) Adenosine triphosphate and inorganic polyphosphate fructoki-nases of *Mycobacterium phlei*. Acta Biochim Pol 13: 129

Szymona O, Szymona M (1978) Multiple forms of polyphosphate-glucose phosphotransferase in various *Mycobacterium* strains. Acta Microbiol Pol 27: 73

Szymona O, Szymona M (1979) Polyphosphate- and ATP-glucose phosphotransferase activities of *Nocardia minima*. Acta Microbiol Pol 28: 153–160

Szymona O, Kowalska H, Szymona M (1969) Search for inducible sugar kinases in *Mycobacte-rium phlei*. Ann Univ Mariae Curie-Sklodowska Sect D Med 24: 1

West MW, Ponnamperuma C (1970) Chemical evolution and the origin of life. A comprehensive bibliography. Space Life Sci 2: 225–295

Woese CR (1987) Bacterial evolution. Microbiol Rev 51: 221–27

Wood HG, Clark JE (1988) Biological aspects of inorganic polyphosphates. Annu Rev Biochem 57: 235–260

Wood HG, Goss NH (1985) Phosphorylating enzymes of the propionic acid bacteria and the roles of ATP, inorganic pyrophosphate, and polyphosphates. Proc Natl Acad Sci USA 82: 312–315

Wurst H, Kornberg A (1994) A soluble exopolyphosphastase of *Saccharomyces cerevisiae*. Puri-fication and characterization. J Biol Chem 269: 10996–11001

Yamagata Y, Watanabe H, Saitoh M, Namba T (1991) Volcanic production of polyphosphates and its relevance to prebiotic evolution. Nature 352: 516–519

Zeng C, Alesin EA, Hardie JB, Harrison RB, Fromm JH (1996) ATP-binding site of human brain hexokinase as studied by molecular modelling and site-directed mutagenesis. J Bio Chem 35: 13157–64

Cytoplasmic Inorganic Pyrophosphatase

A.A. Baykov[1], B.S. Cooperman[2], A. Goldman[3], and R. Lahti[4]

1
Introduction

Pyrophosphate (PP_i) is the smallest member of the polyphosphate family and is formed by two phosphate (P_i) residues linked by a phosphoanhydride bond. A specific enzyme hydrolyzing PP_i to P_i was discovered in animal tissues in 1928 (Kay) and later in a great many other organisms and cell types, in virtually all in which it has been sought. Its initial name was "pyrophosphatase", later elongated with a questionable "inorganic".

The ubiquity of pyrophosphatase (PPase) results from its central role in cell metabolism. A number of vital reactions, such as the syntheses of nucleic acids, proteins, lipids and polysaccharides are driven by nucleoside triphosphate hydrolysis to nucleoside monophosphate and PP_i. In adult man, production of PP_i via these reactions reaches several kilograms per day. Removal of PP_i by PPase drives these reactions in the direction of synthesis (Kornberg 1962). Although PPase is quite active and generally comprises as much as 0.1–0.5% of cell protein, cellular PP_i concentration is maintained at micromolar level (Veech et al. 1980), which exceeds its equilibrium concentration by two orders of magnitude. The value of ΔG^0 for the reaction $PP_i \leftrightarrow 2P_i$ is about -4 kcal/mol under conditions close to physiological (Flodgaard and Fleron 1974; de Meis 1984), and the equilibrium concentration of PP_i in a mixture with 10 mM P_i is only about 0.1 µM. At micromolar level, PP_i affects a number of cell functions, dictating a need for regulating PPase activity (Veech et al. 1980).

Plants and certain bacteria have a membrane-bound PPase, which works as a reversible proton pump. In these organisms, the PP_i level is much higher, and PP_i is used as a phosphoryl and energy donor, instead of or in parallel with ATP. Membrane-bound PPase differ in many respects from sol-

[1] A.N. Belozersky Institute of Physico-Chemical Biology, Moscow State University, Moscow 119899, Russia.
[2] Department of Chemistry, University of Pennsylvania, Pennsylvania 19104-6323, USA.
[3] Department of Biochemistry, University of Turku, Centre for Biotechnology, University of Turku and Åbo Akademi University, 20251 Turku, Finland.
[4] Department of Biochemistry, University of Turku, 20500 Turku, Finland.

Progress in Molecular and Subcellular Biology, Vol. 23
H. C. Schröder, W. E. G. Müller (Eds.)
© Springer-Verlag Berlin Heidelberg 1999

uble cytoplasmic PPases and are beyond the scope of this survey. Mitochondrial PPase appears to occupy an intermediary position – although it contains noncatalytic subunits, which anchor it to the inner mitochondrial membrane (Volk et al. 1983), its catalytic subunits are structurally and functionally similar to cytoplasmic PPase (Lundin et al. 1991) and it does not work as a proton pump. Reviews on membrane PPase have been published recently (Baltscheffsky and Baltscheffsky 1992; Rea and Poole 1993).

Interest in PPase is twofold. First, it provides a simple model for the studies of the chemical mechanism of the enzyme-catalyzed reversible phosphoryl transfer to water. Information obtained in these studies may be pertinent to other enzymes catalyzing similar reactions of longer inorganic polyphosphates or organic polyphosphates, such as ATP. As will be shown below, PPase has clearly become the best-studied enzyme of this group. Second, its central role in PP_i metabolism makes PPase an attractive larget for the treatment of diseases associated with disorders of PP_i metabolism in man (Russell 1976). As shown by Sonnewald (1992), increasing PPase level by gene techniques has a dramatic influence on plant cell metabolism.

The present survey, while providing also some background, focuses on most recent researches in soluble PPase structure and mechanism, with an emphasis on the best-studied PPases, those of *Saccharomyces cerevisiae* (Y-PPase) and *Escherichia coli* (E-PPase). Earlier reviews on cytoplasmic PPase were published by Kunitz and Robbins (1961), Butler (1971), Josse and Wong (1971), Cooperman (1982), Lahti (1983) and Cooperman et al. (1992).

2
Catalytic Properties

2.1
Substrate and Metal Activator Specificity

PPase is active only in the presence of divalent cations, with Mg^{2+} conferring the highest activity and being the physiological activator. Inorganic tri- and tetraphosphates are the only naturally occurring substrates beside PP_i that are converted by PPase in the presence of Mg^{2+} (Josse 1966). The triphosphatase activity of Y-PPase is quite low at pH close to neutrality ($< 0.1\%$ of activity with PP_i) but greatly increases with pH, in contrast to PP_i-hydrolyzing activity, which is maximal at pH 6.5–7 (Shafranskii et al. 1977). The physiological significance, if any, ot these alternative activities of PPase is unkown. It is possible that triphosphate and, probably, other polyphosphates act as PPase inhibitors in vivo (Josse 1966). By contrast, PPase isolated from the archeon *Methanothrix soehngenii* hydrolyzes tri- and tetraphosphate at 44 and 8% of PP_i hydrolysis rate (Jetten et al. 1992) and may therefore be involved in their metabolism. The effectiveness of cations as activators of Y-PPase falls in the order $Mg^{2+} > Zn^{2+} > Mn^{2+} > Co^{2+}$ (Kunitz and Robbins 1961). In the presence of Zn^{2+}, Mn^{2+} or Co^{2+}, but not Mg^{2+}, PPase acquires the ability also to hydrolyze monoesters of pyro- and tri-phosphate, such as ADP and ATP, at a rate of about 1/100 of that measured with $MgPP_i$ (Schlessinger and Coon 1960). An exchange-inert $Co(NH_3)_3$(tri-phosphate)$_2$, complex is hydrolyzed in the presence of Zn^{2+}, but not Mg^{2+} (Knight et al. 1983), suggesting that the substrate specificity is determined by the metal ion(s) bound to protein (see below), rather than to substrate.

2.2
Pyrophosphate Synthesis

The turnover numbers for PP_i hydrolysis exhibited by PPases of different origin are within the range of 100–500 s^{-1}, corresponding to about 10^{10} acceleration compared with noncatalyzed hydrolysis in water. For PP_i synthesis, the turnover numbers are less by a factor of 50–100 (Daley et al. 1986; Baykov et al. 1990; Baykov and Shestakov 1992). The ease with which the synthesis reaction can be measured in solution in the absence of any energy-supplying ion gradient is in sharp contrast to ATPases, for which such measurements have not been possible. In the presence of ATP sulfurylase, which effectively removes the synthesized PP_i, initial rates of PP_i synthesis are measurable by a luminometric or radioisotopic procedure (Nyrén and Lundin 1985; Daley et al. 1986).

2.3
Oxygen Exchange

Another important reaction catalyzed by PPase is oxygen exchange between $[^{18}O]P_i$ and $[^{16}O]$ water, resulting from reversible formation of enzyme-bound PP_i according to Eq. (1) [PP_i:P_i equilibration by PPase ($n = 1$–2)] (Janson et al. 1979):

$$\text{EMg}_n \underset{k_2}{\overset{k_1}{\rightleftharpoons}} \text{EMg}_{n+2}\text{PP}_i \underset{k_4}{\overset{k_3}{\rightleftharpoons}} \text{EMg}_{n+2}(\text{P}_i)_2 \underset{k_6}{\overset{k_5}{\rightleftharpoons}} \text{EMg}_{n+1}\text{P}_i \underset{k_8}{\overset{k_7}{\rightleftharpoons}} \text{EMg}_n. \tag{1}$$

As demonstrated by stereochemical and single-turnover experiments, PPase catalysis proceeds by direct water attack on PP_i, without formation of a phosphorylated enzyme (Gonzalez et al. 1984). $\text{EMg}_{n+2}\text{PP}_i$ may account for up to 20 % of total enzyme equilibrated with MgP_i. Each cycle of PP_i synthesis and hydrolysis results in incorporation of one oxygen from water per two phosphates. Analysis of the distribution of all five P_i species containing from zero to four exchanged oxygens in the resulting P_i allows evaluation of the partition coefficient $P_c = k_4/(k_4 + k_5)$ (Hackney 1980). This parameter is of particular importance for PPase studies, because in combination with the results of steady-state measurements of PP_i hydrolysis and synthesis, as well as the equilibrium measurements of the $EMg_{n+2}PP_i$ species, it allows one to estimate all eight rate constants k_1-k_8 in Eq. (1) without use of rapid kinetic methods (Springs et al. 1981; Fabrichniy et al. 1997).

2.4
Inhibitors

Physiological inhibitors of PPase include phosphate and, probably, polyphosphates (Sect. 2.1). Ca^{2+} exerts a strong inhibition in vitro via the $CaPP_i$ complex (Kurilova et al. 1984), a strong competitive inhibitor with respect to $MgPP_i$, but the inhibiting concentrations exceed by far the physiological Ca^{2+} concentration, with the exception of bovine retinal rod outer segment PPase, which is inhibited by Ca^{2+} at submicromolar levels (Yang and Wensel 1992a). Some phosphoric acid monoesters inhibit PPase in a slowly reversible manner (Sklyankina and Avaeva 1980). Although some of them, for instance phosphoethanolamine, are found in cells, their concentration there appears to be insufficient to inhibit PPase.

Diphosphonates, PP_i analogs in which the bridge O is replaced with C, inhibit mammalian cytoplasmic PPases quite strongly (K_i in the micromolar range) but not mitochondrial and bacterial PPases (Smirnova et al. 1988; Unguryte et al. 1989). Most potent diphosphonate inhibitors, some of which are used in therapy, contain an OH or NH_2 linked to the bridge C. Imidodiphosphate, containing a P-N-P group, is also a strong inhibitor, but is slowly hydrolyzed by PPase (Smirnova et al. 1986).

Eukaryotic PPases are strongly inhibited by fluoride. The inhibition results from stabilization of enzyme-substrate complex, presumably due to substitution of the nucleophilic water molecule in the active site, and occurs in two steps – rapid inhibitor binding and slow isomerization of the resulting complex (Baykov et al. 1992). As isolated by gel filtration, the inactivated Y-PPase contains one PP_i molecule, two Mg^{2+} and one fluoride ions entrapped per active site (Baykov et al. 1979). Prokaryotic PPases are either inhibited by fluoride in a rapidly reversible manner (Kurilova et al. 1984) or not at all (Jetten et al. 1992).

3
Pyrophosphatase Gene

3.1
Cloning the Gene

So far eight prokaryotic (*E. coli*, *Sulfolobus acidocaldarius*, *Thermoplasma acidophilum*, thermophilic bacterium PS-3, *Bartonella bacilliformis*, *Thermus thermophilus*, *Bacillus stearothermophilus*, and cyanobacterium *Synechocystis* sp.) and seven eukaryotic (*S. cerevisiae* cytoplasm and mitochondria, *Kluyveromyces lactis*, *Schizosaccharomyces pombe*, *Arabidopsis tha-*

liana, Solanum tuberosum, and *Bos taurus*/bovine retina) genes encoding soluble PPases have been cloned (Yang and Wensel 1992b; du Jardin et al. 1995; Maryama et al. 1996; Gomez et al. 1997; Mitchell and Minnick 1997; Meyer et al. 1995; Samejima et al. 1997); for further references, see Cooperman et al. 1992. The first two genes cloned were those of *E. coli* and *S. cerevisiae* (cytoplasm). The former was cloned as a 1.3-kb fragment of a partial digest of chromosomal DNA by screening the transformants for high PPase activity in magnesium pyrophosphate overlay test (Lahti et al. 1988), whereas the latter was cloned by hydridization screening using two "long" oligonucleotide probes, designed on the basis of the published amino acid sequence for cytoplasmic Y-PPase (Kolakowski et al. 1988). Interestingly, the *E. coli* genome has, at about 29 kb from the native *E. coli ppa* gene, a pseudogene with an open reading frame that could encode an inactive, "truncated" PPase having the same amino acid sequence as the native E-PPase, but 17 residues shorter (Burland et al. 1995).

3.2
Expressing the Genes and Producing Enzyme Variants

The wild-type E-PPase and its variants have been expressed in *E. coli* under the *ppa* promoter by inserting the intact *ppa* gene into the *EcoRI-HindIII* site of the high-copy number plasmid, pUC19. In this expression system, 200–400 mg of pure protein are produced per liter of cell culture. The first site-directed E-PPase variants produced were contaminated with a small amount (about 0.5 % of total protein) of chromosomal encoded wild type E-PPase, which complicated detailed studies on variants with low activity (Lahti et al. 1990b; 1991; Käpylä et al. 1995). Therefore an *E. coli* strain [MC1061/*YPPAI* (Δ*ppa*)] was constructed in which the chromosomal *ppa* gene was replaced with the *S. cerevisiae* gene for soluble PPasc (Salminen et al. 1995). This strain maintained viability by producing Y-PPase, not E-PPase. Differences in the elution volume during ion-exchange chromatography permitted a facile separation of plasmid-encoded E-PPase variants from the chromosome-encoded Y-PPase (Salminen et al. 1995).

Background-free wild type Y-PPase and its variants have been produced in an expression vector, pKW9, under *tac* promoter in *E. coli* and separated from the chromosome-encoded wild-type E-PPase by ion-exchange chromatography (Heikinheimo et al. 1996b). This expression system produces 150–300 mg of pure protein per 1 of cell culture.

3.3
Conservation of Enzyme Primary Structure

The full E-PPase sequence aligns within residues 28–225 of Y-PPase (Lahti et al. 1990a). Whereas the overall identity between the two sequences is only modest (22–28 %, depending on the choice of alignment parameters), of the

17 polar residues identified by X-ray crystallographic studies of Y-PPase as being at the active site (Terzyan et al. 1984) 14–16 are identical between Y-PPase and E-PPase. PPase thus appears to be an example of enzymes from widely divergent species that conserved common functional elements within the context of substantial overall sequence variation. In what follows, residues with similar functions in both E-PPase and Y-PPase are numbered as *yy/ee*, where *yy* and *ee* are the residue numbers of equivalent residues in E-PPase and Y-PPase (Fig. 1).

As more sequence information has accumulated, the conclusions mentioned above have been confirmed and extended. Out of the 15 complete PPase sequences currently available in the GenBank only 17 residues are conserved in all sequences, and strikingly 13 of these (Glu48/20, Lys56/29, Glu58/31, Arg78/43, Tyr93/55, Asp115/65, Asp117/67, Asp120/70, Asp147/97, Asp152/102, Lys154/104, Tyr192/141 and Lys193/142) belong to the group of the 17 polar active site residues, originally identified by Terzyan et al. (1984; Fig. 1). The four other conserved residues are Asp71/42, Gly94/56, Thr99/61 and Gly141/91. Asp42 points directly to the active site cavity of E-PPase beside Lys29 and Arg43 (Kankare et al. 1994), which together with their Y-PPase counterparts (Lys56 and Arg78) are important for substrate binding (Salminen et al. 1995; Pohjanjoki et al. 1998). Asp42 has been recently shown to be important for substrate binding in E-PPase (Avaeva et al. 1996b), even though in Y-PPase Asp71 does not seen to be important for catalysis (Heikinheimo et al. 1996b). Three residues (Tyr89, Gly132, Phe189) are conserved in 14 out of the 15 sequences. Of these, Tyr89 belongs to the group of the 17 Y-PPase active site residues mentioned above. However, mutational and functional analyses have indicated Tyr89 to be nonessential for catalysis (Pohjanjoki et al. 1998). Of the 17 Y-PPase active site residues, two are somewhat less well conserved (Glu148 and Asp150 are conserved in 7/15 and 11/15 sequences, respectively), whereas Lys198 is completely conserved in the sequence alignments within group I and group II PPases (see below), but not in the three-dimensional structural alignment used in Fig. 1. Mutational

---

Fig. 1. Alignment of the 15 PPase sequences currently available in GenBank (Salminen et al., ▶ unpubl.). Sequences were first aligned with PILEUP (gap weight and length weight values of 2.0 and 0.6, respectively) of the GCG package within prokaryotic and eukaryotic groups separately and then improved using three-dimensional structural alignment of E-PPase with Y-PPase (Heikinheimo et al. 1996a; http://www.btk.utu.fi/xray/PPase/). Residues conserved in all sequences are *boldface* and shown by *letter* below. The 13 conserved active site residues (Kankare et al. 1994) are further emphasized by *underlining*. Other residues conserved in 14 sequences are *boldface*. Inter- and intratrimeric residues of E-PPase (Kankare et al. 1996a) are *underlined* with *boldface* and *italics*, respectively. Residues located at the monomer–monomer interface of Y-PPase (Heikinheimo et al. 1996a) are *underlined* and *boldface*. The four-residue-long insertion specific for mitochondrial (*S. cerevisiae* M.) PPase is *underlined*. Positions of consensus secondary structural elements in E- and Y-PPase are marked as α and β, *numbered* sequentially from the N-terminus. Secondary structural elements found in Y-PPase but not in E-PPase are marked with *negative numbers* (e.g. β-4) if they occur N-terminal to the conserved core

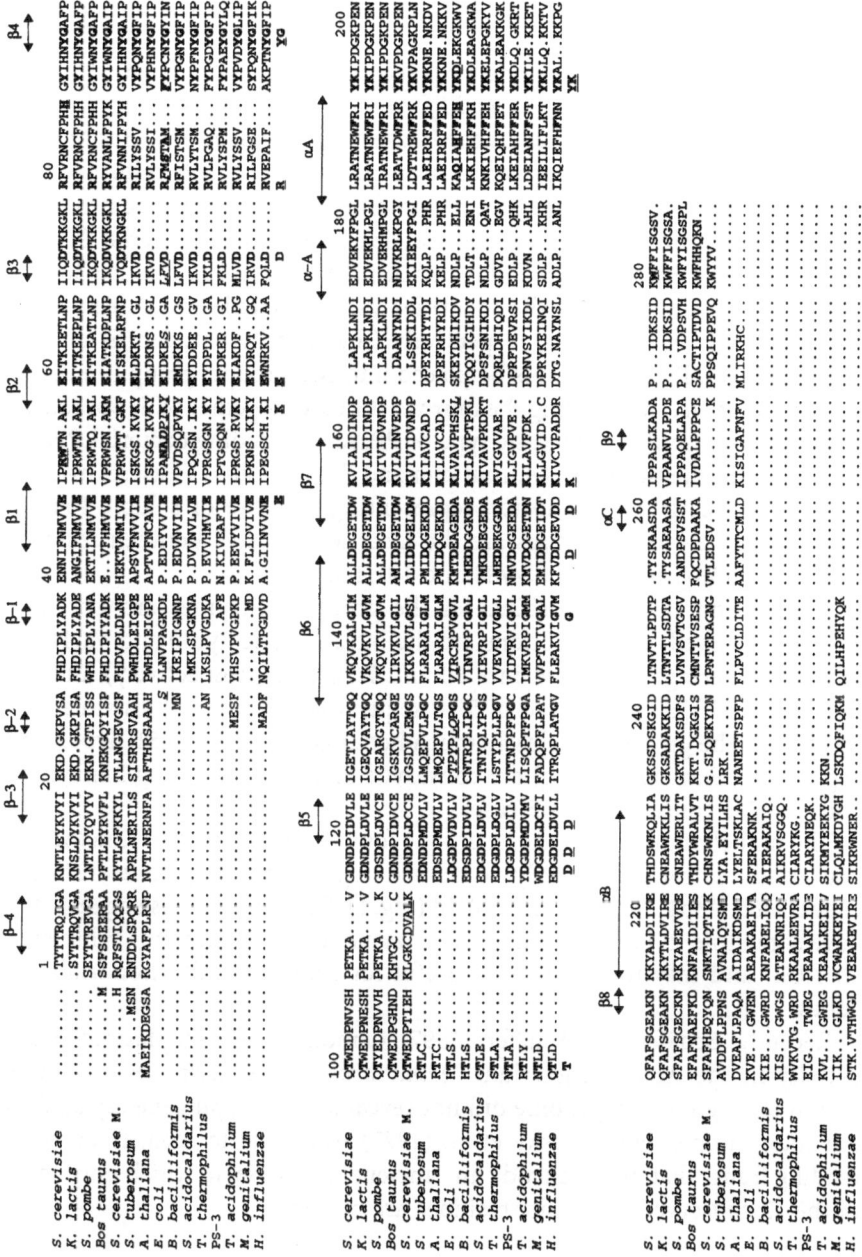

analyses have shown these three residues to be nonessential for catalysis (Heikinheimo et al. 1996b).

Internal identities of the nine prokaryotic PPases vary from 31 % (*H. influenca* vs. *E. coli*) to 53 % (*B. bacilliformis* vs. *E. coli*), whereas in the case of the four yeast PPases as well as the PPase from bovine retina (*Bos taurus*) the range of internal identity varies from 50 % (*S. cerevisiae* mitochondria vs. cytoplasm) to 85 % (*K. lactis* vs. *S. cerevisiae*). Interestingly, the two other known eukaryotic PPase sequences are clearly more like bacterial PPases. The identities between *A. thaliana* and *S. tuberosum* vs. *E. coli* are 36 and 37 %, respectively, whereas the identities vs. the cytoplasmic *S. cerevisiae* PPase are only 23 and 24 %, respectively (the identity between E- and Y-PPase is 28 %). PPases can thus be placed into two groups: group I (eukaryotic type of PPases), including the four yeast PPases together with *Bos taurus* PPase, and group II (prokaryotic type of PPases), including bacterial PPases together with the two plant (*A. thaliana* and *S. tuberosum*) PPases. Group II PPases are clearly much shorter, representing the core of soluble PPases. Typical for the group I PPases are the insertions located between residues (Y-PPase numbering) 71/78, 84/88 and 101/104.

The subunit interface residues of the group II enzymes are generally poorly conserved. The exceptions are the intratrimeric Tyr30 and Val41 (E-PPase numbering: conserved in 9/10 and 7/10 cases, respectively) and the intertrimer His136 (conserved in 5/10 sequences). In contrast, the interface residues of the group I are clearly much better conserved: Arg51, Trp52 and Trp279 (Y-PPase numbering) are conserved in all the five group I enzymes, and His87 is conserved in 4/5 cases.

The mitochondrial PPase has a unique four-residue-long insertion located two residues from the highly conserved region including three residues (Asp115/65, Asp117/67 and Asp120/70 (Fig. 1). This insertion is suggested to have some role in loosely linking the enzyme to the membrane (Vihinen et al. 1992) and may turn out to be useful for screening mitochondrial *PPA* genes.

Recently, Schäfer and Schäfer (1997) demonstrated cytoplasmic PPase activity in the Archaeabacterium *Methanococcus jannaschii*, even though search through its full genome did not reveal a typical *ppa* gene. By analyzing the genome, Baltscheffsky et al. (1977) were able to identify an open reading frame that might encode an "ancient" ur-PPase, in which only some of the 13 conserved active site residues are present.

4
Three-Dimensional Structure

Structural analysis of soluble inorganic pyrophosphatase started with the publication, by Bunick et al. in 1974, of the first Y-PPase diffraction pattern followed by the 6-Å structure of Y-PPase by Harutyunyan and coworkers

(Makhaldiani et al. 1978). The first near-atomic resolution structure came in 1981, of the *apo*-Y-PPase structure (Harutyunyan et al. 1981). The next significant breakthrough was the Y-PPase product complex structure (Chirgadze et al. 1991) which, unfortunately, contained errors, as we shall discuss below (see Sect. 4.2). Since 1994, a veritable deluge of structures has revolutionized discussion of enzyme structure and function in PPases: in 1994, three PPase structures were published, two from *E. coli* (Kankare et al. 1994; Oganessyan et al. 1994) and one at high-resolution from *T. thermophilus* (Teplyakov et al. 1994). In 1996 came high-resolution refined E-PPase structures from the Goldman (Kankare et al. 1996a, b) and Harutyunyan laboratories (Harutyunyan et al. 1996b), a corrected structure of the product complex of Y-PPase (Harutyunyan et al. 1996a) and two high-resolution structures of Y-PPase, the product complex and the resting enzyme (Heikinheimo et al. 1996a). Most recently, another E-PPase crystal structure was published (Avaeva et al. 1997; Harutyunyan et al. 1997).

4.1
Monomeric Structure

The core PPase structure, found in all bacterial PPases and in yeast PPase, consists of 8 β-strands and 2 α-helices (Harutyunyan et al. 1981; Oganessyan et al. 1994; Teplyakov et al. 1994; Kankare et al. 1996b). The molecule is best described as a distorted, highly twisted five-stranded β-barrel with four excursions (Kankare et al. 1994, 1996b). It belongs to the OB (oligonucleotide-binding) fold (Murzin 1994). Strands β1, β4, β5, β7, and β6 form

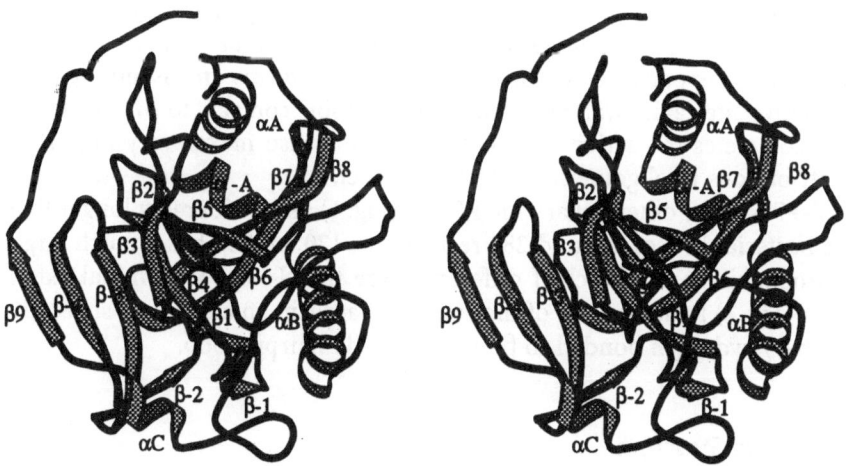

Fig. 2. Overall fold of Y-PPase in stereo. Figure was generated from the Mn_2:Y-PPase:$(MnP_i)_2$ structure. Helices are shown as *spirals*, strands as *arrows*; *numbering* of secondary structural elements is as in Fig. 1. The active site is between helix A and the top of the conserved β-barrel. (Made with MOLSCRIPT; Kraulis 1991)

the barrel, capped on top by α-helix B and on the botton by a loop between β5 and β6 (Fig. 2).

The excursions built onto the barrel create a very large active site. About 15 residues are significant for catalysis (Salminen et al. 1995; Avaeva et al. 1996a; Pohjanjoki et al. 1998), of which 12 (Kankare et al. 1994) are in the excursions (Fig. 2). Consequently, anchoring the excursions to the barrel is important, especially in the smaller, group II PPases (around 170 residues). In these, almost 10 % of the protein is part of the catalytic mechanism, and some 20 % of the protein forms the active site surface.

Lys56/29 and Arg78/43, two of the four catalytically important basic residues (Salminen et al. 1995), are in excursion II (50–91/22–53) between barrel strands β1 and β4. This large excursion is bound to the barrel by hydrophobic contacts between excursions I and II and by a side chain–main chain H-bond between Gly48/20 and I59/32 (Salminen et al. 1995; Kankare et al. 1996a). Interactions between helix A (excursion IV) and the β3–β4 loop form part of the hydrophobic core. Excursion III includes key catalytic residues: residues 115–122/65–72 form a structure/sequence motif for PPases (Kankare et al. 1994; Fig. 1), Asp115/65 and Asp120/70 bind essential cations (Heikinheimo et al. 1996a; Kankare et al. 1996a), and mutating Asp120/70 has the largest effect on catalytic activity (Salminen et al. 1995; Pohjanjoki et al. 1998). The long, highly twisted strand β6 bends by about 90° in the middle so that its N-terminal part contributes to the β-barrel and the hydrophobic core, and its C-terminal part is hydrogen bonded to β7. The loop region 147–152/97–102 is quite flexible. In the (apo)E-PPase structures, it is disordered; in our Y-PPase structures it is ordered, but undergoes relatively large conformational changes, changes that may affect metal-binding and catalytic properties (Sect. 4.4). Interestingly, this region is ordered in the thermophilic T-PPase (Teplyakov et al. 1994) and S. acidocaldarius PPase structures (Leppänen et al., in prep.). α-Helix A forms an essential part of the active site cavity wall and makes hydrophobic contacts to β7, the strand that completes the active site cavity. α-Helix A also makes very important contributions to oligomeric contacts (see Sect. 4.3).

Y-PPase, like other eukaryotic PPases (Fig. 1), is considerably longer than the bacterial PPases, around 280 residues vs. 170, and so has a much larger hydrophobic core. The chief differences are the N- and C-terminal extensions (Fig. 2) folded on top of the enzyme that form an additional sheet (β9/β-4/β-3) hydrogen bonded to β3 in the β2–β3 hairpin loop.

4.2
Active Site Structure

The Y-PPase:Mn$_2$ complex (YP1) structure, refined at 2.2 Å (Heikinheimo et al. 1996a), was the first structure of a "resting" PPase: all other structures had been of apo-enzymes, or E-PPase:Mn complex (Harutyunyan et al.

1996b). Overall, it is very similar to the E-PPase structure: 158 Cα atoms can be superimposed with a root mean square deviation (rmsd) of 1.45 Å. The active sites are very similar but slightly less open in YPI, as the flexible 97–102/147–152 loop is rather closer to the rest of the active site. M1 is coordinated to Asp115, Asp120 and Asp152, through a water to the backbone carbonyl of Trp153, and to two more H_2O. M2 is coordinated to Asp120, through a water to Glu48/20 and four additional waters. M1 and M2 are separated by two H_2O, 4.9 Å apart.

Our 2.0-Å Mn_2:Y-PPase:$(MnP_i)_2$ (YP2) structure is in the same crystal form as reported earlier (Chirgadze et al. 1991). In that preliminary report, the authors describe an Mn_3P_{i2} complex, where one of the P_i was misplaced. The authors later corrected that structure, and both we (at 2.0 Å; Heikinheimo et al. 1996a) and they (at 2.4 Å; Harutyunyan et al. 1996a) reported a product (Mn_4P_{i2}) complex structure, in which the positions of all four Mn^{2+} and positions and orientations of the two P_i are clearly visible (Fig. 3). Changes between the YP1 and YP2 structures involve movements of loops Glu101-Asp115 and Met143-Trp153, with the movement of the latter being more pronounced. M3 and M4 are on opposite sides of the phosphates

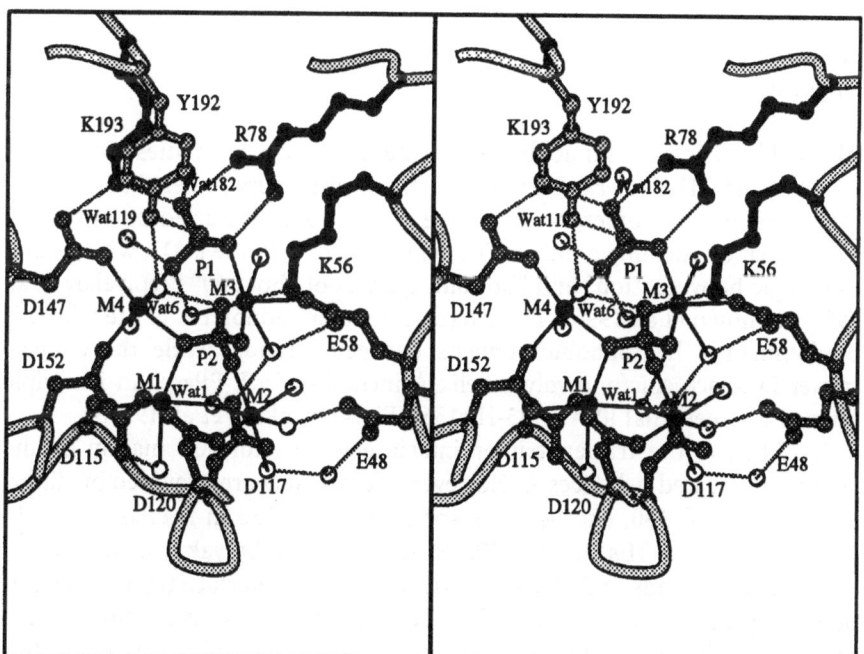

Fig. 3. Structure of the Y-PPase active site in stereo. Figure shows the Y-PPase [Mn_2:Y-PPase:$(MnP_i)_2$] active site, and indicates the proposed nucleophile (*Wat1*), and three potential general acids (*182, 119* and *6*). Hydrogen bonds to substrate or water molecules are shown as *gray lines*, metal coordination (to Mn1–4) as *black lines*; P1 and P2 are labelled as are conserved active site residues. View is from inside the molecule, looking out, and some residues (e.g. Y93, K154) have been omitted for clarity. (Made with MOLSCRIPT; Kraulis 1991)

approximately in the center of the active site, above the lower phosphate (P2) but below the upper one (P1). M3 is coordinated to P1, P2 and Glu58/31; M4 is coordinated to P1, P2, Asp152/102, and Asp147/97. The motion of the active site loops has three effects on the metal binding: M1 and M2 are now separated by a single H_2O (Wat1), at a distance of 3.7 Å; Asp152/102 coordinates both M1 and M4; and Asp147/97 binds M4 and accepts an H-bond from Lys142/193. Wat1 is also hydrogen-bonded to Asp117/67. Viewed from the phosphates, P1 is coordinated to Arg78/56, Tyr192/141 and Lys193/142, M3 and M4; P2 is coordinated to Lys56/29, Tyr93/55 and all four Mn^{2+}. Lys154/104 functions in a support role, by positioning Asp120/70 and Tyr93/55 for M1, M2 and P1 binding. As suggested by the dual numbering scheme, we believe that this mechanism is in essence preserved in E-PPase.

Although there is no three-dimensional structural similarity between PPase and other phosphoryl transfer enzymes, there is a considerable chemical similarity to exonucleases, polymerases and integrases, which share a common "two metal ion mechanism" (Beese and Steitz 1991). For example, metals M1 and M4 and Y1PPase align with the metal ions in the 3',5' exonuclease of Klenow fragment, and the putative hydroxide ions overlap, as does the electrophilic phosphate P2.

4.3
Oligomeric Structure

All soluble PPases are homooligomers – hexamers in prokaryotes and dimers in eukaryotes. Hexameric E-PPase is a dimer of trimers rather than a trimer of dimers: per monomer, a much larger surface area is buried on forming a trimer that on forming a dimer. The symmetry of E-PPase is D3, with a dihedral angle between "top" and "bottom" trimers of about 60° (Teplyakov et al. 1994; Salminen et al. 1996). The interface between monomers in the trimer is predominantly hydrophobic contacts between strands, while the trimer–trimer interface chiefly involves helix A including, in E-PPase, an ion-triple formed between His140-Asp143-His136' (Fig. 4; Kankare et al. 1996b).

In T-PPase, the trimer–trimer interface is also formed mainly by the symmetry-related α-helices A. However, the contacts are provided by different residues, Gly130, His134 (corresponding to His136 in E-PPase), Thr138 (corresponding to His140 in E-PPase) and Leu142 (Teplyakov et al. 1994).

When Mg^{2+} binds to E-PPase, the trimer–trimer interface tightens, as evidenced by an increase in the surface area buried upon forming dimers from monomers. This is due to Mg^{2+} binding to a site at the trimer–trimer interface, demonstrated by soaking "long c-axis" (Heikinheimo et al. 1995) crystals in decimolar Mg^{2+} (Kankare et al. 1996b; Harutyunyan et al. 1997). The Mg^{2+}, present at a stoichiometry of 0.5/monomer, is octahedrally coordinated to six water molecules, which in turn H-bond to the side chains of Asn24/24' and Asp26/26', as well as to the backbone carbonyls of Asn24 and

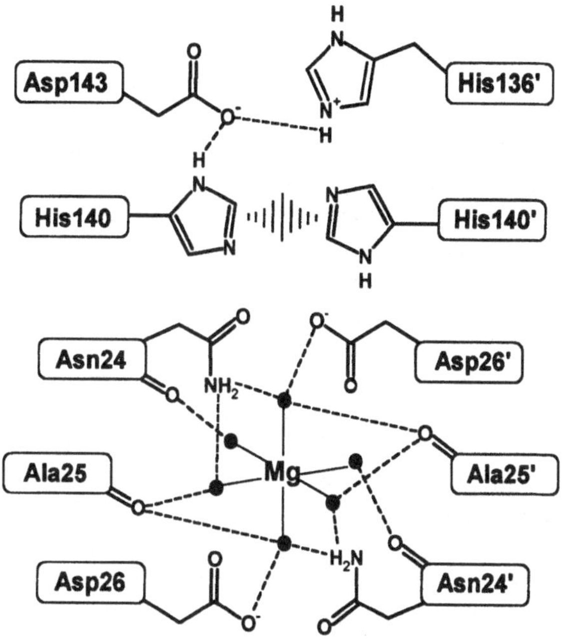

Fig. 4. Scheme of major contacts in the trimer–trimer interface of E-PPase. Unprimed and primed amino acid residue numbers refer to two different subunits. *Black circles* are water molecules, *dashed lines* hydrogen bonds. Also shown is a hydrophobic contact between His140 and His140'. For clarity, symmetry-related interactions His136-Asp143'-His140' are not shown. [Made with Chem-Sketch (ACD Labs)]

Ala25 (Fig. 4). D26N and D26S substitutions eliminate Mg^{2+} binding to the intertrimer cavity and eliminate or reverse the effect of Mg^{2+} on the rates of hexamer dissociation into trimers at pH 5 and trimer association at pH 7.2 (Salminen and Velichko, unpubl.). The latter substitutions also increase hexamer stability in acidic medium, consistent with Asp26 being the residue whose protonation leads to hexamer dissociation pH < 6 (Borshchick et al. 1986). This interfacial metal-binding site, unique to E-PPase, is not a significant factor in catalysis.

The dimeric Y-PPase has a completely different oligomerization interface, involving the formation of a three-layer "base-stack" of aromatic residues containing Trp52, His87/His87' and Trp52', buttressed by charged hydrogen bonds from Arg51 on one end and Arg51' at the other (Heikinheimo et al. 1996a). This dimer interface is likely to be conserved in other eukaryotic PPases, since Arg51 and Trp52 (and Trp279) are conserved in bovine and all known yeast sequences. His87 is conserved in all yeast sequences but is replaced by a Lys in the bovine PPase (see Sect. 3.3).

4.4
The Role of Oligomeric Structure in Catalysis and Termostability

The trimer–trimer interaction is quite strong in E-PPase – no dissociation into trimers is observed down to 10^{-4} mg/ml enzyme concentration at pH 7.2 (Volk et al. 1996). Dimeric Y-PPase is similarly stable, and its monomeric form could be only obtained by covalent immobilization (Plaksina et al. 1981). Replacing either His136 or His140 by Gln destabilizes E-PPase hexamer (Baykov et al. 1995), whereas replacing both makes trimer the dominant species in solution at protein concentrations as high as 26 mg/ml (Velichko et al. 1998). The hexameric variant enzyme is, however, readily formed in the presence of MgP_i (Velichko et al. 1998). Compared to hexamer, the (H136, 140Q)-PPase trimer exhibits a greatly decreased rate constant for substrate binding (k_1). By contrast, the value of the catalytic constant is similar for trimer and hexamer, indicating that the structures of the corresponding enzyme–substrate complexes are similar.

A plausible structural explanation for the reciprocal effect of MgP_i binding on hexamer stability can be found in the role of α-helix A. The disruption of the intertrimer contacts upon hexamer dissociation would lead to destabilization of α-helix A in the sense that it would not be positioned accurately along one face of the active site cavity. This would lead to mispositioning of, for instance, Lys142, which binds MgP_i and Mg_2PP_i. Conversely, correct positioning of α-helix A can be achieved in two ways, by forming hexamer or by binding substrate/product to the active site. This mechanism also explains why some active-site mutations cause dissociation of the hexamer into trimers (Volk et al. 1996; Fabrichniy et al. 1997).

Preliminary data suggest that monomeric E-PPase prepared by mutating Gln80 and Tyr77 in the intratrimer contact exhibits still poorer substrate-binding affinities (Salminen and Velichko, unpubl.). Oligomerization is also critical in thermostability: the chief difference between the thermophilic T-PPase and E-PPase involves an "oligomeric shift". Each "top" monomer (Salminen et al. 1996) skews with respect to its adjacent "bottom" monomer by about 1 Å in the xy plane, is 0.3-Å closer to the center of gravity of the hexamer in the z-direction, and rotates by about 7° about its center of gravity in the xy plane. Therefore, although an optimized monomer–monomer alignment of E- and T-PPase has an rmsd per Cα of 0.83 Å, the hexamer–hexamer superposition yields, per monomer, an rmsd per Cα of 2.22 Å. T-PPase thus has tighter interfaces than E-PPase, as shown by the surface area buried upon oligomerization. In T-PPase, each monomer makes contact with all four nearest neighbors (including both twofold related neighbors), whereas in E-PPase, each monomer only makes contracts to one twofold related neighbor (Salminen et al. 1996). In T-PPase, but not E-PPase, multi-center ionic interactions occur across the interfaces between monomers. Similar oligomeric changes occur in S. acidocaldarius PPase, but the stabilizing interactions are different (Leppänen et al. in prep.).

5
Mechanism of Catalysis of Water Attack on PP_i

The $Mn_2:Y\text{-PPase}:(MnP_i)_2$ structure, in combination with earlier functional studies of both Y-PPase and E-PPase, and model studies of phosphoryl transfer, leads us to propose the detailed mechanism for PPase catalysis of

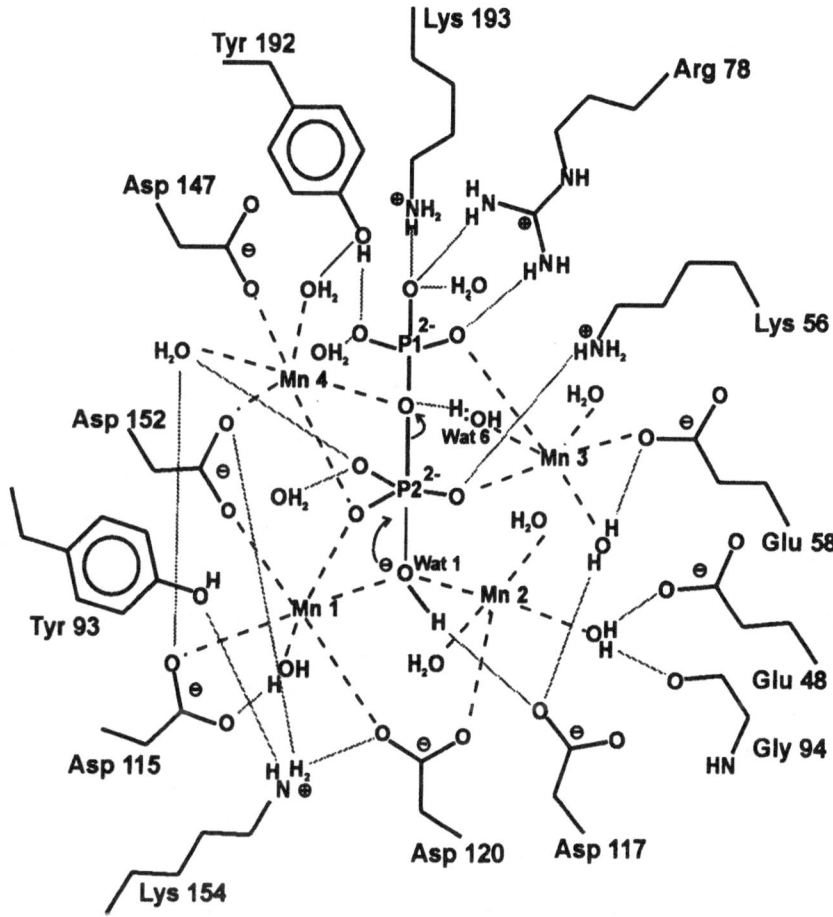

Fig. 5. Proposed PPase mechanism. Active site is shown in two dimensions. Hydrogen bonding is shown with *dotted* and metal coordination with *dashed lines*. Flow of electrons is indicated with *arrows*. The reaction involes an hydroxide ion (*Wat1*) coordinated to Mn1 and Mn2, which hydrolyzes the substrate $Mn_2P_2O_7$. The required general acid is a water molecule (*Wat6*) coordinated to Mn3 and hydrogen bonded to P1 in the Mn_2:Y-PPase:$(MnP_i)_2$ structure (Fig. 3). (Heikinheimo et al. 1996a)

PP$_i$ hydrolysis shown in Fig. 5. According to this proposal, PPase catalyzes PP$_i$ hydrolysis by lowering the pK_a of the leaving group in site P1, by forming an incipient hydroxide ion that functions as a stronger nucleophile than water in attacking the electrophilic phosphorus in site P2, and by shielding the charge on the electrophilic phosphorus, thus permitting attack by the hydroxide anion.

5.1
Assignment of P1 and P2

PPases bind the two P_i at the active site with substantially different affinities (Cooperman et al. 1981; Baykov and Shestakov 1992; Käpylä et al. 1995; Baykov et al. 1996). The assignment of P1 as the higher affinity site is based on the following three observations. First, in the presence of three Mn^{2+} ions, P_i bound in the high affinity site protects Arg78 from chemical modification, with little additional protection conferred by P2 site binding (Cooperman et al. 1981). Second, only the P1 site is occupied by sulfate ion in *T. thermophilus* PPase (Teplyakov et al. 1994) and E-PPase (Avaeva et al. 1997). Third, one P_i ion binds to Y-PPase in the absence of divalent ion; its dissociation constant decreases only threefold when one or two Mn^{2+} bind, but 30- to 80-fold when a third Mn^{2+} binds (Cooperman et al. 1981). This pattern is fully consistent with binding to P1; at this site, two positively charged residues and a tyrosine hydroxyl offer a total of four side-chain hydrogen bond donors for binding (Arg78 is bidentate) and the P_i in P1 interacts with M3 and M4 but with neither M1 nor M2. By contrast, P2 has only one positively charged residue, Lys56, and a total of two side-chain hydrogen bond donors available for binding, and interacts directly with M1 and M2.

In our proposed mechanism (Fig. 5) the phosphoryl groups in P1 and P2 are assigned as the leaving and electrophilic groups, respectively. This assignment is based on the observations that P2 is closer to the most reasonable candidate for the nucleophilic water, Wat1, and that the electrophilic P_i dissociates before the leaving group P_i (Springs et al. 1981), and our expectation that the more weakly bound P_i in P2 should dissociate first. Placing the leaving group in the high affinity site also makes sense mechanistically, since tighter binding should correlate with higher conjugate acidity, and thus a faster hydrolysis rate. The pK_a of P_i in P1 should be lower than that of P_i in P2, as strong hydrogen bonding should lower pK_a more effectively than coordination to metal ions, especially coordination to relatively weak acidic metal ions like Mg^{2+} and Mn^{2+} (Breinig and Jones 1963; Spiro 1971; Cooperman 1976). The P_i in P2 is well-positioned to dissociate with the exposed M3 metal ion, via a deep groove that leads down to the P2 binding pocket in the active site (Heikinheimo et al. 1996a).

5.2
Consistency with Model Studies

Model studies (Benkovic and Schray 1973) have shown that phosphoryl transfer from phosphomonoester dianions is exquisitely sensitive to the conjugate acidity of the leaving group, and is characterized by a Brønsted β of −1.2. According to the proposed mechanism, the conjugate acidity of the leaving phosphoryl group (at P1) is increased by coordination to Arg78,

Tyr192, Lys193, M3, M4 and Wat6. Although phosphoryl transfer from phosphomonoesters is much less sensitive to base strength of the nucleophile, there is evidence that strong, oxyanion nucleophiles can attack phosphomonoester dianions at faster rates than the corresponding protonated neutral nucleophiles, providing that the charge repulsion between the nucleophile and electrophile can be overcome, as for instance in the presence of divalent metal ion cations (Hsu and Cooperman 1976; Herschlag and Jencks 1990). In the proposed mechanism, the nucleophile is an hydroxide ion (Wat1), the conjugate base of a water molecule having an abnormally low pK_a (~6) by virtue of its coordination to M1 and M2 and hydrogen bonding to Asp117, and the charge on the electrophilic phosphoryl group is neutralized through coordination to Lys56, Tyr93, and M1–M4.

5.3
The General Acid

The pH dependence of k_{cat} for PPases indicates the presence of an activating general acid of apparent pK_a 7.9–8.5 (Y-PPase; Knight et al. 1981, Pohjanjoki et al. 1998) to 9.7 (E-PPase; Käpylä et al. 1995, Avaeva et al. 1996a). To what group in Y-PPase does the activating general acid correspond? Conservative mutation (D/E, K/R, Y/F) of some 11 active site residues raises the pK_a of the activating general acid by > 1.5 pH units. Variants of amino acid sice chains that hydrogen bond to P1 (Arg78, Lys193, Tyr192) retain appreciable activity, do not show atypically large changes in the apparent pK_a of the general acid, and are thus unlikely candidates (Pohjanjoki et al. 1998). We therefore believe that the general acid is a water molecule hydrogen bonded to P1. A strong candidate is Wat6 (Fig. 5). The acidity of Wat6 would be expected to increase by coordination to M3, and the general rise in pK_a with mutation could reflect perturbation of the M3–Wat6 bond, and of H-bonding interactions in which Wat6 participates, within the tightly integrated PPase active site.

5.4
The Nucleophilic Water

The pH dependence of k_3 for variant forms of E-PPase (Käpylä et al. 1995; Baykov et al. 1996, Fabrichniy et al. 1997) provides evidence for an essential base, with an apparent pK_a between 7.8 and 8.7 for the active site variants D97E, Y55F and K104R (corresponding to D147E, Y93F, and K154R in Y-PPase). The corresponding pK_a in wild type E-PPase and Y-PPase is < 6.5.

Conservative mutation of virtually every conserved active site residue in either E-PPase or Y-PPase increases the apparent pK_a of the essential base (Salminen et al. 1995, Pohjanjoki et al. 1998), with the large majority of such changes exceeding 1.6 pH units. As in the case of the general acid, this generalized response to mutation must reflect a group with an abnormally low pK_a that depends on the integrity of the active site structure, such that any perturbation of that structure leads to an increase in pK_a. This well describes Wat1, the nucleophile in our proposed mechanism, the pK_a of which is lowered by binding to M1 and M2, and hydrogen binding to Asp117. The fluoride inhibition of PPase referred to above presumably reflects fluoride replacement of Wat1.

5.5
Relaxation at the Active Site: Alternative Mechanisms

Our proposed mechanism requires that the $Mn_2:Y$-PPase:$(MnP_i)_2$ structure (Fig. 3) has relaxed since the chemical catalytic step such that the oxygen from the putative nucleophilic hydroxide ion (Wat1 in Fig. 5) has rotated away to become coordinated to M2 alone in Fig. 3 and another water molecule has replaced it in bridging M1 and M2. The mechanism in Fig. 5 was generated by docking a model of the transition state into the structure of the product complex without making any changes to the protein and trying to overlap the phosphorus oxygens in the transition state and the product as much as possible. Total overlap was not possible, because an inversion of configuration is in progress at P2, and because no exygen on P1 in the product complex points directly towards P2. Nonetheless, the only adjusted parameters, all in the transition state, were: the axial O-P bond lengths on the trigonal bipyramidal P2 (range 1.9–2.4 Å), the P-O-P bend angle (range 130–180°) and the torsion angles in the transition state.

An alternative mechanism we considered had Wat1, the putative bridging hydroxide, acting as a general base to activate a second water as the incoming nucleophile, placed between O1 and O2 in the product complex (Fig. 3), and corresponding to one of them. This mechanism has the advantage of not requiring a relaxation step following hydrolysis, but suffers, vis-à-vis the proposed mechanism, in that it is much harder to satisfy the geometrical requirements for catalysis without either distorting the transition state beyond acceptable limits or invoking conformational change on the part of the protein.

Harutyunyan et al. (1997) have recently proposed yet a third mechanism, in which the two P sites have the same identities as above, but a different water molecule, bound between M2 and Tyr93, is the nucleophile. Their mechanism also has the advantage of not requiring a relaxation following hydrolysis, but suffers from two major problems. First, it does not explain the substantial (10–20%) residual activity in the Y93F variant of Y-PPase and the corresponding Y55F variant of E-PPase (Fabrichniy et al. 1997; Prohjanjoki et al. 1998). Second, it does not account for the proximity of M1 and M2 – why are they brought so close together, at what must represent a considerable free energy penalty, if not to activate Wat1?

Current attempts to determine the structure of Y-PPase complexed with either substrate or substrate analog, if successful, may allow a definitive choice to be made among these alternatives.

Two reports have been published recently (Young et al. 1998; Shintani et al. 1998) showing that *Bacillus subtilis* has a unique type of soluble PPase, which is completely different, by its amino acid sequence, from the other PPases studied so far. Accordingly, *B. subtilis* PPase is a first example of a new family of soluble PPases, which was named a "Bs family". Putative representatives of the Bs family were also observed in four other bacterial strains, at least two of which (*M. jannaschii* and *Archaeoglobus fulgidues*), like *B. subtilis*, do not seem to have a typical *ppa* gene in their genomes.

Acknowledgments. The author's work cited in this review was supported by grants from the Russian Foundation for Basic Research, the Russian State Project Bioengineering/Enzyme Engineering, the United States National Institutes of Health and the Finnish Academy of Sciences.

References

Avaeva S, Ignatov P, Kurilova S, Nazarova T, Rodina E, Vorobyeva N, Oganessyan V, Harutyun-yan E (1996a) *Escherichia coli* inorganic pyrophosphatase: site-directed mutagenesis of the metal binding sites. FEBS Lett 399: 99–102

Avaeva SM, Rodina EV, Kurilova SA, Nazarove TI, Vorobyeva NN (1996b) Effect of D42N substitution in *Escherichia coli* inorganic pyrophosphatase on catalytic activity and Mg^{2+} binding. FEBS Lett 392: 91–94

Avaeva S, Kurilova S, Nazarova T, Rodina E, Vorobyeva N, Sklyankina V, Grigorjeva O, Harutyunyan E, Oganessyan V, Wilson K, Dauter Z, Huber R, Mather T (1997) Crystal structure of *Escherichia coli* inorganic pyrophosphatase complexed with SO_4^{2-}. Ligand-induced molecular asymmetry. FEBS Lett 410: 502–508

Baltscheffsky H, Baltscheffsky M, Nadanaciva S, Persson B, Schultz A (1997) Possible origin and evolution of inorganic pyrophosphatases. In: Lahti R (ed) Ist Int Meet on Inorganic pyrophosphatases, University of Turku, Turku, pp 1–3

Baltscheffsky M, Baltscheffsky H (1992) Inorganic pyrophosphate and inorganic pyrophosphatase. In: Ernster L (ed) Molecular mechanisms in bioenergetics. Elsevier, Amsterdam, pp 331–348

Baykov AA, Shestakov AS (1992) Two pathways of pyrophosphate hydrolysis and synthesis by yeast inorganic pyrophosphatase. Eur J Biochem 206: 463–470

Baykov AA, Tam-Villoslado JJ, Avaeva SM (1979) Fluoride inhibition of inorganic pyrophosphatase. IV Evidence for metal participation in the active center and a four-site model of metal effect on catalysis. Biochim Biophys Acta 569: 228–238

Baykov AA, Shestakov AS, Kasho VN, Vener AV, Ivanov AH (1990) Kinetics and thermodynamics of catalysis by the inorganic pyrophosphatase of *Escherichia coli* in both directions. Eur J Biochem 194: 879–887

Baykov AA, Alexandrov AP, Smirnova IN (1992) A two-step mechanism of fluoride inhibition of rat liver inorganic pyrophosphatase. Arch Biochem Biophys 294: 238–243

Baykov AA, Dudarenkov VY, Käpylä J, Salminen T, Hyytiä T, Kasho VN, Husgafvel S, Cooperman BS, Goldman A, Lahti R (1995) Dissociation of hexameric *Escherichia coli* inorganic pyrophosphatase into trimers on His-136→Gln or His-140→Gln substitutions and its effect on enzyme catalytic properties. J Biol Chem 270: 30804–30812

Baykov AA, Hyytiä T, Volk SE, Kasho VN, Vener AV, Goldman A, Lahti R, Cooperman BS (1996) Catalysis by *Escherichia coli* inorganic pyrophosphatase: pH and Mg^{2+} dependence. Biochemistry 35: 4655–4661

Beese LS, Steitz TA (1991) Structural basis for the 3'-5' exonuclease activity of *Escherichia coli* DNA polymerase I: a two metal ion mechanism. EMBO J 10: 25–33

Benkovic SJ, Schray KJ (1973) Chemical basis of biological phosphoryl transfer. In: Boyer PD (ed) The enzymes vol 8, 3rd edn. Academic Press, New York, pp 201–237

Borshchik IB, Magretova NN, Chernyak VY, Sklyankina VA, Avaeva SM (1986) Structural organization of *E. coli* inorganic pyrophosphatase. Biokhimiya 51: 1484–1489

Breinig JB, Jones MM (1963) The effect of coordination of the reactivity of aromatic ligands. VII. Specific ion effects on diazo coupling rates. J Org Chem 28: 852–854

Bunick G, McKenna GP, Colton R, Voet D (1974) The X-ray structure of yeast inorganic pyrophosphatase, crystal properties. J Biol Chem 249: 4647–4649

Burland V, Plunkett G II, Sofia HJ, Daniels DC, Blattner FR (1995) Analysis of the *E. coli* genome VI: DNA sequence of the region from 92.8 through 100 minutes. Nucleic Acids Res 23: 2105–2119

Butler LG (1971) Yeast and other inorganic pyrophosphatases. In: Boyer PD (ed) The enzymes, vol 4, 3rd edn. Academic Press, New York, pp 529–541

Chirgadze NY, Kuranova IP, Nevskaya NA, Teplyakov AV, Wilson K, Strokopytov BV, Arutyunyan G, Khene V (1991) Crystal structure of MnP complex of inorganic pyrophosphatase of yeast at resolution of 2.35 Å. Sov Phys Crystallogr 36: 128–132

Cooperman BS (1976) The role of divalent metal ions in phosphoryl and nucleotidyl transfer. In: Sigel H (ed) Metal ions in biological systems, vol 5. Dekker, New York, pp 79–125

Cooperman BS (1982) The mechanism of action of yeast inorganic pyrophosphatase. Methods Enzymol 87: 526: 548

Cooperman BS, Panackal A, Springs B, Hamm DJ (1981) Divalent metal ion, inorganic phosphate, and inorganic phosphate analogue binding to yeast inorganic pyrophosphatase. Biochemistry 20: 6051–6060

Cooperman BS, Baykov AA, Lahti R (1992) Evolutionary conservation of the active site of soluble inorganic pyrophosphatase. TIBS 17: 262–266

Daley LA, Renosto F, Segel IH (1986) ATP sulfurylase-dependent assays for inorganic pyrophosphate: application to determining the equilibrium constant and reverse direction kinetics of the pyrophosphatase reaction, magnesium binding to orthophosphate, and unknown concentrations of pyrophosphate. Anal Biochem 157: 385–395

De Meis L (1984) Pyrophosphate of high and low energy. Contributions of pH, Ca^{2+}, Mg^{2+}, and water to free energy of hydrolysis. J Biol Chem 259: 6090–6097

Du Jardin P, Rojas-Beltran J, Gebhardt, C, Brasseur R (1995) Molecular cloning and characterization of a soluble inorganic pyrophosphatase in potato. Plant Physiol 109: 853–860

Fabrichniy IP, Kasho VN, Hyytiä T, Salminen T, Halonen P, Dudarenkov VY, Heikinheimo P, Chernyak VY, Goldman A, Lahti R, Cooperman BS, Baykov AA (1997) Structural and functional consequences of substitutions at the tyrosine 55-lysine 104 hydrogen bond in *Escherichia coli* inorganic pyrophosphatase. Biochemistry 36: 7746–7753

Flodgaard H, Fleron P (1974) Thermodynamic parameters for the hydrolysis of inorganic pyrophosphate at pH 7.4 as a function of $[Mg^{2+}]$, $[K^+]$ and ionic strength determined from equilibrium studies of the reaction. J Biol Chem 249: 3465–3474

Gomez R, Losada M, Serrano A (1997) Cyanobacterial and algal inorganic pyrophosphatases and the molecular phylogeny of the higher plant enzymes. In: Lahti R (ed) Proc Ist Int Meet on Inorganic pyrophosphatases, University of Turku, Turku, pp 6–9

Gonzalez MA, Webb MR, Welsh KM, Cooperman BS (1984) Evidence that catalysis by yeast inorganic pyrophosphatase proceeds by direct phosphoryl transfer to water and not via a phosphoryl enzyme intermediate. Biochemistry 23: 797–801

Hackney DD (1980) Theoretical analysis of distribution of $[^{18}O]P_i$ species during exchange with water. Application to exchanges catalyzed by yeast inorganic pyrophosphatase. J Biol Chem 255: 5320–5328

Harutyunyan EH, Terzyan SS, Voronova AA, Kuranova IP, Smirnova EA, Vainshtein BK, Höhne W, Hansen G (1981) An X-ray study of yeast inorganic pyrophosphatase at 3 Å resolution. Dokl Akad Nauk SSSR 258: 1481–1485

Harutyunyan EH, Kuranova IP, Vainshtein BK, Höhne WE, Lamzin VS, Dauter Z, Teplyakov AV, Wilson KS (1996a) X-ray structure of yeast inorganic pyrophosphatase complexed with manganese and phosphate. Eur J Biochem 239: 220–228

Harutyunyan EH, Oganessyan VY, Oganessyan NN, Terzyan SS, Popov AN, Rubinsky SB, Vainstein BK, Nazarova TI, Kurilova SA, Vorobjeva NN, Avaeva SM (1996b) The structure of E. coli inorganic pyrophosphatase in its Mn^{2+} complex at a 2.2 Å resolution, Kristallografiya 41: 84–96

Harutyunyan EH, Oganessyan VV, Oganessyan NN, Avaeva SM, Nazarova TI, Vorobyeva NN, Kurilova SA, Huber R, Mather T (1997) Crystal structuce of holo inorganic pyrophosphatase from Escherichia coli at 1.9 Å resolution. Mechanism of hydrolysis. Biochemistry 36: 7754–7760

Heikinheimo P, Salminen T, Lahti R, Cooperman BS, Goldman A (1995) New crystal forms of Escherichia coli and Saccharomyces cerevisiae soluble inorganic pyrophosphatases. Acta Crystallogr D 51: 399–401

Heikinheimo P, Lehtonen J, Baykov AA, Lahti R, Cooperman BS, Goldman A (1996a) The structural basis for pyrophosphatase catalysis. Structure 4: 1491–1508

Heikinheimo P, Pohjanjoki P, Helminen A, Tasanen M, Cooperman BS, Goldman A, Baykov AA, Lahti R (1996b) A site-directed mutagenesis study of Saccharomyces cerevisiae pyrophosphatase. Functional conservatioin of the active site of soluble inorganic pyrophosphatases. Eur J Biochem 239: 138–143

Herschlag D, Jencks WP (1990) Catalysis of the hydrolysis of phosphorylated pyridines by $Mg(OH)^+$: a possible model for enzymatic phosphoryl transfer. Biochemistry 29: 5172–5179

Hsu CH, Cooperman BS (1976) Metal ion catalysis of phosphoryl transfer via a ternary complex. The effects of changes in leaving group, metal ion, and attacking nucleophile. J Am Chem Soc 98: 5657

Janson CA, Degani C, Boyer PD (1979) The formation of enzyme-bound and medium pyrophosphate and the molecular basis of the oxygen exchange reaction of yeast inorganic pyrophosphatase. J Biol Chem 254: 3743: 3749

Jetten MSM, Fluit TJ, Stams AJM, Zehnder AJB (1992) A fluoride-insensitive inorganic pyrophosphatase isolated from Methanothrix soehngenii. Arch Microbiol 157: 284–289

Josse J (1966) Constitutive inorganic pyrophosphatase of Escherichia coli. II. Nature and binding of active substrate and the role of magnesium. J Biol Chem 261: 1948–1957

Josse J, Wong SCK (1971) Inorganic pyrophosphatase of Escherichia coli. In: Boyer PD (ed) The enzymes, vol 4, 3rd edn. Academic Press, New York, pp 499–527

Kay HD (1928) Phosphatases of mammalian tissues. II. Pyrophosphatase. Biochem J. 22: 1446–1448

Kankare J, Neal G, Salminen T, Glumoff T, Cooperman BS, Lahti R, Goldman A (1994) The structure of E. coli soluble inorganic pyrophosphatase at 2.7 Å resolution. Protein Eng 7: 823–830

Kankare J, Salminen T, Lahti R, Cooperman BS, Baykov AA, Goldman A (1996a) Structure of Escherichia coli inorganic pyrophosphatase at 2.2 Å resolution. Acta Crystallogr D52: 551–563

Kankare J, Salminen T, Lahti R, Cooperman B, Baykov AA, Goldman A (1996b) Crystallographic identification of metal binding sites in Escherichia coli inorganic pyrophosphatase. Biochemistry 35: 4670–4677

Käpylä J, Hyytiä T, Lahti R, Goldman A, Baykov AA, Cooperman BS (1995) Effect of D97E substitution on the kinetic and thermodynamic properties of Escherichia coli inorganic pyrophosphatase. Biochemistry 34: 792–800

Knight WB, Fitts SW, Dunanway-Mariano D (1981) Investigation of the catalytic mechanism of yeast inorganic pyrophosphatase. Biochemistry 20: 4079–4086

Knight WB, Ting S-J, Chuang S, Dunaway-Mariano D, Haromy T, Sundaralingam M (1983) Yeast inorganic pyrophosphatase substrate recognition. Arch Biochem Biophys 227: 302–309

Kolakowski LF, Schloesser MG, Cooperman BS (1988) Cloning, molecular characterization, and chromosome localization of the inorganic pyrophosphatase (PPA) gene from S. cerevisiae. Nucleic Acids Res 22: 10441–10452

Kornberg A (1962) On the metabolic significance of phosphorolytic and pyrophosphorolytic reactions. In: Kasha M, Pullman D (eds) Horizons in biochemistry. Academic Press, New York, pp 251–264

Kraulis PJ (1991) MOLSCRIPT: a program to produce both detailed and schematic plots of protein structures. J Appl Crystallogr 24: 946–950

Kunitz M, Robbins PW (1961) Inorganic pyrophosphatases. In: Boyer PD, Lardy H, Myrbäck K (eds) The enzymes, vol 5, 2nd edn. Academic Press, New York, pp 169–178

Kurilova SA, Bogdanova AV, Nazarova TI, Avaeva SM (1984) Changes in E. coli inorganic pyrophosphatase activity on its interaction with magnesium, zinc, calcium and fluoride ions. Bioorg Khim 10: 1153–1160

Lahti R (1983) Microbial inorganic pyrophosphatases. Microbiol Rev 47: 169–179

Lahti R, Pitkäranta T, Valve E, Ilta I, Kukko-Kalske E, Heinonen J (1988) Cloning and characterization of the gene encoding inorganic pyrophosphatase of Escherichia coli K-12. J Bacteriol 170: 5901–5907

Lahti R, Kolakowski LF, Heinonen J, Vihinen M, Pohjanoksa K, Cooperman BS (1990a) Conservation of functional residues between yeast and E. coli inorganic pyrophosphatases. Biochim Biophys Acta 1038: 338–345

Lahti R, Pohjanoksa K, Pitkäranta T, Heikinheimo P, Salminen T, Meyer P, Heinonen J (1990b) A site-directed mutagenesis study on Escherichia coli inorganic pyrophosphatase. Glutamic acid-98 and lysine-104 are important for structural integrity, whereas aspartic acids-97 and -102 are essential for catalytic activity. Biochemistry 29: 5761–5766

Lahti R, Salminen T, Latonen S, Heikinheimo P, Pohjanoksa K, Heinonen J (1991a) Genetic engineering of Escherichia coli inorganic pyrophosphatase. Tyr55 and Tyr141 are important for the structural integrity. Eur J Biochem 198: 293–297

Lundin M, Baltscheffsky H, Ronne H (1991) Yeast PPA2 gene encodes a mitochondrial inorganic pyrophosphatase that is essential for mitochondrial function. J Biol Chem 266: 12168–12172

Makhaldiani VV, Smirnova EA, Voronova AA, Kuranova IP, Arutyunyan EG, Vainshtein BK, Höhne W, Binwald B, Hansen G (1978) X-ray diffraction study of inorganic pyrophosphatase from baker's yeast at a resolution of 6 Å. Dokl Akad Nauk SSSR 240: 1478–1481

Maryama S, Maeshima M, Nishimura M, Aoki M, Ichiba T, Sekiguchi J, Hachimori A (1996) Cloning and expression of the inorganic pyrophosphatase gene from thermophilic bacterium PS-3. Biochem Mol Biol Int 40: 679–688

Meyer W, Moll R, Kath T, Schäfer G (1995) Purification, cloning and sequencing of Archaebacterial pyrophosphatase from the extreme thermoacidophile Sulfolobus acidocaldarius. Arch Biochem Biophys 319: 149–156

Mitchell SJ, Minnick MF (1997) Cloning, functional expression, and complementation analysis of an inorganic pyrophosphatase from Bartonella bacilliformis. Can J Microbiol 43: 734–743

Murzin AG (1994) New protein folds. Curr Opin Struct Biol 4: 441–449

Nyrén P, Lundin A (1985) Enzymatic method for continuous monitoring of inorganic pyrophosphate synthesis. Anal Biochem 151: 504–509

Oganessyan VY, Kurilova SA, Vorobyeva NN, Nazarova TI, Popov AN, Lebedev AA, Avaeva SM, Harutyunyan EH (1994) X-ray crystallographic studies of recombinant inorganic pyrophosphatase from Escherichia coli. FEBS Lett 348: 301–304

Plaksina EA, Sergienko OV, Sklyankina VA, Avaeva SM (1981) Preparation of immobilized dimer and monomer of inorganic pyrophosphatase and evidence for catalytic activity of the monomer. Boorg Khim 7: 357–364

Pohjanjoki P, Lahti R, Goldman A, Cooperman S (1998) Evolutionary conservation of enzymatic catalysis: quantitative comparison of the effects of mutation of aligned residues in *Saccharomyces cerevisiae* and *Escherichia coli* inorganic pyrophosphatases on enzymatic activity. Biochemistry (in press)

Rea PA, Poole RJ (1993) Vacuolar H^+-translocating pyrophosphatase. Annu Rev Plant Physiol Plant Mol Biol 44: 157–180

Russell RGG (1976) Metabolism of inorganic pyrophosphate (PP_i). Arthritis Reum 19: 465–478

Salminen T, Käpylä J, Heikinheimo P, Goldman A, Heinonen J, Baykov AA, Cooperman BS, Lahti R (1995) Structure and function analysis of *Escherichia coli* inorganic pyrophosphatase: is a hydroxide ion the key to catalysis? Biochemistry 34: 782–791

Salminen T, Teplyakov A, Kankare J, Cooperman BS, Lahti R, Goldman A (1996) An unusual route to thermostability disclosed by the comparison of *Thermus thermophilus* and *Escherichia coli* inorganic pyrophosphatases. Protein Sci 5: 1014–1025

Samejima T, Takahashi Y, Shinoda H, Satoh T (1997) Protein engineering study on inorganic pyrophosphatases from *Bacillus stearothermophilus*, *Thermus thermophilus* and *Escherichia coli*. In: Lahti R (ed) Proc. Ist Int Meet on Inorganic pyrophosphatases, University of Turku, Turku, pp 24–26

Schäfer T, Schäfer G (1997) Pyrophosphatases from *Sulfolobus* and other Archae. In: Lahti R (ed) Proceedings of the First International Meeting on Inorganic Pyrophosphatases. Universtiy of Turku, Turku, pp 9–12

Schlessinger MJ, Coon MJ (1960) Hydrolysis of nucleoside di and triphosphates by crystalline preparations of yeast inorganic pyrophosphatase. Biochim Biophys Acta 41: 30–36

Shafranskii YuA, Baykov AA, Andrukovich PF, Avaeva SM (1977) Comparative kinetic studies of the Mg^{2+}-activated hydrolysis of tripolyphosphate and pyrophosphate by inorganic pyrophosphatase. Biokhimiya 42: 1244–1251

Shintani T, Uchiumi T, Yonezawa T, Salminen A, Baykov AA, Lahti R, Hachimori A (1998) Cloning and expression of a unique inorganic pyrophosphatase from *Bacillus subtilis*. Evidence for a new family of enzyme. FEBS Lett 439: 263–266

Sklyankina VA, Avaeva SM (1980) The quaternary structure of *Escherichia coli* inorganic pyrophosphatase is essential for phosphorylation. Eur J Biochem 191: 195–201

Smirnova IN, Baykov AA, Avaeva SM (1986) Studies on inorganic pyrophosphatase using imidodiphosphate as substrate. FEBS Lett 206: 121–124

Smirnova IN, Kudryavtseva NA, Komissarenko SV, Tarusova NB, Baykov AA (1988) Diphosponates are potent inhibitors of mammalian inorganic pyrophosphatase. Arch Biochem Biophys 267: 280–284

Sonnewald U (1992) Expression of *E. coli* inorganic pyrophosphatase in transgenic plants alters photoassimilate partitioning. Plant J 2: 571–581

Spiro TG (1971) Phosphate transfer and its activation by metal ions: alkaline phosphatase. In: Eichhorn GL (ed) Inorganic biochemistry. Elsevier, Amsterdam, pp 578–581

Springs B, Welsh KM, Cooperman BS (1981) Thermodynamics, kinetics and mechanism in yeast inorganic pyrophosphatase catalysis of inorganic pyrophosphate: inorganic phosphate equilibration. Biochemistry 20: 6384–6391

Teplyakov A, Obmolova G, Wilson KS, Ishii K, Kaji H. Samejima T, Kuranova I (1994) Crystal structure of inorganic pyrophosphatase from *Thermus thermophilus*. Protein Sci 3: 1098–1107

Terzyan SS, Voronova AA, Smirnova EA, Kuranova IP, Nekrasov YV, Arutyunyan EG, Vainstein BK, Höhne W, Hansen G (1984) Spatial structure of yeast inorganic pyrophosphatase at a resolution of 3 Å. Bioorg Khim 10: 1469–1482

Unguryte AL, Smirnova IN, Kasho VN, Baykov AA (1989) Comparison of the catalytic properties of mitochondrial and cytosolic inorganic pyrophosphatases of rat liver. Biol Membr 6: 356–361

Veech RL, Cook GA, King MT (1980) Relationship of free cytoplasmic pyrophosphate to liver glucose content and total pyrophosphate to cytoplasmic phosphorylation potential. FEBS Lett 117 Suppl: K65–K72

Velichko IS, Mikalahti K, Kasho VN, Dudarenkov VY, Hyytiä T, Goldman A, Cooperman BS, Lahti R, Baykov AA (1998) Trimeric inorganic pyrophosphatase of *Escherichia coli* obtained by directed mutagenesis. Biochemistry 37: 734–740

Vihinen M, Lundin M, Baltscheffsky H (1992) Computer modeling of two inorganic pyrophosphatases. Biochem Biophys Res Commun 186: 122–128

Volk SE, Baykov AA, Kostenko EB, Avaeva SM (1983) Isolation, subunit structure and localization of inorganic pyrophosphatase of heart and liver mitochondria. Biochim Biophys Acta 744: 127–134

Volk SE, Dudarenkov VY, Käpylä J, Kasho VN, Voloshina OA, Salminen T, Goldman A, Lahti R, Baykov AA, Cooperman BS (1996) Effect of E20D substitution in the active site of *Escherichia coli* inorganic pyrophosphatase on its quaternary structure and catalytic properties. Biochemistry 35: 4662–4669

Yang Z, Wensel TG (1992a) Inorganic pyrophosphatase from bovine retinal rod outer segments. J Biol Chem 267: 24634–24640

Yang Z, Wensel TG (1992b) Molecular cloning and functional expression of cDNA encoding a mammalian inorganic pyrophosphatase. J Biol Chem 267: 24641–24647

Young TW, Kuhn NJ, Wadeson A, Ward S, Burges D, Cooke GD (1998) *Bacillus subtilis* ORF *yybQ* encodes a manganese-dependent inorganic pyrophosphatase with distinctive properties: the first of a new class of soluble pyrophosphatase? Microbiology 144: 2563–2571

Polyphosphate/Poly-(R)-3-Hydroxybutyrate) Ion Channels in Cell Membranes

R. N. Reusch[1]

1
Introduction

Among the most intriguing structures formed by inorganic polyphosphates (polyP) are their complexes with poly-(R)-3-hydroxybutyrates) (PHB) that form ion channels in lipid bilayers. All peptide and protein ion channels, as well as synthetic ion channels, are amphiphilic structures with an outer coat of non-polar residues and a lining of polar and charged residues (Urry 1985; Christensen et al. 1988; Nakano et al. 1990; Sansom 1991; Kobuke et al. 1992; Epand 1993; Fyles et al. 1993). These attributes are provided in a cooperative fashion by the two structurally distinct homopolymers, polyP and PHB. The polymeric anion, polyP, forms a ladder of cation binding sites that stretches across the bilayer, shielded from the hydrophobic environment by the amphiphilic solvating polyester, PHB. Complexes of polyP and PHB, located in the plasma membranes of diverse bacteria (Reusch and Sadoff 1983; Reusch et al. 1986, 1987), are the first nonproteinaceous, ion-selective channels discovered in biological cells. Their capacity to form voltage-activated, calcium-selective channels in planar lipid bilayers has been established (Reusch et al. 1995; Das et al. 1997). As yet, there is no direct evidence of their in vivo functions; however, a number of studies point to calcium involvement in important cellular functions in bacteria, such as chemotaxis (Matsushita et al. 1989; Tisa and Adler 1992) and cell division (Chang 1986; Smith 1995; Norris 1996). In addition, there is substantial evidence that polyP/PHB complexes may serve as calcium pumps and DNA channels. Here we will discuss the singular molecular characteristics of polyP and PHB that relate to their roles in ion transport, and consider how the two polymers act in synergy to form these interesting transmembrane ion channels.

[1] Department of Microbiology, Giltner Hall, Michigan State University, East Lansing, Michigan 48824, USA.

Progress in Molecular and Subcellular Biology, Vol. 23
H. C. Schröder, W. E. G. Müller (Eds.)
© Springer-Verlag Berlin Heidelberg 1999

2
Inorganic Polyphosphates:
Cation Attraction, Selection, and Transport

Inorganic polyphosphates have distinctive molecular features that make them useful vehicles for the regulation of ion movement across cell membranes. They are linear chains of tetrahedral phosphate anions, linked through common oxygen atoms by phosphoanhydride bonds (Fig. 1). Although polyP are rapidly hydrolyzed at low pH and elevated temperatures,

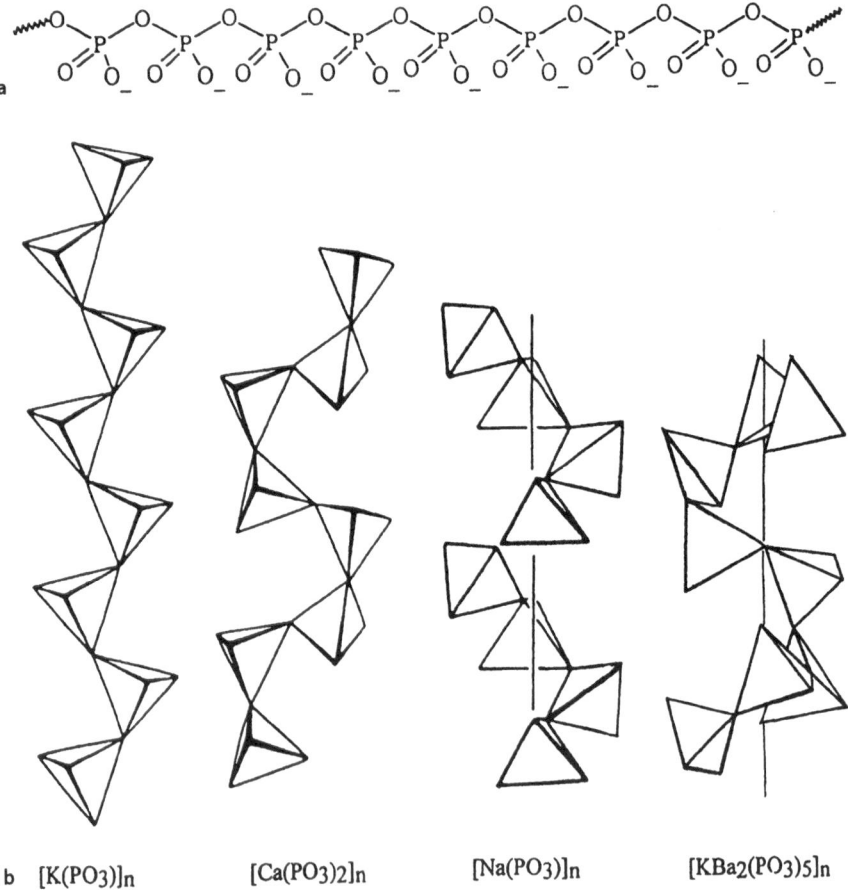

b   [K(PO₃)]n   [Ca(PO₃)₂]n   [Na(PO₃)]n   [KBa₂(PO₃)₅]n

Fig. 1. a Chain of polyphosphate residues. **b** Chain configurations of polyphosphate salts. Drawings show a few of many possible conformation taken from crystal structures (Matheja and Degens 1971; Corbridge 1995). Individual phosphate residues, each with a monovalent negative charge, are represented by a *tetrahedron*. Rotational flexibility of P-O-P bonds joining the tetrahedra allows the chains to twist into a large variety of conformations, depending on associated cations. (Reusch et al. 1995)

they are reasonably stable under physiological conditions. At physiological pH, each residue bears a monovalent negative charge. The repulsive forces between neighboring tetrahedra favor backbone rigidity; however, the P-O-P bonds linking these units have considerable rotational flexibility, allowing the tetrahedra to twist into a number of different conformations. As a result, the inorganic polyphosphates are polymorphic (Matheja and Degens 1971; Majling and Hanic 1980; Corbridge 1995) and their arrangement depends on the coordination preferences of the associated cations. In this way, polyP differs from more rigid anionic complexing agents which are specifically suited to the coordination requirements of particular cations. An important consequence of this flexibility is that it enables adjacent tetrahedra to approach within a few angstroms to form a binding site for divalent cations (Eisenman and Horn 1983). The stronger electrostatic interaction of divalent cations with polyP leads to their preferential selection over monovalent cations. It is binding energy rather than ion size or coordination geometry that is the rationale for the sequestering of Ca^{2+} and Mg^{2+} by polyP, and makes them effective agents in water softening (Corbridge 1995).

A polyP chain of negatively charged residues joined by flexible bonds can efficiently attract cations, select for divalent cations and, in response to voltage or concentration gradients, move them along its backbone of closely spaced binding sites. However, polyP is too polar to penetrate a lipid bilayer and it does not effectively discriminate among divalent cations. In order to form an effective ion channel, polyP must associate with a molecule that can provide these capabilities. That molecule is poly-(R)-3-hydroxybutyrate) (PHB).

3
Poly-(R)-3-Hydroxybutyrate)

Poly-(R)-3-hydroxybutyrate) (PHB) is a linear homopolymer of R-(3-hydroxybutyric acid), that is formed from acetyl-CoA in bacteria under growth-limiting conditions (Fig. 2). PHB was discovered by Lemoigne in 1925 in *Bacillus megaterium*, and since then has been found in many different microorganisms (Dawes and Senior 1973; Doi 1990; Brandl et al. 1995). It is best known as a storage material of high molecular weight (10^5–10^6 daltons) that can account for up to 90% of the cell dry weight. Frequently, the inclusion granules contain not only PHB but also its homologues, collectively known as polyhydroxyalkanoates (PHAs). The PHAs have assumed considerable commercial importance as biodegradable plastics because of their polypropylene-like material properties (Anderson and Dawes 1990). More recently, a low molecular weight form of PHB was found in bacterial membranes (Reusch and Sadoff 1983; Reusch et al. 1986). Subsequently, this form of PHB has been found in other cell fractions of prokaryotes and also in diverse plant and animal tissues (Reusch 1989, 1992; Reusch et al. 1992;

Fig. 2. a Chain of poly-(R)-3-hydroxybutyrate) residues. b Common pathway for synthesis of poly-(R)-3-hydroxy-butyrate) in bacteria. (Dawes and Senior 1973)

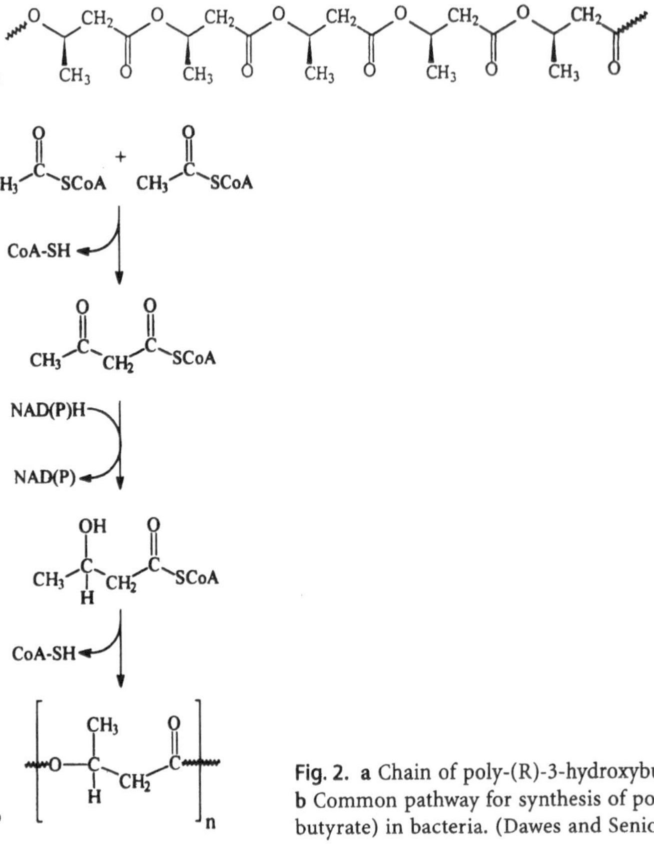

Müller and Seebach 1994; Seebach et al. 1994a; Huang and Reusch 1996). Apparently, PHB like polyP is a ubiquitous and perhaps universal cell constituent. Low molecular weight PHB is always found complexed to other macromolecules, such as proteins and polynucleotides, as well as polyphosphates. It is this form of the polyester that is of interest to us here.

3.1
Low Molecular Weight Poly-(R)-3-Hydroxybutyrate) in Bacterial Membranes

In 1983, Reusch and Sadoff discovered an 'ordered gel structure' of PHB in the membranes of *Azotobacter vinelandii* and *Bacillus subtilis*. The structure was discovered fortuitously during an investigation of membrane organization in genetically competent bacteria. Since competence is a transient and labile phenomenon, it was essential to use minimally intrusive procedures. Hence, modifications in the membrane environment were examined directly in whole cell cultures using the hydrophobic fluorescent probe N-phenyl-1-

naphthylamine (NPN). At low concentrations ($< 2 \times 10^{-5}$M), NPN does not influence the temperatures of the lipid phase transitions and does not interact significantly with protein. The probe partitions into the hydrocarbon region of the cell membranes where it responds to changes in the viscosity or polarity of its environment by a change in fluorescence intensity. This procedure, developed by Traüble and Overath (1973), is rapid, results in minimal disturbance of cell processes, and reports lipid transitions specifically and with reasonable agreement with transitions determined by light scattering and X-ray diffraction (Overath and Traüble 1973).

Genetic competence develops in the Gram-negative soil bacterium *A. vinelandii*, when log-phase cells are transferred to an iron-deficient medium (Page and von Tigerstrom 1979). Transformation becomes measurable after ca. 12 h, reaches a maximum at ca. 18 h and then slowly diminishes. At stages during this process, the NPN fluorescence spectra were observed as temperatures were increased slowly from 15 to 65 °C. In addition to the broad and reversible gel to liquid-crystalline phase transition of the membrane phospholipids, which begins below 0 °C and ends at about 25 °C, the spectra of normal log-phase cells showed a small and relatively sharp, irreversible fluorescence peak at ca. 56 °C (Reusch and Sadoff 1983; Fig. 3A). After transfer of the cells to iron-deficient medium, the intensity of the 56 °C peak increased considerably with time, reached a maximum at the time of highest competence, and then diminished.

No significant changes could be found in the phospholipid or neutral lipid composition of the cell membranes throughout this period, but the plasma membranes showed a steady increase in PHB content (Law and Slepecky 1961; Fig. 3B). There was a corresponding increase in cytoplasmic PHB granules, but the granules do not display appreciable fluorescence with NPN and could not be responsible for the 56 °C peak. The outer membranes were not analyzed because they cosediment with the PHB granules when the membranes are separated on sucrose gradients. The molecular weight of the plasma membrane PHB was estimated at 13 000 ± 15 % daltons by viscosity measurements. Similar studies of competence in the Gram-positive organism *B. subtilus* showed a comparable correlation between competence, plasma membrane PHB, granule PHB, and the intensity of a fluorescence peak at 56 °C. The relatively steep slope and high temperature of the fluorescence curve led to the inference that membrane PHB was part of an organized gel structure. Presumably, the ordered form of PHB has little influence on NPN environment, but release of the polyester into the bilayer when the structure dissociates effects a significant increase in bilayer viscosity, and consequently an increase in NPN fluorescence.

Unlike *A. vinelandii* and *B. subtilis*, *Escherichia coli* does not accumulate PHB in cytoplasmic inclusion bodies, and in fact were believed incapable of synthesizing the polyester. It was also generally accepted that the deveopment of genetic transformability in *E. coli* was physiochemically mediated,

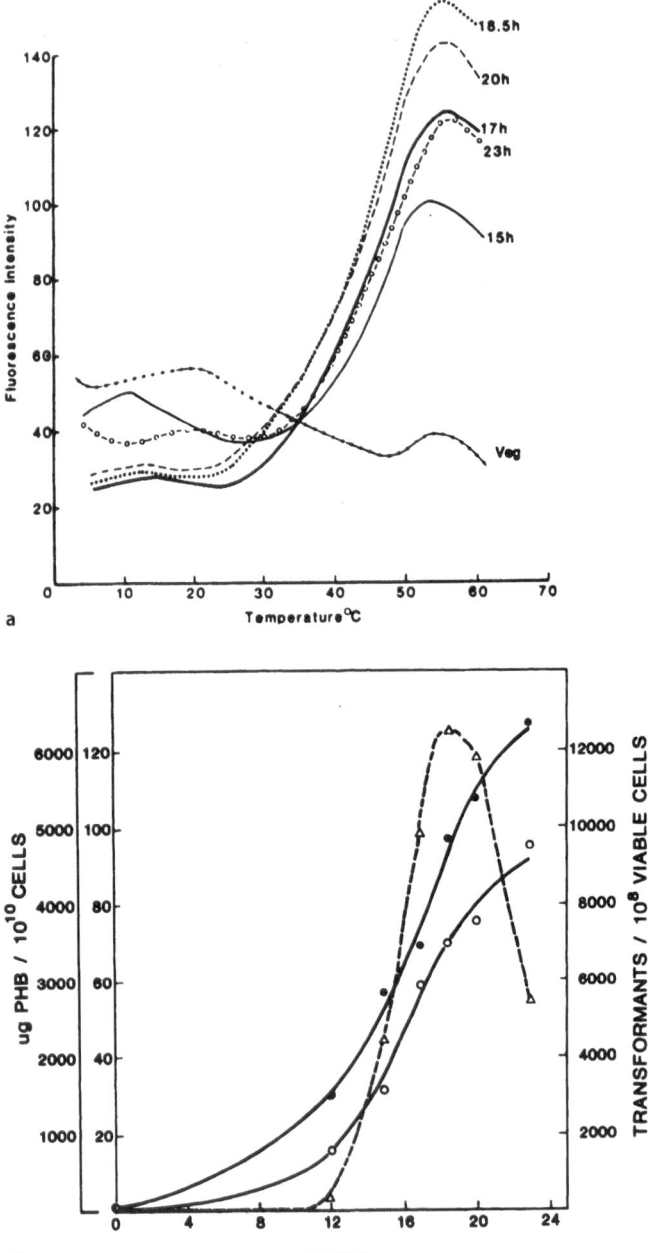

a

b

◀ **Fig. 3. a** Fluorescence spectra of N-phenyl-1-naphthylamine (NPN) in *A. vinelandii* UW1 cells during log-phase vegetative growth and at various stages of transformability. Log-phase growth was in Burk nitrogen-free buffer (Wilson and Knight 1952), pH 7.2, plus ammonium acetate (1.1 mg/ml) and glucose (1%). Cells were cultured at 30 °C with moderate aeration. For development of transformability, cells were transferred to the same medium, except that Fe^{2+} was omitted from the buffer, and cultured as above. At indicated times, NPN was added to 4 ml cell culture to a final concentration of 10^{-5} M, and thermotropic fluorescence spectra were recorded. Fluorescence intensities are relative. Excitation: 360 nm; emission: 410 nm. Measurements were made at increasing temperature (ca. 2 °C per min) **b** Relationship between concentrations of total PHB (*solid circles*), plasma membrane PHB (*open circles*) and transformation efficiency (*triangles*). Membranes were isolated and separated on linear sucrose gradients (25 to 65%). PHB concentrations were determined by the method of Law and Slepecky (1961). Transformation efficiency was determined by the method of Page and von Tigerstrom (1979). (Reusch and Sadoff 1983)

and not under physiological control. The exposure of log-phase cells to cold calcium-containing buffers was presumed to modify membrane structure and render the membranes more permeable to DNA entry (Smith et al. 1981). Reusch et al. (1986) observed the thermotropic fluorescence spectra of NPN in *E. coli* cells before and after the development of competence. Normal log-phase cells did not show any notable fluorescence at temperatures above the gel to liquid-crystalline phase transition, but a strong, sharp, irreversible peak was observed at 56 °C, after the cells were suspended in competence-inducing buffers (Fig. 4). The peak developed slowly in the buffer of Mandel and Higa (1970), and rapidly in the buffer recommended by Hanahan (1983). This corresponded to the time required for the development of transformability in the two procedures. Analysis of the membranes indicated that competent cells, but not log-phase cells, contained low molecular weight PHB, and PHB concentrations correlated with peak intensities and transformation efficiencies. Since there was no interfering cytoplasmic PHB, it was possible to establish that PHB was confined to the plasma membranes by analysis of plasma and outer membrane fractions.

Observation of the NPN thermotropic fluorescence spectra in cells has proven to be a very useful and sensitive technique for detection of these PHB complexes in bacterial membranes. The concentration of complexed PHB can be estimated by comparing the intensity of the 56 °C peak to that of the phospholipid phase transition, the latter being proportional to cell density. The integrity of the complexes can be assessed from the shape and temperature range of the high temperature peak; the peak broadens and lessens in intensity as the complexes dissociate.

Freeze fracture electron microscopy studies of the membranes of *E. coli* and *A. vinelandii* support the fluorescence data (Reusch et al. 1987). The micrographs show that, as cells incorporate PHB into the plasma membranes, small semiregular plaques which possess shallow particles are formed. These plaques grow in size and frequency as the concentration of membrane PHB increases (Fig. 5).

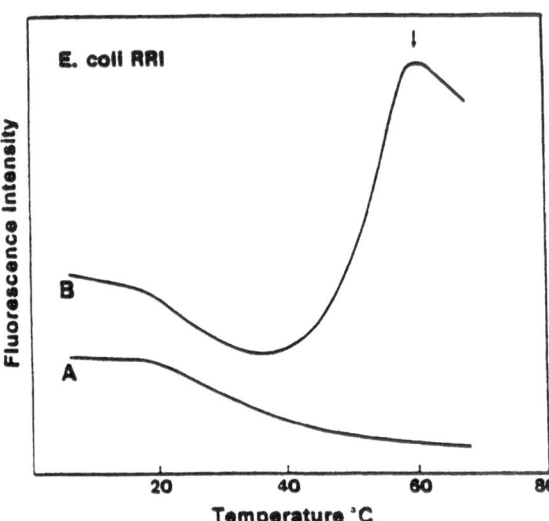

Fig. 4. Fluorescence spectra of N-phenyl-1-naphthylamine (NPN) in *E. coli* RR1 cells before and after induction of competence. Log-phase growth and competence protocol were as described by Hanahan (1983). Spectra were measured as in Fig. 3. Final concentration of NPN was 1 µM. Trace *A* log-phase cells; trace *B* competent cells. *Arrow* marks fluorescence peak of the membrane PHB complex. (Reusch 1992)

3.2
Poly-(R)-3-Hydroxybutyrate) as an Amphiphilic Salt Solvent

The properties of PHB most fundamental to its role in forming ion channels with polyP are its amphiphilic nature and its ability to form complexes with salts. The repeating unit of PHB contains both hydrophobic methyl groups and hydrophilic ester carbonyl oxygens, enabling the polyester to act as an intermediary between polyP and the bilayer. Perhaps even more important is the capacity of PHB to 'dissolve' salts. Indeed, PHB may be unique among biological polymers in having the structural features common to a small

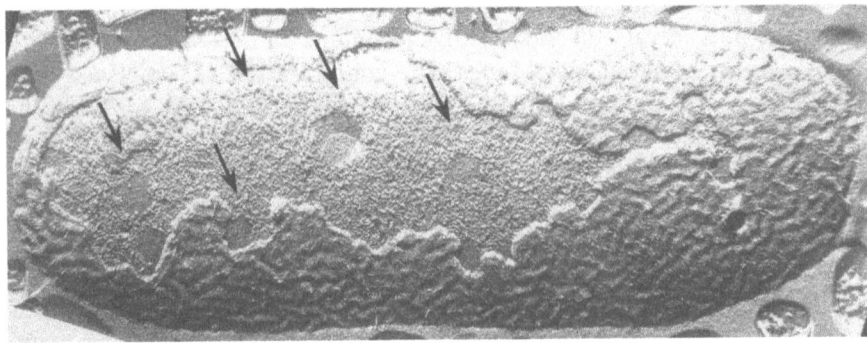

Fig. 5. Representative freeze-fracture electron micrograph of competent *E. coli* DH1. Micrograph shows typical appearance of small semiregular plaques (*arrows*) in plasma membranes of *E. coli* DH1 cells exposed to cold calcium-containing competence buffer. (Reusch et al. 1987)

group of polymers that form ion-conducting salt complexes (Armand 1987; Watanabe and Ogatu 1987; Gray 1992). Important characteristics of this class are:

1. flexible backbones with low barriers to bond rotation to ease segmental motions of the polymer chain;
2. heteroatoms that have sufficient electron donor power to form coordinate bonds with cations; and
3. a suitable distance between heteroatoms to permit the formation of multiple intrapolymer coordinate bonds to cations.

The stability of these polymer salt complexes is enhanced, as compared to complexes formed by small molecules, by the entropic advantage known as the 'polymer effect', attributable to the cooperative effect of neighboring ligands attached to a common backbone.

The most well-known member of this class of polymers is the polyether, polyethylene oxide, whose complexes with lithium perchlorate are used commercially in lithium batteries. The good solvating power of polyethylene oxide is attributed to an optimal spacing of the electron-donating ether oxygens along a flexible backbone. When this distance is decreased, as in polymethylene oxide, chain flexibility is greatly reduced; when it is increased, as in 1,3-polypropylene oxide, the distance between oxygens is too great to allow the polymer to assume the low energy conformations that maximize polymercation coordinations. In both cases. the ability to solvate cations is thus lost or greatly diminished. Substituents pendant to the backbone, as in 1,2-polypropylene oxide (1,2 PPO), have a lesser effect, but they may attenuate the solvating ability of the oxygens by increasing steric hindrance.

The carbonyl oxygens of esters are weak electron donors as compared to ether oxygens: nevertheless, polyesters with suitable spacing between ester groups also form ion-conducting salt complexes. The polyester poly-β-propiolactone forms ion-conducting complexes with lithium perchlorate (Watanabe et al. 1984). This polyester and PHB have an identical backbone structure and differ only in that PHB has a methyl group in the β position of the repeating unit (Fig. 6). It may be inferred from the parallel situation in the polyethers that this substitution may lessen the solvating ability but not abolish it.

The capacity of PHB to 'dissolve' salts was demonstrated experimentally by Bürger and Seebach (1993) who showed that oligomers of R-3-hydroxybutanoic acid could transport alkali and alkaline earth salts across chloroform layers, and by Reusch and Reusch (1993) who prepared conducting complexes from PHB and its homologue, poly-(R)-3-hydroxyvalerate), with lithium perchlorate.

It is also important to consider the types of salts that form complexes with PHB. As an aprotic polymer, PHB does not have the hydrogen-bond donating groups needed to solvate anions. Accordingly, it 'dissolves' salts of large anions with diffused charge that require little solvation (MacCallum

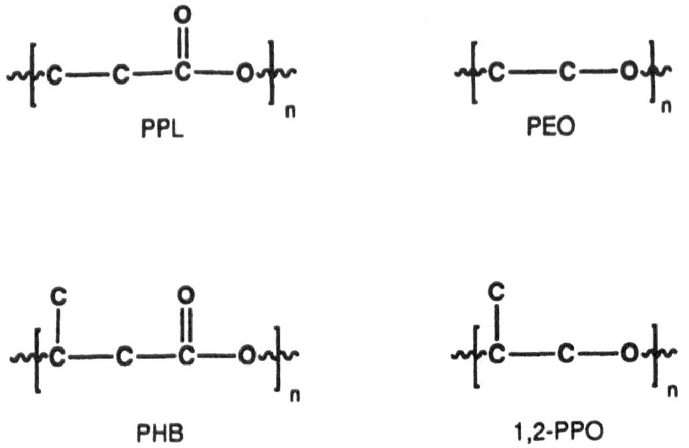

Fig. 6. Backbone structures of polyesters and polyethers of salt-solvating polymers. Figure shows the similarity of backbone structures, with optimal spacing between electron-donating oxygens, of polymers that form ion-conducting salt complexes. *PPL* Poly-β-propiolactone; *PEO* polyethylene oxide; 1,2-PPO 1,2-polypropylene oxide. (Reusch 1992)

and Vincent 1987). Its capacity to solvate cations is also limited. Ester carbonyl oxygens are weak Lewis bases of low polarity and can form coordinate bonds only with hard cations that have large solvation energies. The calcium salt of polyP satisfies both these requirements; the single negative charge carried by each polyP residue is shared by two oxygens, and Ca^{2+} has a high charge density and large solvation energy (Table 1).

Table 1. Ionic radii, hydration numbers, free energies of solvation, surface charge densities and coordination geometry of alkali and alkaline earth cations of interest

Cation	Ionic radius[a] (Å)	Hydration number[a]	$-\Delta G^{oa}$ (kcal/mol)	Coordination geometry
Na^+	0.98	6	98.5	Octahedral
K^+	1.33	6	80.5	Octahedral
Mg^{2+}	0.72	6	454	Regular octahedral
Ca^{2+}	1.06	7, 8	379	Irregular cubic
Sr^{2+}	1.27	7, 8	340	Irregular cubic
Ba^{2+}	1.43	7, 8	314	Irregular cubic

3.3
Poly-(R)-3-Hydroxybutyrate) Ion Channels in Planar Lipid Bilayers

Since PHB can transport salts across chloroform layers, its ability to conduct salts across bilayers was examined (Seebach et al. 1996b) in a planar bilayer voltage-clamp setup (Miller 1983; Alvarez et al. 1985). In this system, a bilayer is formed between two aqueous solutions by 'painting' a decane solution of phospholipids across a small aperture (ca. 0.2 mm in diameter) in a partition separating two chambers containing salt solutions. The hydrocarbon drains away and the phosphilipids spontaneously arrange themselves into a bilayer (black lipid membrane or BLM). The aqueous solution on one side (*cis*) of the bilayer represents the cell cytoplasm and the solution on the other side (*trans*) represents the aqueous environment outside the cell. The *trans side* is maintained as ground and external voltage steps are applied to the *cis* side (Fig 7).

In this manner, monodisperse and polydisperse oligomers of R-3-hydroxybutanoic acid (3-HB) ranging in size from 8 to 96 monomer units were incorporated into bilayers of synthetic 1-palmitoyl, 2-oleoyl, phosphatidyl choline (POPC) by premixing the PHB oligomers and phospholipids before painting the bilayer. Bilayers formed from pure phospholipids or phospholipids with low concentrations of the oligomers were not conductive. However, when the concentrations of the oligomers were 0.1–5 % of the phospholipid concentration, discrete current fluctuations were observed for all chain lengths $\geq$ 16 monomer units. The conductance at a given potential was reasonably constant for a given preparation, but varied for different preparations of the same oligomer, so that current–voltage relationships could not be established. As expected, the 3-HB oligomers did not discriminate well among ions. The current records were similar to those observed with ionic and non-ionic detergents such as Triton-X-100, octyl glucoside, or sodium dodecyl sulfate (van Zutphen et al. 1972; Tanaka et al. 1986), and for pure phosphatidylcholine bilayers at temperatures of the gel to liquid-crystalline phase transition (Marsh et al. 1976; Antonov et al. 1980).

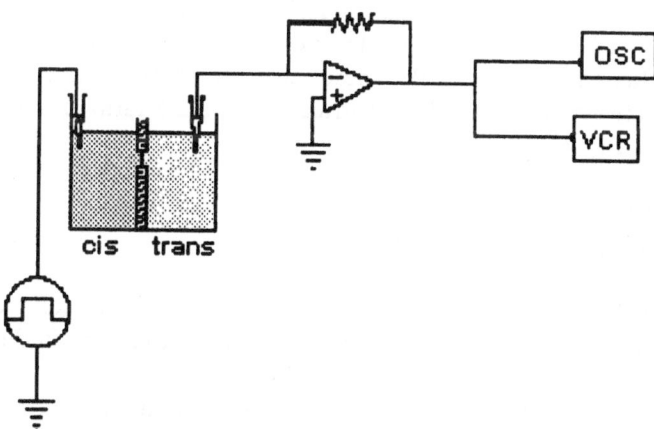

Fig. 7. Planar bilayer setup. The system consists two aqueous solutions, labeled *cis* and *trans*, separated by a planar bilayer. External voltage commands are applied to the *cis* side, with the *trans* side maintained at ground (defined as zero voltage). *OSC* oscilloscope; *VCR* recording tape system

The arrangement of the oligomers in the bilayer is uncertain but it is of interest that the length of a 16mer in the crystalline state is ca. 50 Å (Barham et al. 1984; Welland et al. 1989) and the thickness of the hydrophobic region of a planar lipid bilayer formed from a decane solution of phosphatidylcholine, as determined from electrical capacitance measurements, is ca. 48 Å (Fettiplace and Haydon 1980). From this, Seebach et al. (1996b) suggest that the uncomplexed oligomers form islands of lamellar crystallites in the bilayer, 16 residues per turn, and that ion permeability results from areas of mismatch at the interfacial regions between phospholipids and the oligomers.

These studies demonstrate the ability of PHB to serve as a solvating agent for ions in the bilayer; however, the PHB 'channels' are neither voltage-dependent nor selective. Like polyP, PHB has some of the essential properties of ion channels but not others. Fortunately, the two polymers complement each other, and together possess all the characteristics needed to form well-regulated ion-selective channels in lipid bilayers.

4
Polyphosphate/Poly-(R)-3-Hydroxybutyrate) Complexes in Bacterial Membranes

When the PHB gel structure in *E. coli* membranes was isolated and analyzed, the component complexed to PHB was identified as polyP, and Ca^{2+} was found to be the predominant neutralizing cation (Reusch and Sadoff 1988). The best estimates of the polymer lengths are 130–150 residues for PHB (MW ca. 12 000 daltons), as measured by non-aqueous size-exclusion chromatography (Seebach et al. 1994a), and 55–70 residues for polyP (MW ca. 5000 daltons), as determined by acrylamide gel electrophoresis (Castuma et al. 1995). The molecular weight of the channel complex is estimated as 17 000 ± 4000 daltons by nonaqueous size-exclusion chromatography (Reusch et al. 1995). These measurements indicate a 1:1 relationship between the two polymer strands, and a 2:1 ratio of monomer residues for PHB:polyP.

4.1
Selectivity for Calcium

Although Ca^{2+} is the primary cation in polyP/PHB complexes isolated from *E. coli* competent cells, one might argue that this finding is inconclusive since the competence-inducing buffers contain high concentrations of this cation. Huang and Reusch (1995) examined the cation composition of polyP/PHB complexes formed when *E. coli* cells were suspended in cold MES buffer, pH 6.3, with no added salts or with high concentrations of chloride salts of alkali metals or alkaline earth cations (other than calcium). In all cases, Ca^{2+} was the predominant cation. Evidently, polyP/PHB sequestered

Ca^{2+} from other celluar sites, demonstrating their strong avidity for this cation. From this selectivity, we may conclude that the geometry of the binding cavity formed by the association of the two polymers must be optimal for distinguishing Ca^{2+} from other cations. This suggests that each binding cavity typically has seven or eight oxygen ligands arranged in irregular geometry (Lehn 1973; Simon et al. 1973). The ability to tolerate variant bond angles and bond lengths makes Ca^{2+} the cation of choice for cross-linking macromolecules. Such a binding cavity would discriminate well against the other major physiological divalent cation, Mg^{2+}, which accommodates only six ligands and has a strong preference for regular octahedral geometry (Martin 1990).

4.2
Models for the Structure of PolyP/Poly-(R)-3-Hydroxybutyrate) Membrane Complexes

The detailed structure of polyP/PHB complexes is unknown. However, some assumptions can be made concerning the general organization of the complexes from the physical properties and sizes of the polymers, and the low dielectric environment they inhabit. It is clear that the highly polar polyanionic polyP must be shielded from the hydrophobic region of the bilayer by the amphiphilic PHB. Two structures with this general arrangement have been suggested.

Reusch et al. (Reusch and Sadoff 1988; Reusch et al. 1995) assume that polyP/PHB is basically a polymer electrolyte complex. Ion conduction only occurs in these complexes when the solvating polymer is in the elastomeric state; prevailing theories suggest that the ions are 'carried' from one binding site to the next by segmental motions of the polymer backbone (Gray 1992). They propose that PHB is in the amorphous state and forms a helical structure in the membrane (Marchessault et al. 1970), turning its lipophilic methyl groups to face the fatty acyl chains of the phospholipid lattice and its hydrophilic ester carbonyl oxygens towards CapolyP (Fig. 8A,B). Computer-modeling studies indicate that the molecule may arrange itself in a large variety of such exolipophilic-endopolarophilic configurations (Reusch et al. 1995). The helical shape is determined primarily by four backbone dihedral angles; the ester groups prefer an antiperiplanar arrangement, but small changes in one or more of the three less constrained backbone dihedrals have large effects on helical rise and diameter. The argument is not that PHB itself retains such conformations, but rather that the flexibility of the molecule enables it to wrap around the CapolyP framework helix and conform to its contours. In the idealized model, each Ca^{2+} is held in an ionophoretic cage solvated by eight oxygen atoms in distorted cubic geometry – four phosphoryl oxygens from two adjoining units of polyP (total charge of −2) and four ester carbonyl oxygens (two neighboring PHB units from one turn and two contiguous units from the turn directly above or below; Fig. 8C).

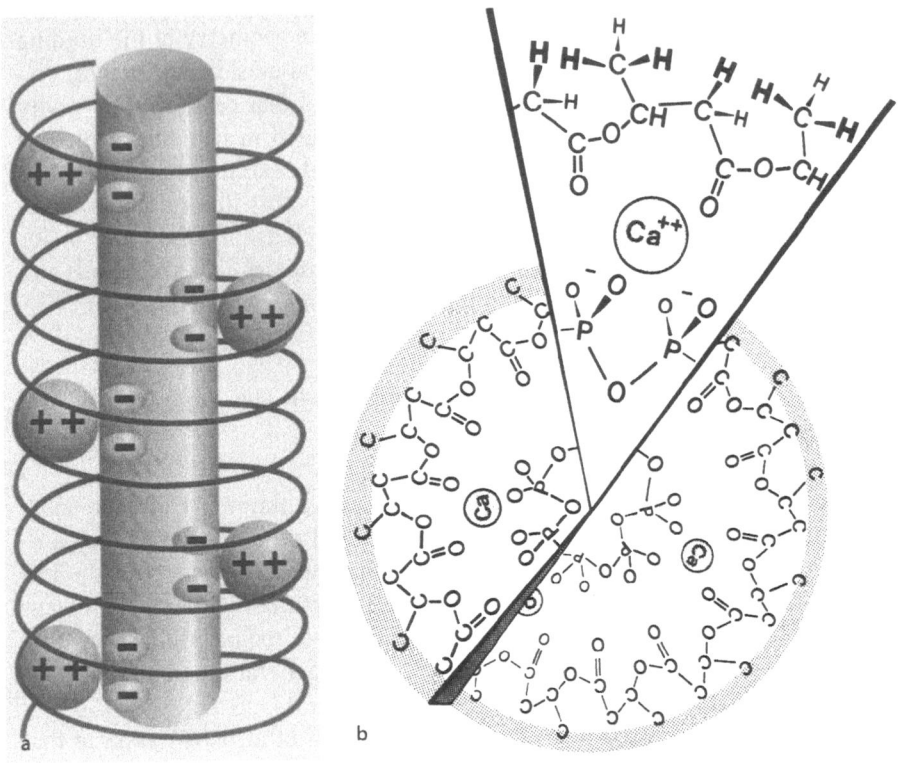

The calcium ions, held by ionic bonds to the core polyP helix, form weak ion–dipole bonds to ester carbonyl oxygens that crosslink the turns of the PHB helix.

Seebach et al. (1994b, 1996a) disagree with this view of membrane PHB structure. They propose that the molecules preserve the 2_1 helicity, 6-Å pitch, determined from fiber X-ray scattering for the crystalline state of the polymer (Cornibert and Marchessault 1972, 1975; Yokouchi et al. 1973; Brückner et al. 1988), and consider that the polyester chain must contain the stable dihedral angles determined from X-ray structures for the solid state cyclic oligomers by Plattner et al. (1993). Transmission electron microscopy, wide-angle X-ray scattering and atomic force microscopy show that the crystallites formed by solid monodisperse synthetic oligomers of (R)-3-hydroxybutyrate have the same lamellar morphology as the high molecular weight polymer (Seebach et al. 1996a). The crystal thickness, determined from small angle X-ray scattering (Welland et al. 1989; Sykes et al. 1995) as 52 Å, corresponds closely to the length of 16 3-HB units. From these considerations, they suggest a model in which the PHB molecules fold in the same lamellar morphology as PHB crystallites, 16 residues per turn across the bilayer. Eight consecutive antiparallel helices then form an ellipsoid arrangement, reminiscent of a β-sheet, surrounding the CapolyP (Fig. 9). There are too few carbonyl oxygens available in this model to fully coordinate the Ca^{2+} ions. To compensate, it is suggested that water molecules are incorporated (Seebach et al. 1996a) or that the PHB helix is distorted within the lipid lattice to allow the carbonyl oxygens to lie inside (Seebach et al. 1994b).

It may not be possible to resolve the structure of the complexes with certainty. In the Reusch model, the complexes have the liquidic properties of polymer electrolytes and this suggests a family of conformations rather than a single defined structure. In the Seebach model, several PHB molecules are involved in surrounding polyP. The individual PHB chains are free to adopt various positions in the phospholipid lattice, hence a well-defined structure is again unlikely. Nevertheless, further studies may help to decide between the two views of the arrangement of PHB molecules in the complexes and delineate the more probable conformations.

◀ **Fig. 8.** Models of the polyP/PHB channel structure proposed by Reusch et al. **a** The central cylinder represents the polyP helix which contains pairs of closely spaced monovalent negative charges that provide a ladder of binding sites for Ca^{2+}. Ca(polyP) is surrounded and solvated by an exolipophilic-endopolarophilic helix of PHB. (Das et al. 1997.) **b** Cross section. (Reusch 1989.) **c** Putative coordination geometry of calcium ions in polyP/PHB. Calcium forms ionic bonds with four phosphoryl oxygens of polyP and ion-dipole bonds with four ester carbonyl oxygens of PHB to form a neutral complex with distorted cubic geometry. (Reusch and Sadoff 1988)

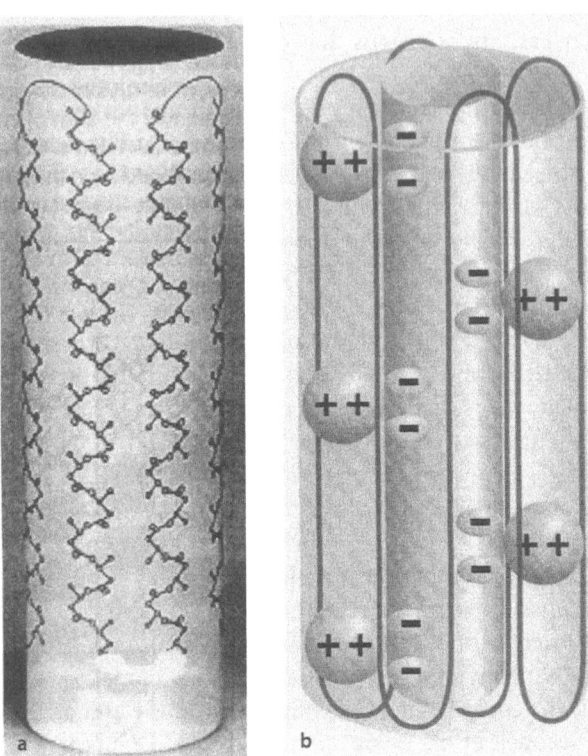

Fig. 9. Models of the polyP/PHB channel structure proposed by Seebach et al. **a** The tube-like arrangement of PHB helices. A chain of 3-HB units in a 6-Å pitch, 2_1 helical conformation, folded as in the lamellar crystallites, runs up and down parallel to the fatty acyl chains in the bilayer. Neighboring helices form a cylinder surrounding CapolyP. (Seebach et al. 1994b, 1996a.) **b** The central cylinder represents the polyP helix which contains pairs of closely spaced monovalent negative charges that provide a ladder of binding sites for Ca^{2+}. Ca(polyP) is surrounded and solvated by an outer cylinder composed of a β-like sheet of PHB formed as described above. (Das et al. 1997)

5
Polyphosphate/Poly-(R)-3-Hydroxybutyrate) Complexes in Planar Lipid Bilayers

5.1
E. Coli PolyP/Poly-(R)-3-Hydroxybutyrate) Complexes as Ion Channels

The ability of *E. Coli* polyP/PHB complexes to form calcium-selective channels in planar bilayers was investigated in the system described above (Fig. 7). *Escherichia coli* DH5α cells were made competent to increase the concentration of polyP/PHB. Then vesicles were prepared from the cell enve-

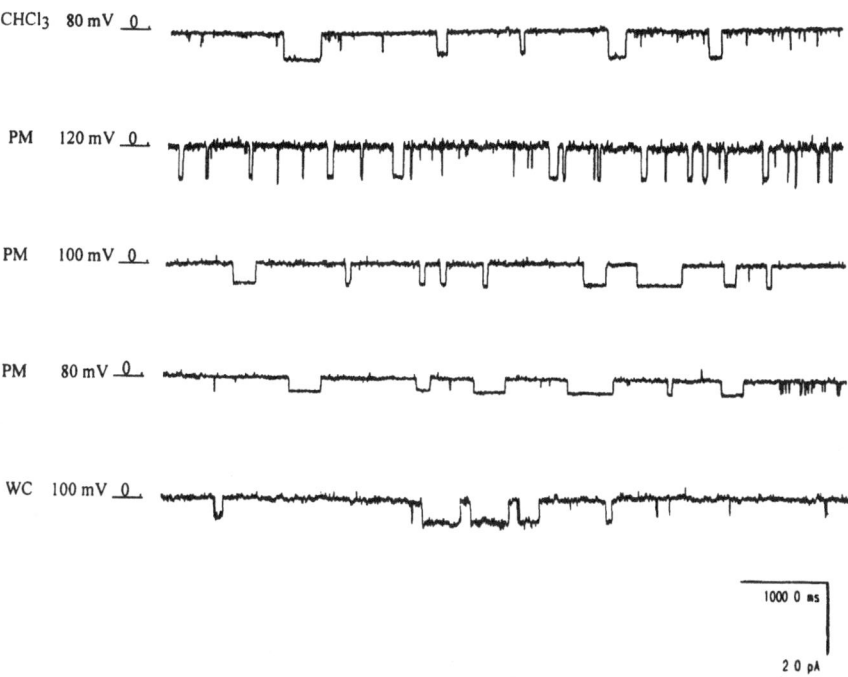

Fig. 10. Calcium channels from *E. coli* in planar bilayers. Single-channel currents observed in membrane vesicles and extracts of *E. coli*-competent cells when incorporated into planar bilayers of synthetic 1-palmitoyl, 2-oleoyl, phosphatidylcholine (POPC) between aqueous bathing solutions of 250 mM CaCl$_2$, 5 mM MgCl$_2$, 10 mM Tris Hepes, pH 73. *Line 1* Membrane vesicles of whole cells (*WC*) were added to *cis* bathing solution and the solution was gently stirred to allow spontaneous incorporation of vesicles into the bilayer. Single-channel currents were activated by a voltage step ≥ 60 mV. A representative current record at +100 mV is shown; *Lines 2, 3 and 4* plasma membrane vesicles (*PM*) were incorporated into the bilayer and single-channel currents were activated as above. Representative records at +80 mV, +100 mV and +120 mV are shown; *Line 5* a chloroform extract of genetically competent cells was added to a decane solution of POPC. The chloroform was removed by evaporation, and the remaining lipid mixture was used to form the bilayer. Single-channel currents were activated as above. A representative current record at +80 mV is shown. (Reusch et al. 1995)

lopes and added to the *cis* side of a planar bilayer formed by synthetic 1-palmitoyl, 2-oleoyl, phosphatidylcholine (POPC) between symmetric bathing solutions of 250 mM CaCl$_2$, 5 mM MgCl$_2$, 10 mM Tris Hepes, pH 7.3, and allowed to insert spontaneously into the bilayer (Reusch et al. 1995). No activity was observed in the absence of an applied voltage, but when a potential ≥ 60 mV was held for several minutes, single-channel currents were observed signifying the presence of Ca^{2+}-permeant channels (Fig. 10, line 1). The envelopes were then separated on sucrose gradients into plasma and outer membranes. Each was tested in a similar manner, but channel activity was only observed with the plasma membrane vesicles (Fig. 10, lines 2 and 3). This is consistent with earlier findings, by chemical analysis and freeze fracture electron microscopy, that polyP/PHB complexes are confined to the plasma membranes (Reusch et al. 1986, 1987).

PolyP/PHB complexes were also extracted from the cell membranes into chloroform and premixed with the phospholipid solution before painting the bilayer. Single-channel currents were again observed with voltage steps of 60 mV or greater (Fig. 10, line 4). When the complexes are extracted from membranes or cells, the chloroform solution contains protein and lipopolysaccharides in addition to polyP/PHB. To remove these components and evaluate their influence on channel activity, the complexes were further purified by size-exclusion column chromatography. This eliminated all detectable contaminants and also provided an estimate of the molecular weight of the complexes as $17\,000 \pm 4000$. The purified complexes were found to be more labile, but the single-channel activity they produced closely resembled that observed for the membrane complexes (Reusch et al. 1995).

To still further establish the composition of the channels, the PHB/CapolyP complexes were reconstituted. PHB was recovered from *E. coli* and carefully purified, and CapolyP was prepared from commercial sodium polyphosphate and calcium chloride. Three procedures were used:

1. PHB and excess CapolyP were mixed intimately, heated briefly in a microwave oven, and then sonicated in dry cold chloroform. CapolyP is very insoluble in chloroform so only PHB-complexed CapolyP dissolves. This solution was then premixed with phospholipid and used to form a bilayer;
2. PHB/CapolyP complexes, prepared as in (1), were incorporated into liposomes and added to the aqueous bathing solutions for spontaneous insertion into a preformed bilayer of phospholipid;
3. PHB was premixed with phospholipids before painting the bilayer. CapolyP was then added to the aqueous bathing solutions, and a potential of 60 mV was applied to induce the formation of the complexes within the bilayer. Single-channel currents similar to those described above were obtained with all three procedures (Reusch et al. 1995).

It is, of course, important to distinguish polyP/PHB channels from the conductive structures composed only of PHB (Sect. 3.3). In all the above channel preparations, the concentration of PHB in the bilayer was restricted to one-hundredth or less of that required to form PHB channels. This avoids having to distinguish the two activities in the current records. They differ in many respects, e.g. polyP/PHB channels are voltage-dependent and hence current–voltage relationships can be determined. Other differences such as ion selectivity and blockage will be discussed further below.

5.2
Completely Synthetic PolyP/Poly-(R)-3-Hydroxybutyrate) Complexes as Ion Channels

Despite the accumulation of evidence indicating that the observed channel activity was produced by polyP/PHB complexes, doubt remained as to whether the activity was actually effected ty trace protein contaminants that had been overlooked. To resolve this uncertainty once and for all, Das et al. (1997) performed a total synthesis of the channel complex from (R)-3-

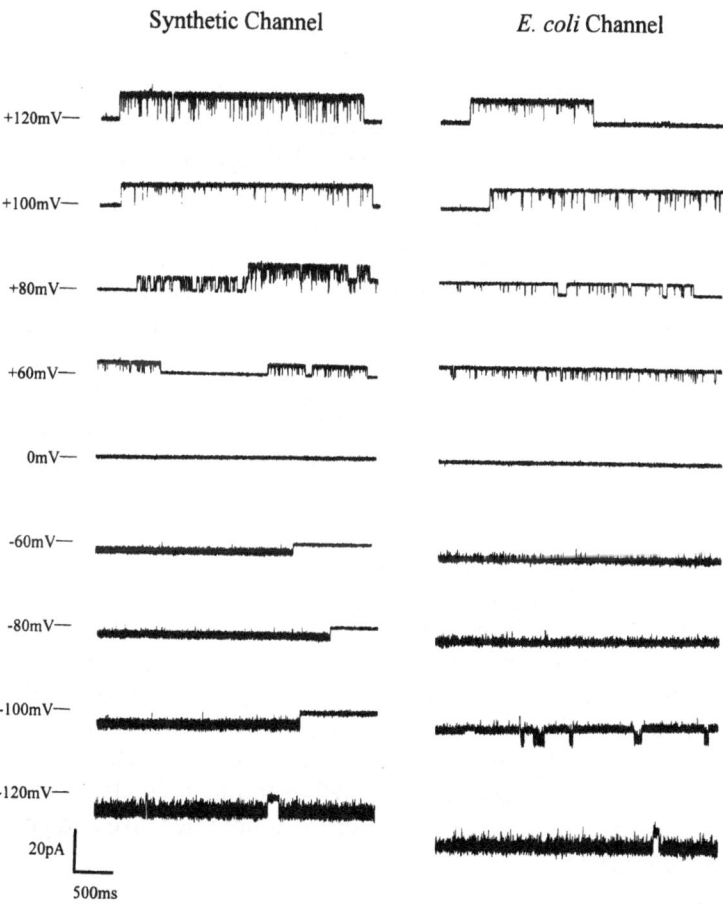

Fig. 11. Representative single-channel current fluctuations of synthetic and *E. coli*-derived polyP/PHB complexes at various clamping potentials. *Left* Synthetic HB$_{128}$/polyP: *right* channels extracted from competent cells of *E. coli* DH5α. Complexes were incorporated into planar lipid bilayers composed of synthetic 1-palmitoyl, 2-oleoyl, phosphatidylcholine and cholesterol (5:1; w/w) between aqueous bathing solutions of 200 mM CaCl$_2$, 5 mM MgCl$_2$, 10 mM Tris Hepes, pH 7.4 at 22 °C. *Bars* at side of each profile indicate fully closed state of channel. Clamping potentials with respect to ground are indicated at *left*. (Das et al. 1997)

Fig. 12. Current–voltage relations for synthetic HB$_{128}$/polyP (*squares*) and *E. coli* PHB/polyP channel complexes (*triangles*). Conductance of the channel for Ca^{2+} in symmetric solutions, under experimental conditions described in Fig. 11, is 101 ± 6 pS for synthetic channels and 104 ± 12 pS for *E. coli* channels. *Data points* represent mean values of ten observations. (Das et al. 1997)

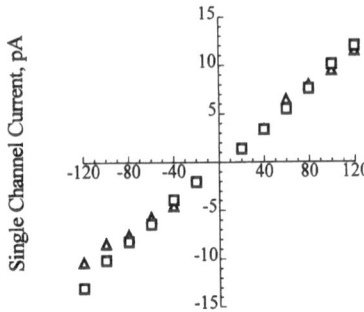

Holding Potential, mV

hydroxybutanoic acid, sodium polyphosphate and calcium chloride. An exponential fragment-coupling strategy was employed to prepare the 128mer of 3-hydroxybutyrate (HB$_{128}$) (Lengweiler et al. 1996). This synthetic polymer has an MW of 11.4 KDa, which is close to that of *E. coli* PHB (ca. 12 KDa). Calcium polyphosphate was prepared from sodium polyphosphate glass (Av. residue number 65) and calcium chloride. Complexes were formed from a mixture of the dry polymers as described above (Sect. 5.1). The size of complexed polyP was determined by acrylamide gel electrophoresis to be in the same range (55–65 residues) as that in the *E. coli* complexes (Castuma et al. 1995).

The channels formed in planar bilayers by the synthetic complexes were virtually identical to those formed by polyP/PHB complexes extracted from *E. coli* (Fig. 11). The conductances of the synthetic and *E. coli* channels were equivalent (Fig. 12). The conductance of the fully open synthetic channels was 101 ± 6 pS, and that of the *E. coli* channel was 104 ± 12 pS.

Both synthetic and *E. coli* polyP/PHB channels exhibited multiple subconductance states, and complex gating kinetics. As shown in Fig. 11, the channel conductances fluctuate between two major modes of gating. In mode 1, illustrated at positive clamping potentials, the channel openings are several seconds long with few complete closures, occasionally interspersed by long closed states. In mode 2, shown at negative clamping potentials, the channels open in long bursts separated by brief complete closures, with fast flickering closures to an intermediate closed state as well as to the main closed state. Switching between mode 1 and 2 can be observed at both negative and positive clamping potentials.

5.3
Cation Selectivity of PolyP/Poly-(R)-3-Hydroxybutyrate) Channels

The channels formed by polyP/PHB complexes, *E. coli* or synthetic, show strong selectivity for divalent over monovalent cations. The selectivity of an ion channel is usually investigated by determining the reversal potential

Fig. 13. Selectivity of synthetic HB_{128}/polyP complexes for Ca^{2+} over Na^+. Single-channel current–voltage relations of HB_{128}/polyP complexes, incorporated in planar lipid bilayers composed of synthetic 1-palmitoyl, 2-oleoyl, phosphatidylcholine and cholesterol (5:1; w/w) between unequal solutions; *cis* 65 mM $CaCl_2$, 10 mM NaCl, 5 mM $MgCl_2$, 10 mM Trips Hepes, pH 7.4, and *trans* (ground) 200 mM NaCl, 0.1 mM $CaCl_2$, 5 mM $MgCl_2$, 10 mM Tris Hepes, pH 7.4. Equilibrium potentials (calculated from concentration) were $E_{Ca} = -82$ mV, $E_{Cl} = +9$ mV, and E_{Na} is +76 mV. *Error bars* indicate standard deviation from mean ($n = 3$). (Das et al. 1997)

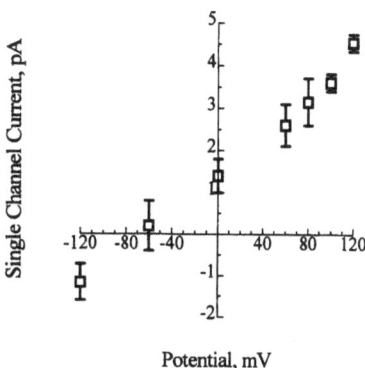

Potential, mV

(zero-current potential) when the channel is in a bilayer formed between aqueous solutions of unequal ion composition. Typically, the reversal potential is estimated graphically from the single-channel current–voltage relationships. This value can be compared with the theoretical reversal potential for each cation in a perfectly selective channel, calculated from the Nernst equation, e.g. for cation A of charge z

$$E_A = \frac{RT}{zF} \ln \frac{[A]_o}{[A]_i}.$$

The single-channel current–voltage relations for synthetic polyP/PHB complexes in bilayers between unequal solutions of Ca^{2+} and Na^+ are shown in Fig. 13. With solutions of 65 mM $CaCl_2$, 10 mM NaCl, 5 mM $MgCl_2$, 10 mM Tris Hepes, pH 7.4 *cis* and an isotonic solution of 200 mM NaCl, 0.1 mM $CaCl_2$, 5 mM $MgCl_2$, 10 mM Tris Hepes, pH 7.4 *trans*, the reversal potential as estimated from the graph is –67 mV. This is close to the equilibrium potential (calculated from concentrations) for Ca^{2+} of –82 mV, whereas the Cl^- equilibrium potential is +9 mV and the equilibrium potential for Na^+ is +76 mV. Similar studies for the *E. coli* complexes showed strong selectivity for Sr^{2+} over Na^+ (Reusch et al. 1995).

5.4
Block of PolyP/Poly-(R)-3-Hydroxybutyrate) Channels

One of the characteristics of protein calcium channels is their sensitivity to block by transition metal cations (Hille 1992). Lanthanum is a particularly potent blocker. It is suggested that permeant and blocking ions compete for common binding sites in the channel. The polyP/PHB channel complexes are also blocked by transition metal cations in a concentration-dependent manner. Nearly complete block of single-channel currents was observed in the synthetic complexes at concentrations > 0.1 mM La^{3+} (0.1 % of Ca^{2+}) (Das et al. 1997; Fig. 14). The *E. coli*-derived complexes were completely blocked

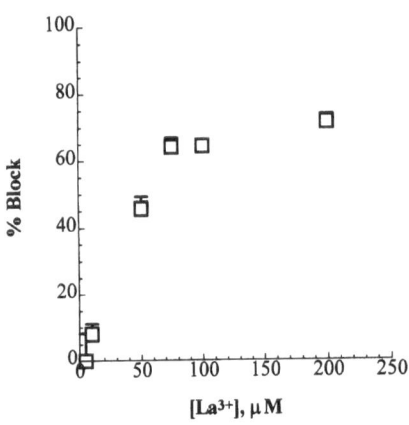

Fig. 14. Block of synthetic HB_{128}/polyP channel by transition metal cation, La^{3+}. The bilayer composed of synthetic 1-palmitoyl, 2-oleoyl, phosphatidylcholine and cholesterol (5:1; w/w) was formed between aqueous bathing solutions of 200 mM $CaCl_2$, 5 mM $MgCl_2$, 10 mM Tris Hepes, pH 7.4. After incorporation of the channel, activities were recorded for 5 min at a clamping potential of -80 mV. Then, $LaCl_3$ was added to the *trans* compartment to achieve the indicated concentrations. The bath was stirred and activities were recorded after 1 min of addition of La^{3+}. *Data points* represent mean values of amplitude histograms; *error bars* show standard deviation from mean. (Das et al. 1997)

by 0.6 mM La^{3+}, 1.5 mM Co^{2+} or 8 mM Cd^{2+} (Reusch et al. 1995). Mg^{2+} also has an effect on the currents. At low concentrations (1 % of Ca^{2+} concentration), it appears to increase the stability of the channel complexes and acts as a mild blocker, thus making it easier to observe channel openings and closures. For this reason, low concentrations of Mg^{2+} were customarily included in the buffers. Raising Mg^{2+} concentrations to higher levels significantly diminished single-current magnitudes, and nearly complete blocking of Ca^{2+} currents was observed in the *E. coli* channels when Mg^{2+} was present in amounts in excess of 10 % of $[Ca^{2+}]$.

5.5
How Do PolyP/Poly-(R)-3-Hydroxybutyrate) Channels Work?

The mechanism of ion conduction by polyP/PHB channel complexes can be rationalized in terms of the structures and properties of the component polymers. One view of how the channel may operate in the cell membrane or planar bilayer follows. CapolyP, surrounded and solvated by PHB, forms a salt bridge extending from the cytoplasm to the periplasm. A multi-lane channel is formed between the two polymers; the outer wall is lined with solvating oxygens, the inner wall is girdled by monovalent, phosphoryl anions (Figs. 8, 9). At the outer interface, cations are drawn to the mouth of the channel by polyP, and divalent cations are preferentially bound. Most of the binding sites within the channel are occupied by Ca^{2+} and the strong bonds between Ca^{2+} and polyP prevent ion movement, and so the channel is 'closed'.

The polyP 'wire' of negative-charges across the bilayer acts as a sensor of membrane potential. PolyP reacts to membrane depolarization (or a voltage-step of sufficient strength) by stretching or sliding within the PHB pore, thus dislodging the resident Ca^{2+} and initiating an ion flow. Ca^{2+} at the interface then preferentially permeate into the binding cavities at the end of the channel by virtue of their well-suited coordination geometry and the relatively rapid rate at which they undergo replacement of water of hydration. Sr^{2+} and Ba^{2+} are also permeant, but they are not normally found in physiological systems. These cations have the same coordination geometry as Ca^{2+} (Table 1) and, evidently, the flexible PHB envelope can adjust to accommodate the larger ion size. Since binding sites on polyP are identical and spaced at frequent intervals, there is no net potential energy cost to cation movement within the channel. Segmental motions of the PHB backbone and librational movements of ester carbonyl oxygens carry Ca^{2+} from site to site in parallel single-file lanes until internal concentrations rise to an appropriate level or the membrane is again polarized. Transition metal cations, particularly trivalent cations like La^{3+}, bind tightly to polyP at the interface but have difficulty entering because of their unsuitable coordination preferences, and consequently they block the ion flow.

No channel is perfectly selective. The reversal potential indicates that Na^+ passes through the channel, even though infrequently, and undoubtedly other cations are also occasionally transported. The conformations of polyP and PHB are less rigid and uniform than is suggested by the diagrammatic models. Within the channel there may be regions of atypical structure and liquidic character as the two polymers adjust in synchrony to the size, charge and coordination requirements of individual cations passing through.

6
Polyphosphate/Poly-(R)-3-Hydroxybutyrate) Complexes as Calcium Pumps

Bacterial cells, like those of eukaryotes, maintain low internal Ca^{2+} concentrations. The means by which they do this is still uncertain, although secondary systems for Ca^{2+} export have been identified in a number of bacteria (Ambudkar et al. 1984, Lynn and Rosen 1987). PolyP/PHB complexes have many structural features that make them likely agents for Ca^{2+} extrusion. The ionic bond between Ca^{2+} and polyP is sufficiently strong to withstand the considerable Ca^{2+} gradient across the membrane, estimated as $2\,mM$ outside and $0.1\,\mu M$ inside for *E. coli* (Gangola and Rosen 1987). At the same time, Ca^{2+} is held to PHB only by weak ion–dipole bonds. Consequently, the CapolyP helix is postulated to slide within the PHB channel, as was suggested may be the response to a voltage change (Sect. 5.5). This organization implies that Ca^{2+} could be transported out of the cell by extending the polyP chain on the cytoplasmic side of the membrane and transporting it through the PHB pore. As the appended phosphate units move into the PHB channel, Ca^{2+} is sequestered from the cytoplasm; at the outer face of the membrane, CapolyP is exported (Fig. 15).

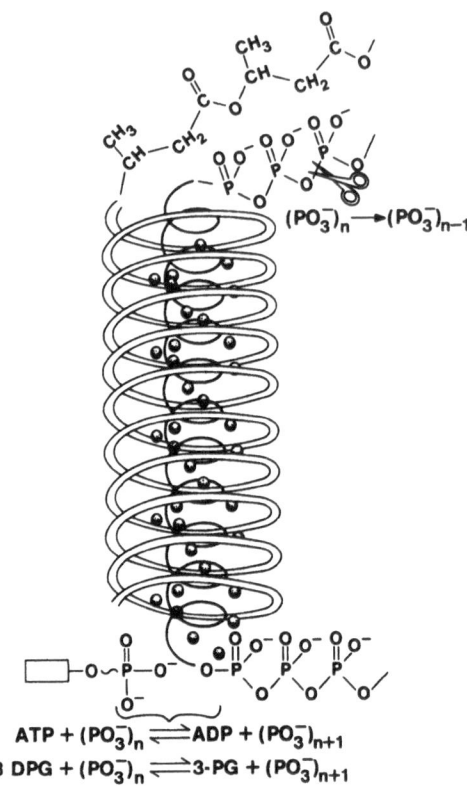

Fig. 15. Model of putative polyP/ PHB calcium pump. Schematic drawing of the putative polyP/ PHB complex in the membranes, indicating a hypothetical mechanism for coexport of Ca^{2+} and polyP from the cytoplasm. (Reusch 1992)

$$ATP + (PO_3^-)_n \rightleftharpoons ADP + (PO_3^-)_{n+1}$$
$$1,3\ DPG + (PO_3^-)_n \rightleftharpoons 3\text{-}PG + (PO_3^-)_{n+1}$$

The enzymes required to carry out this mechanism are known in many organisms (Kulaev and Vagabov 1983; Wood and Clark 1988; Kornberg 1995). PolyP occupies an intermediate position in the free energy scale of phosphorylated compounds, and can act as both a donor and acceptor of phosphate groups. It has a free energy of hydrolysis similar to that of ATP, and in bacterial cells is formed enzymatically from ATP and other high energy phosphates. Polyphosphate kinase, discovered by Kornberg et al. (1956) in *E. coli*, catalyzes the reversible transfer of the γ phosphate of ATP to the end of a polyphosphate chain. Similar enzymes have since been reported in *Neisseria* and coryneform bacteria (Bark et al. 1993; Tinsley et al. 1993). Phosphorylated compounds with still higher free energy of hydrolysis, such as 1,3-diphosphoglycerate, can also donate high-energy phosphates to polyP; the enzyme 3-phosphoglyceroyl phosphate-polyphosphate-phosphotransferase forms polyphosphates in both prokaryotes and eukaryotes (Kulaev and Vagabov 1983). At the periplasmic face, polyP may be degraded by polyphosphatases (Kulaev and Vagabov 1983; Kornberg 1995).

There is at present no experimental evidence for this alleged calcium-export function of polyP/PHB in *E. coli*; however, it is significant that polyP and PHB were recently found to be components of a Ca^2-ATPase pump in human erythrocyte membranes. The Ca^{2+}-ATPase is the sole transporter of Ca^{2+} in red blood cells and was the first plasma membrane enzyme demonstrated to work as a Ca^{2+} pump (Dunham and Glynn 1961; Schatzmann 1966). Reusch et al. (1997) purified the erythrocyte Ca^{2+}-ATPase to homogeneity by calmodulin affinity chromatography (Niggli et al. 1979; Kosk-Kosicka et al. 1986). The presence of polyP in the protein was signaled by its characteristic red-violet color on staining with o-toluidine blue (Griffin et al. 1965). PolyP was then measured by an enzyme assay in which polyphosphate kinase of *E. coli* is used to catalyze the transfer of phosphoryl groups from polyP to ADP (Crooke et al. 1994). The presence of PHB was demonstrated by a positive reaction to anti-PHB IgG on a Western blot; PHB was then measured by a chemical assay in which the polyester is degraded to crotonic acid (Karr et al. 1983; Reusch 1989). Though only minute quantities of the polymers were found (0.5 µg polyP/mg protein and 1.3 µg PHB/mg protein), the amounts are likely to be greatly understated, since both PHB and polyP are subject to extensive chemical and enzymatic hydrolysis during the isolation and purification procedures.

The Ca^{2+}-ATPase protein also showed some remarkable enzymatic activities. The ability of polyP to directly phosphorylate the Ca^{2+}-ATPase was examined using $[^{32}P](polyP)$. Then the capacity of the Ca^{2+}-ATPase to act as a polyphosphate kinase was investigated by radioactive assays, i.e. the ability to catalyze the transfer of the γ-phosphate of ATP to polyP (Ahn and Kornberg 1990), and to utilize polyP to phosphorylate ADP and generate ATP (Crooke et al. 1994):

$$[\gamma\text{-}^{32}P]\,ATP + (polyP)_n \rightarrow ADP + [^{32}P](polyP)_{n+1}\,;$$

$$[^{32}P](polyP)_n + ADP \rightarrow [^{32}P](polyP)_{n-1} + [^{32}P]ATP.$$

The results have importance in that they place these two polymers, with proven capacities to facilitate ion transport, in a human calcium-transporting structure. However, the manner in which the PHB and polyP interact with the protein is not known and it remains uncertain whether they play an active role in the export of Ca^{2+}.

7
Polyphosphate/Poly-(R)-3-Hydroxybutyrate) Complexes as DNA Channels?

PolyP/PHB complexes were first discovered in genetically competent cells. Although the complexes are present in log-phase cells, their concentrations in most organisms during exponential growth are very small. Had it not been for the remarkable increase in their numbers when the cells are made genetically transformable and the sensitivity of the fluorescence assay with NPN, it is likely they would have remained undetected for many years to come. It was the striking correlation between polyP/PHB concentrations and transformation efficiencies in *A. vinelandii*, *B. subtilis* and *E. coli* that led Reusch and Sadoff (1988) to postulate that the complexes play a role in DNA transmembrane transport. Here, we will consider the experimental and structural evidence in support of this hypothesis, and speculate on a mechanism by which polyP/PHB may be used for DNA transmembrane transport.

For the purpose of this discussion, the processes leading to DNA transmembrane transfer can be divided into three stages:

1. development of competence;
2. DNA binding; and
3. DNA uptake.

The details of the procedures used to carry out these steps vary for each organism. Nevertheless, we have seen that regardless of the method used to develop competence, the result is a conspicuous increase in the concentration of polyP/PHB complexes in the plasma membranes. When formation of the complexes is prevented by any means, transformation is inhibited (Reusch et al. 1986; Huang and Reusch 1995). DNA binding requires divalent cations, and only certain ones will do – Mg, Ca, Mn, Sr. All of these cations form strong ionic bonds to phosphate and can crossbridge the phosphate residues of DNA and polyP. For DNA uptake to occur, the cells must be returned to normal growth medium. If the thermotropic fluorescence spectrum of the cells is examined at this time, a rapid decrease in the intensity of the 56 °C fluorescence peak is observed, indicating that polyP/PHB complexes are being removed from the membrane. This suggests a mechanism for DNA transmembrane transfer. As polyP is retrieved by cytoplasmic enzymes, it may draw the bound DNA molecule into and through the PHB channel. In the case of *E. coli*, this stage is usually preceded by applying a brief heat pulse. The glass transition temperature of PHB is 0 °C (Holmes 1987), and the maximum temperature of the heat pulse, 42 °C, is just below the point at which dissociation of polyP/PHB begins (Fig. 4). The heat pulse would have the effect of making PHB more flexible so that it can more easily make necessary conformational adjustments, while avoiding dissociation of polyP/PHB.

From this viewpoint, the various procedures for competence development and DNA transformation are simply resourceful methods for changing the direction of movement of polyP within the PHB pore from outward to inward. The cells are first placed in an environment that leads to a substantial increase in polyP/PHB, with sufficient divalent cations to bind DNA to polyP. Then they are transferred to a medium in which they ordinarily sustain much lower levels of the membrane complexes, thus inducing an inward flow of polyP. In support of this hypothesis, single-stranded donor DNA was found complexed to PHB when DNA uptake in *E. coli* RR1 was interrupted in the first few minutes (Fig. 16).

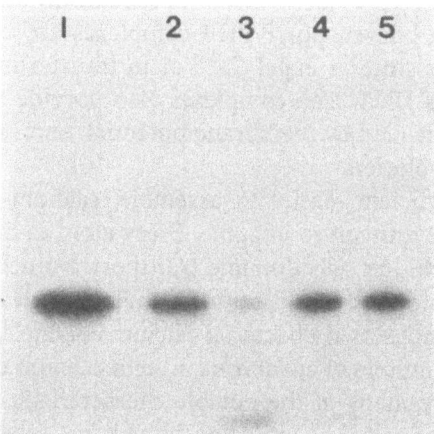

Fig. 16. Evidence for a PHB/DNA complex. Chloroform-soluble donor ssDNA extracted from *E. coli* RR1 cells in early stages of DNA uptake. Cells were made genetically competent by the method of Hanahan (1983), and transformed with ^{32}P-DNA prepared from the same organism. DNA uptake was interrupted after 5 min. The cell pellet was washed sequentially with methanol, methanol:acetone (1:1), acetone and the complex was extracted from the dry residue with dry cold chloroform. Solvated DNA was recovered by washing the chloroform extract with 10 mM EDTA, pH 7.5. Electrophoresis was carried out on 15% acrylamide gels. *Lanes 1 and 2* Untreated extract; *lane 3* digestion with mung bean nuclease; *lane 4* digestion with λ-exonuclease; *lane 5* digestion with RNase 1A. (Reusch 1992)

8
Evolutionary Aspects and Concluding Remarks

PolyP and PHB are ubiquitous cell constituents; both polymers have been found in representatives of nearly all classes of organisms, including bacteria, fungi, plants and mammals (Kulaev 1979; Kulaev and Vagabov 1983; Reusch 1989; Reusch et al. 1992; Seebach et al. 1994a; Kumble and Kornberg 1995). This alone suggests they have a physiological role(s) that are fundamental to life. The studies here suggest that one of these roles is ion transport.

It is intriguing to speculate that polyP/PHB channels date from primordial times. PolyP and PHB are composed of rudimentary molecules available to the earliest cells. In fact, inorganic polyphosphates are prebiotic molecules. In nature they have been found in volcanic condensates and thermal vents at the bottom of the ocean (Kulaev 1979; Yamagata et al. 1991). The synthesis of PHB is more demanding but requires only acetate and a reducing agent, both available in the primordial soup. The need for a system to control Ca^{2+} must have arisen early in cell development, as Ca^{2+} would tend to bind to and precipitate organic phosphates which were essential to the cell. PolyP could be employed to seclude divalent cations, but this would also remove Mg^{2+}, a critical cation for many biosynthetic reactions. The association of polyP with PHB provides a means to specifically sequester Ca^{2+}. Once positioned across a bilayer, polyP/PHB complexes are versatile structures and may be used to store or expel Ca^{2+}, or to transfer informational polyphosphates, such as DNA. The complexes also provide a means for rapid import of Ca^{2+}, controlled by membrane potential, and set the stage for cell signaling later in evolution.

Simpler, more efficient, easier to assemble, calcium-selective channels than polyP/PHB are difficult to imagine. Every atom in both polymers contributes to the task of ion selection and transport. No deletions or substitutions are conceivable that would improve their effectiveness. If we accept that polyP/PHB complexes are bacterial calcium channels, we must consider them to be the progenitors of eukaryotic protein calcium channels. Certainly the complexes share many of the notable characteristics reported for proteinaceous Ca^{2+} channels, i.e. activation by voltage, selectivity for divalent over monovalent cations, permeance to Ca^{2+}, Sr^{2+} and Ba^{2+}, and concentration-dependent blocking by transition metal cations. Moreover, the structures postulated for the complexes bear a striking resemblance to the single-file, multiple-site Ca^{2+} channel structures described by Hagiwara and Byerly (1981), Hess and Tsien (1984), McCleskey and Almers (1985) and Tsien et al. (1987). The process by which protein channels evolved from polyP/PHB complexes is unknown and needs to be explored. Most provoking is the presence of polyP and PHB in a human calcium pump. Do other calcium pumps and channels also harbor these simple polymers? Perhaps the channels of higher eukaryotes are not just protein but supramolecular structures in which protein, polyP and PHB have joined together for more efficient regulation of ion transport.

References

Ahn K, Kornberg A (1990) Polyphoshate kinase from *Escherichia coli*. J Biol Chem 265: 11734–11739

Alvarez O, Benos D, Latorre R (1985) The study of ion channels in planar lipid bilayer membranes. J Electrophysiol Tech 12: 159–177

Ambudkar SV, Zlotnick EW, Rosen BP (1984) Calcium efflux from *Escherichia coli*s: evidence for two systems. J Biol Chem 259: 6142–6146

Anderson AJ, Dawes EA (1990) Occurrence, metabolism, metabolic role, and industrial uses of bacterial polyhydroxyalkanoates. Microbiol Rev 54: 450–472

Antonov VF, Petrov VV, Molnar AA, Predvoditdelev DA, Ivanov AS (1980) The appearance of single-ion channels in unmodified lipid bilayer membranse at the phase transition temperature. Nature 283: 585–586

Armand MB (1987) Current state of PEO-based electrolyte. In: MacCallum JR, Vincent CA, (eds) Polymer electrolyte reviews 1. Elsevier, New York, pp 1–37

Barham PJ, Keller A, Otun EL, Holmes PA (1984) Crystallization and morphology of a bacterial thermoplastic: poly-3-hydroxybutyrate. M Mater Sci 19: 2781–2794

Bark K, Kampfer P, Sponner A, Dott W (1993) Polyphosphate-dependent enzymes in some coryneform bacteria isolated from sewage sludge. FEMS Microbiol Lett 107: 133–138

Brandl H, Aeberli B, Bachofen R, Schwegler I, Müller H-M, Bürger MH, Hoffmann T, Lengweiler UD, Seebach D (1995) Biodegradation of cyclic and substituted linear oligomers of poly(3-hydroxybutyrate). Can J Microbiol 41: 180–186

Brückner S, Meille AV, Malpezzi L, Cesaro A, Navarini L, Tombolini R (1988) The structure of poly(D-(-)-β-hydroxybutyrate). A refinement based on the *Rietveld* method. Macromolecules 21: 967–971

Bürger HM, Seebach D (1993) Cation transport across bulk liquid organic membranes with oligomers of (R)-3-hydroxybutanoic acid. Helv Chim Acta 76: 2570–2580

Castuma CE, Huang R, Kornberg A, Reusch RN (1995) Inorganic polyphosphates in the acquisition of competence. J Biol Chem 270: 12980–12983

Chang CF (1986) Electron probe analysis, X-ray mapping, and electron energy-loss spectroscopy of calcium, magnesium, and monovalent ions in log-phase and in dividing *Escherichia coli* B cells. J Bacteriol 167: 935–939

Christensen B, Fink J, Merrifield RB, Mauzerall D (1988) Channel-forming properties of cecropins and related model compounds incorporated into planar lipid membranes. Proc Natl Acad Sci USA 85: 5072–5076

Corbridge DEC (1995) Phosphorus. An outline of its chemistry, biochemistry and technology. Stud Inorg Chem 20: 224–243

Cornibert J, Marchessault RH (1972) Physical properties of poly-β-hydroxybutyrate. IV. Conformational analysis and crystal structure. J Mol Biol 71: 735–756

Cornibert J, Marchessault H (1975) Conformational isomorphism. A general 2_1 helical conformation for poly(β-alkanoates). Macromolecules 8: 296–305

Crooke E, Akiyama M, Rao NN, Kornberg A (1994) Genetically altered levels of inorganic polyphosphate in *Escherichia coli*. J Biol Chem 269: 6290–6295

Dawes EA, Senior PJ (1973) The role and regulation of energy reserve polymers in microorganisms. Adv Microb Physiol 10: 135–206

Das S, Lengweiler UD, Seebach D, Reusch RN (1997) Proof for a nonproteinaceous calcium-selective channel by total synthesis from (R)-3-hydroxybutanoic acid and inorganic polyphosphate. Proc Natl Acad Sci USA 94: 9075–9079

Doi Y (1990) Microbial polyesters. VCH, New York

Dunham ET, Glynn IM (1961) Adenosinetriphosphate activity and the active movement of alkali metal ions. J Physiol 156: 274–293

Eisenman G, Horn R (1983) Ionic selectivity revisited: the role of kinetic and equilibrium processes in ion permeation through channels. J Membr Biol 76: 197–225

Epand RM (1993) The amphiatic helix. CRC Press, Ann Arbor, M, pp 222–246

Fettiplace D, Haydon A (1980) Water permeability of lipid membranes. Physiol Rev 60: 510–550

Fyles TM, James TD, Kaye KC (1993) Activities and modes of action of artificial ion channel mimics. J Am Chem Soc 115: 12315–12321

Gangola P, Rosen BP (1987) Maintenance of intracellular calcium in *Escherichia coli*. J Biol Chem 26: 12570–12574

Gray FM (1992) Solid polymer electrolytes. VCH, New York, pp 1–4

Griffin JB, Davidian NM, Penniall R (1965) Studies of phosphorus metabolism by isolated nuclei. J Biol Chem 240: 4427–4434

Hagiwara S, Byerly L (1981) Calcium channel. Annu Rev Neurosci 4: 69–125

Hanahan D (1983) Studies on transformation of *Escherichia coli* with plasmids. J Mol Biol 166: 557–580

Hanahan D, Bloom FR (1996) Mechanisms of DNA transformation. In: Neidhardt FC (ed) *Escherichia coli* and *Salmonella*: cellular and molecular Biology. ASM Press, Washington, DC, pp 2449–2459

Hess P, Tsien RW (1984) Mechanism of ion permeation through ion channels. Nature 309: 453–456

Hille B (1992) Ionic channels of excitable Membranes. Sinauer, Sunderland, MA

Holmes PA (1987) Biologically produced (R)-3-hydroxyalkanoate polymers and copolymers. In: Bassett DC (ed) Developments in crystalline polymers-2. Elsevier, New York, pp 1-65

Huang R, Reusch RN (1995) Genetic competence in *Escherichia coli* requires poly-β-hydroxybutyrate/calcium polyphosphate membrane complexes and certain divalent cations. J Bacteriol 177: 486-490

Huang R, Reusch RN (1996) Poly-(R)-3-hydroxybutyrate) is associated with specific protein in the cytoplasm and membranes of *Escherichia coli*. J Biol Chem 271: 22196-22202

Karr DB, Waters JK, Emerich DW (1983) Analysis of poly-β-hydroxybutyrate in *Rhizobium japonicum* bacteroids by ion-exclusion high-pressure liquid chromatography and UV detection. Appl Environ Microbiol 46: 1339-1344

Kobuke Y, Ueda K, Sokabe M (1992) Artificial non-peptide sinlge ion channels. J Am Chem Soc 114: 7618-7622

Kornberg A (1995) Inorganic polyphosphate: toward making a forgotten polymer unforgettable. J Bacteriol 177: 491-495

Kornberg A, Kornberg S, Simms E (1956) Metaphosphate synthesis by an enzyme from *Escherichia coli*. Biochim Biophys Acta 20: 215-227

Kosk-Kosicka D, Scaillet S, Inesi G (1986) The partial reactions in the catalytic cycle of the calcium-dependent adenosine triphosphatase purified from erythrocyte membranes. J Biol Chem 261: 3333-3338

Kulaev IS (1979) The biochemistry of inorganic polyphosphates. Wiley, New York, pp 1-248

Kulaev IS, Vagabov VM (1983) Polyphosphate metabolism in micro-organisms. Adv Microb Physiol 24: 83-171

Kumble KD, Kornberg A (1995) Inorganic polyphosphate in mammalian cells and tissues. J Biol Chem 270: 5818-5822

Law JH, Slepecky RA (1961) Assay of poly-β-hydroxybutyric acid. J Bacteriol 82: 33-42

Lehn JM (1973) Design of organic complexing agents; strategies towards properties. Struct Bonding 16: 1-69

Lengweiler UD, Fritz MG, Seebach D (1996) Sythesis of monodisperse linear and cyclic oligomers of (R)-3-hydroxybutyric acid with 128 residues. Helv Chim Acta 79: 670-701

Lynn AR, Rosen BP (1987) Calcium transport in prokaryotes. In: Rosen BP, Silver S (eds) Ion transport in prokaryotes. Academic Press, New York, pp 181-201

MacCallum JR, Vincent CA (1987) Ion–molecule and ion–ion interactions. In: MacCallum JR, Vincent CA (eds) Polymer electrolyte reviews 1. Elsevier, New York, pp 23-37

Majling J, Hanic F (1980) Phase chemistry of condensed phosphates. Top Phosphorus Chem 10: 341-502

Mandel H, Higa A (1970) Calcium-dependent bacteriophage DNA infection. J Mol Biol 53: 159-162

Marchessault RH, Okamura K, Su CJ (1970) Physical properties of poly(β-hydroxybutyrate). II. Conformational aspects in solution. Macromolecules 3-735-740

Marsh D, Watts A, Knowles PF (1976) Evidence for phase boundary lipid. Permeability of tempo-choline into dimyristoylphosphatidylcholine vesicles at the phase transition. Biochemistry 15: 3570-3578

Martin RB (1990) Bioinorganic chemistry of magnesium. In: Sigel H, Sigel A (eds) Metal ions in biological systems vol 26. Dekker, New York pp 1-13

Matheja J, Degens ET (1971) Structural molecular Biology of phosphates. Fischer, Stuttgart, pp 78-90

Matsushita T, Hirata H, Kusaka I (1989) Calcium channels in bacteria. Ann N Y Acad Sci 560: 276-278

McCleskey EW, Almers W (1985) The Ca channel in skeletal muscle is a large pore. Proc Natl Acad Sci USA 82: 7149-7153

Miller C (1983) Integral membrane channels: studies in model membranes. Physiol Rev 63: 1209-1242

Müller HM, Seebach D (1994) Poly(hydroxyalkanoates): a fifth class of physiologically important organic biopolymers? Angew Chem 32: 477–502

Nakano A, Xie Q, Mallen JV, Echegoyen L, Gokel GW (1990) Synthesis of a membrane-insertable, sodium cation-conducting channel: kinetic analysis by dynamic 23Na NMR. J Am Chem Soc 112: 1287–1289

Niggli V, Penniston J, Carafoli E (1979) Purification of the $(Ca^{2+}\text{-}Mg^{2+})$-ATPase from human erythrocyte membranes using a calmodulin affinity column. J Biol Chem 254: 9955–9958

Norris V, Grant S, Freestone P, Canvin J, Sheikh FN, Toth I, Triner M, Modha K, Norman R (1996) Calcium signalling in bacteria. J Bacteriol 178: 3677–3682

Overath P, Träuble H (1973) Phase transitions in cell membranes, and lipids of *Escherichia coli*. Detection by fluorescent probes, light scattering and dilatometry. Biochemistry 12: 2625–2634

Page WJ, von Tigerstrom M (1979) Optimal conditions for transformation of *Azotobacter vinelandii*. J Bacteriol 139: 1058–1061

Plattner DA, Brunner A, Dobler M, Müller HM, Petter W, Zbinden P, Seebach D (1993) Cyclic oligomers of (R)-3-hydroxybutanoic acid: preparation and structural aspects. Helv Chim Acta 76: 2004–2033

Reusch RN (1989) Poly-β-hydroxybutyrate/calcium polyphosphate complexes in eukaryotic membranes. Proc Soc Exp Biol Med 191: 377–381

Reusch RN (1992) Biological complexes of poly-β-hydroxybutyrate. FEMS Rev 103: 119–130

Reuch RN, Reusch WH (1993) Branched polyhydroxyalkanoate polymer salt compositions and method of preparation. US Patent no 5,266,422

Reusch RN, Sadoff HL (1983) D-(–)-poly-β-hydroxybutyrate in membranes of genetically competent bacteria. J Bacteriol 156: 778–788

Reusch RN, Sadoff HL (1988) Putative structure and function of a poly-β-hydroxybutyrate calcium polyphosphate channel in bacterial plasma membranes. Proc Natl Acad Sci USA

Reusch RN, Hiske TW, Sadoff HL (1986) Poly-β-hydroxybutyrate membrane structure and its relationship to genetic transformability in *Escherichia coli*. J Bacteriol 168: 553–562

Reusch R, Hiske T, Sadoff H, Harris R, Beveridge T (1987) Cellular incorporation of poly-β-hydroxybutyrate into plasma membranes of *Escherichia coli* and *Azotobacter vinelandii* alters native membrane structure. Can J Microbiol 33: 435–444

Reusch RN, Sparrow AW, Gardiner J (1992) Transport of poly-β-hydroxybutyrate in human plasma. Biochim Biophys Acta 1123: 33–40

Reusch RN, Huang R, Bramble LL (1995) Poly-3-hydroxybutyrate/polyphosphate complexes form voltage-activated/Ca^{2+} channels in the plasma membranes of *Escherichia coli*. Biophys J 69: 754–766

Reusch RN, Huang R, Kosk-Kosicka D (1997) Novel components and enzymatic activities of the human erythrocyte plasma membrane calcium pump. FEBS Lett 412: 592–596

Rosen BP, McClees JS (1974) Active transport of calcium in inverted membrane vesicles of *Escherichia coli*. Proc Natl Acad Sci USA 71: 5042–5046

Sansom MSP (1991) The biophysics of peptide models of ion channels. Prog Biophys Mol Biol 55: 139–151

Schatzmann HJ (1966) ATP-dependent Ca^{++}-extrusion from human red cells. Experientia 22: 364–368

Seebach D, Brunner A, Bürger HM, Schneider J, Reusch RN (1994a) Isolation and ¹H-NMR spectroscopic identification of poly-(R)-3-hydroxybutanoate) from prokaryotic and eukaryotic organisms. Eur J Biochem 224: 317–328

Seebach D, Bürger M, Müller HM, Lengweiler UD, Beck AK, Sykes KE, Barker PA, Barham PJ (1994b) Synthesis of linear oligomers of (R)-3-hydroxybutyrate and solid-state structural investigations by electron microscopy and x-ray scattering. Helv Chim Acta 77: 1099–1123

Seebach D, Brunner A, Bachmann BM, Hoffmann T, Kühnle FN, Lengweiler UD (1996a) Biopolymers and -oligomers of (R)-3-hydroxyalkanoic acids – contributions of synthetic organic chemists. Ernst Shering Res Found 28: 1–105

Seebach D, Brunner A, Bürger HM, Reusch RN, Bramble LL (1996b) Channel-forming activity of 3-hydroxybutanoic-acid oligomers in planar lipid bilayers. Helv Chim Acta 79: 507–517

Simon W, Morf WE, Meier PC (1973) Specificity for alkali and alkaline earth cations in membranes, Struct. Bonding 16: 113–160

Smith HO, Danner DB, Deich RA (1981) Genetic transformation. Annu Rev Biochem 50: 41–68

Smith RJ (1995) Calcium and bacteria. Adv Microb Physiol 37: 83–103

Sykes KE, McMaster TJ, Miles MJ, Barker PA, Barham PJ, Seebach D, Müller H-M, Lengweiler UD (1995) Direct imaging of the surfaces of poly-β-hydroxybutyrate and hydroxybutyrate oligomers by atomic force microscopy. J Mater Sci 30: 623–627

Tanaka JC, Furman RE, Barchi RL (1986) Skeletal muscle sodium channels. Isolation and reconstitution. In: Miller C (ed) Ion channel reconstitution. Plenum Press, New York, pp 277–305

Tinsley CR, Manjula BNK, Gotschlich EC (1993) Purification and characterization of polyphosphate kinase from *Neisseria meningitidis*. Infect Immun 61: 3703–3710

Tisa LS, Adler J (1992) Calcium ions are involved in *Escherichia coli* chemotaxis. Proc Natl Acad Sci USA 89: 11804–11808

Träuble, H. Overath P (1973) Membrane structure of *Escherichia coli*. Biochim Biophys Acta 307: 491–512

Tsien RW, Hess P, McCleskey EW, Rosenberg RL (1987) Calcium channels: mechanisms of selectivity, permeation, and block. Annu Rev Biophys Biophys Chem 16: 265–290

Urry DW (1985) Chemical basis of ion transport specificity in biological membranes. Top Curr Chem 128: 175–218

Van Zutphen H, Merola AJ, Brierley GP, Cornwell DG (1972) The interaction of nonionic detergents with lipid bilayer membranes. Arch Biochem Biophys 152: 755–766

Watanabe M, Ogatu N (1987) In: MacCallum JR, Vincent CA (eds) Polymer electrolyte reviews 1. Elsevier, New York, pp 39–68

Watanabe M, Togo M, Sanui K, Ogatu N, Kobayashi T, Ohtaki Z (1984) Ionic conductivity of polymer complexes formed by poly(β-propiolactone) and lithium perchlorate. Macromol Rev 17: 2908–2912

Welland EL, Stejny J, Halter A, Keller A (1989) Selective degradation of chain folded single crystals of poly)β-hydroxybutyrate). Polym Commun 30: 302–304

Wilson PW, Knight SG (1952) Experiments in bacterial physiology. Burgess, Minneapolis, MJ

Wood HG, Clark JE (1988) Biological aspects of inorganic polyphosphates. Annu Rev Biochem 57: 235–260

Yamagata Y, Watanabe H, Saitoh M, Namba T (1991) Volcanic production of polyphosphates and its relevance to prebiotic evolution. Nature 352: 516–519

Yokouchi M, Chaitani Y, Tadokoror H, Teranishi K, Tani H (1973) Structural studies of polyesters: 5. Molecular and crystal structure of optically active and racemic poly(β-hydroxybutyrate). Polymer 14: 267–272

Inorganic Polyphosphate Regulates Responses of *Escherichia coli* to Nutritional Stringencies, Environmental Stresses and Survival in the Stationary Phase

N. N. Rao and A. Kornberg[1]

1
Introduction

The extraordinary conservation of inorganic polyphosphate (polyP) in all cells – bacteria, fungi, plants and animals – and the various uses to which it might be put are reminiscent of the many ways in which other highly conserved molecules, such as ATP, NAD, fatty acids and certain polypeptides, have been exploited in evolution. A serious difficulty in understanding the physiological role of polyP has been the inadequacy of quantitative methods. With polyP-specific enzymatic assays and facile enrichment of polyP from crude extracts (see Chap. 12), it was observed that the accumulation and disappearance of polyP in *Escherichia coli* (and other organisms) are dynamic; fluctuations of 100- to 1000-fold are observed in response to nutritional and environmental stresses. These fluctuations depend on a global network of metabolic pathways. Insight into the metabolic roles for polyP in *E. coli* has been derived from mutant cells lacking PPK, the enzyme responsible for the synthesis of polyP. These roles establish an essential place for polyP in the regulation of responses to nutritional deficiencies, environmental stresses and survival in the stationary phase.

2
Nutritional Stringencies and Environmental Stress

The influence of nutritional limitation (P_i, nitrogen and amino acids) on polyP accumulation in batch cultures has demonstrated that fluctuations in the level of cellular polyP are both rapid and enormous (Ault-Riché et al. 1998; Rao et al. 1998; Fig. 1).

[1] Department of Biochemistry, Stanford University School of Medicine, Stanford, California 94305-5307, USA.

Progress in Molecular and Subcellular Biology, Vol. 23
H. C. Schröder, W. E. G. Müller (Eds.)
© Springer-Verlag Berlin Heidelberg 1999

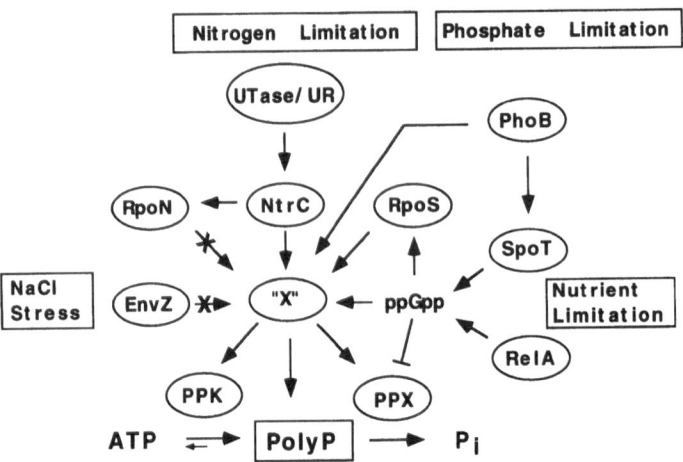

Fig. 1. Model for stress-induced polyP accumulation. NtrC (a member of the signal cascade for nitrogen metabolism), together with RpoS and PhoB, is needed for polyP accumulation in response to nitrogen limitation. Involvement of a sigma factor (RpoS) implies activation of an additional factor ("X") which could lead to polyP accumulation by direct interaction with polyP, inhibition of PPX, stimulation of PPK, or a combination of all three. Under nutrient limitation, ppGpp accumulated by RelA and SpoT actions, can lead to polyP accumulation by PPX inhibition and/or RpoS activation. Failure to accumulate polyP even when ppGpp and RpoS levels are high (as in carbon starvation) implies the presence of additional regulator(s). Also, osmotic stress triggers polyP accumulation through a mechanism that does not involve EnvZ, the osmotic sensor

2.1
Phosphate Limitation: *phoB* and the Pho Regulon

Escherichia coli responds to stresses, acute and prolonged, in a variety of ways. One of the best understood nutritional stresses is depletion of P_i in the medium, sensed by the phosphate (Pho) regulon; very low levels of extracellular P_i (4 µM) induce the Pho regulon (Wanner 1994). PhoB, the response regulator, turns on several genes, among them those for alkaline phosphatase (*phoA*) and proteins involved in P_i uptake and assimilation. Pho-regulon mutants affected in PhoB synthesis were tested for the accumulation of polyP in a MOPS-buffered minimal medium containing low levels of P_i (0.1 mM) and an insufficient amount of amino acids (2 µg/ml). Large amounts of polyP (48 nmol/mg protein) accumulated in a narrow growth range (OD_{540} about 0.3) in wild type cells (Park and Kornberg, unpubl. results). A lack of PhoB due to either missense mutation (*phoB63, phoB23* and *phoB62*) or the insertional inactivation of *phoB* (*phoB*:Tn 5) resulted in low levels of polyP (0.3–1.9 nmol/mg protein). Inactivation of the protein kinases, PhoR and CreC, which activate PhoB by phosphorylation (Wanner 1995), led to barely detectable levels (0.1 nmol/mg protein). Mutations in *phoU*, (*ΔphoU, phoU35*) and in *phoR* (*phoR68, phoU* (*ΔphoR69*) which lead to constitutive expression of the Pho regulon (Shinagawa et al. 1987; Torriani-Gorini 1994; Wanner 1996) resulted in polyP accumulations comparable to those in wild type cells.

Complementation of a *phoB* mutant with a multicopy *phoB* plasmid restored the capacity to accumulate polyP, indicative of a direct role in polyP accumulation. Synthesis of alkaline phosphatase in wild type cells depends solely on the P_i concentration in the growth medium and is under *phoB* control (Wanner 1996). Evidence that an imbalanced supply of amino acids (2 μg/ml) and low P_i (0.1 mM) induced PhoB synthesis was demonstrated by a significant increase in alkaline phosphatase activity which coincided with the accumulation of polyP.

2.2
Amino Acid and Phosphate Limitation

PolyP did not accumulate in the wild type cells grown in a low P_i medium supplemented with large amounts of the amino acids (20 μg/ml), nor did cells grown in a high P_i medium accumulate a significant amount of polyP. However, as mentioned above, large amounts of polyP do accumulate in media deficient both in P_i and amino acids. The accumulation is transient and preceded by appearance of guanosine nucleotides, ppGpp and pppGpp, the stringent response signals formed by ribosome-bound RelA (Cashel et al. 1996). These nucleotides are signals for the transcriptional repression of ribosomal RNA and proteins and for activating the expression of certain biosynthetic genes. Changes in the cellular levels of (p)ppGpp have also been observed in responses to deficiencies of P_i (Spira et al. 1995), carbon (Lazzarini et al. 1971) and nitrogen (Irr 1972). It is noteworthy that mutants in *phoB* do not accumulate polyP even though they develop wild type levels of (p)ppGpp upon amino acid starvation.

2.3
(p)ppGpp and in Vivo Synthesis of PolyP

PolyP accumulation requires high levels of (p)ppGpp, independent of whether they are generated by RelA (active during the stringent response) or SpoT (expressed during P_i starvation). A double mutant lacking both *relA* and *spoT* accumulated neither (p)ppGpp nor polyP. Additional evidence for the role of (p)ppGpp in polyP accumulation was demonstrated by the accumulation of both (p)ppGpp and polyP upon induction of an inducible multicopy plasmid which synthesizes a RelA peptide encoded by a truncated *relA* gene (Svitil et al. 1993). The massive production of both polyP and (p)ppGpp was not observed in a control plasmid lacking the truncated *relA* gene.

2.4
Accumulation of PolyP is Due to Inhibition of Exopolyphosphatase by (p)ppGpp

Inasmuch as the activities (measured in extracts) of the enzymes that synthesize and degrade polyP fluctuate only marginally during polyP accumulation, the accumulation must have another basis. A singular and profound inhibition by pppGpp and/or ppGpp of the hydrolytic breakdown of polyP by exopolyphosphatase (PPX), but without effect on the polyP synthetic enzyme (PPK), could explain blockage of the turnover of polyP and its accumulation (Kuroda et al. 1997). Turnover of polyP (resulting from its cyclic synthesis by PPK and hydrolysis by PPX) was found to be 12 min or less (Park and Kornberg, unpubl. results). This result was simulated by the accumulation of polyP in a reaction mixture containing purified PPK and PPX in the presence of pppGpp and ppGpp (Kuroda et al. 1997). Although this mechanism can account for polyP accumulation in some instances, it does not explain the lack of polyP accumulation in *phoB* mutants which produce (p)ppGpp.

2.5
Nitrogen Starvation: *rpoN*

In wild tpye cells growing in a medium with a limited amount of nitrogen, stoppage of growth due to exhausting the nitrogen coincided with a rapid accumulation of polyP. However, no accumulation occurred when the supply was exhausted in mutants defective in *glnG* (UTase/UR) or in *glnD*, the regulatory NtrC in the nitrogen regulon. Surprisingly, mutants in *rpoN*, the gene that encodes σ^{54} and is activated by NtrC, did accumulate polyP. Thus, a gene other than *rpoN* is the target for NtrC action in polyP accumulation (Fig. 1).

2.6
Nutritional Shift from Rich to Minimal Medium and the Reverse

Growing wild type cells accumulate massive amounts of polyP in a nutritional shift-down from a rich medium [Luria-Bertani broth (LB)] to minimal medium (MOPS-buffered medium with low P_i). Reversal of these conditions resulted in a very rapid disappearance of the polyP. A *relA* mutant (defective in the stringent response) failed to accumulate polyP in a nutritional shift-down.

2.7
Osmotic Stress

Cells grown to mid-log phase in LB (osmolarity equivalent of 0.17 M NaCl) when resuspended in fresh medium containing 1 M NaCl accumulated polyP rapidly. Shifts into LB without added NaCl did not lead to the accumulation of polyP. The pathway of salt-induced polyP accumulation was, however, independent of the osmotic response protein EnvZ; stimulation of EnvZ by the addition of procaine to cells had no effect on polyP levels. Osmotic stress imposed on strains with deletions in *relA* or *rpoN* resulted in significant polyP accumulations (10–20 nmol/mg protein). Strains with deletions in *rpoS* or *phoB* failed to accumulate polyP in response to osmotic stress (Fig. 1).

3
Stationary Phase Adaptations

That the accumulation of polyP of *E. coli* in the stationary phase provides a significant reservoir of ATP energy is implausible because the turnover of ATP consumes only a fraction of a second (Chapman and Atkinson 1977). Even a polyP level (expressed in phosphoanhydride bonds) five times or more that of ATP could supply ATP for only a second or two. Thus, other functions for polyP need to be considered, among them a regulatory role. It is noteworthy that very low levels of polyP (1 nmol/mg protein) in stationary phase cells suffice for the crucial adaptations to stresses and for suvival (Rao and Kornberg 1996). PolyP-deficient cells exhibit a striking phenotype in their failure to develop resistance to heat, oxidants, UV, and osmotic stress, induction of stationary phase-specific genes, and in their death within a few days.

3.1
Heat Sensitivity

During survival in the stationary phase or upon starvation, cells acquire *rpoS*-mediated multiple resistances (Hengge-Aronis 1993; Loewen and Hengge-Aronis 1994). When polyP-deficient cells were held at 55 °C for 2 min, only 2 % survived compared with about 90 % of the wild type cells. The mutant complemented with multiple copies of the plasmid-bearing *rpoS*, the stationary phase-specific RNA polymerase sigma factor, proved to be as heat resistant as wild type cells. Similar results were obtained when the mutant was complemented with a plasmid containing *ppk* (Rao and Kornberg 1996; Fig. 2).

3.2
Response to Oxidative Stress and Sensitivity to Menadione

PolyP-deficient cells in the stationary phase were more sensitive than wild type cells to 42 mM H_2O_2 (Crooke et al. 1994). Complementation of the mutant in the stationary phase with a plasmid bearing the *ppk* gene restored H_2O_2 resistance to near wild type levels (Rao and Kornberg 1996). Exposure of cells to the quinone menadione leads to oxidative stress due to the pro-

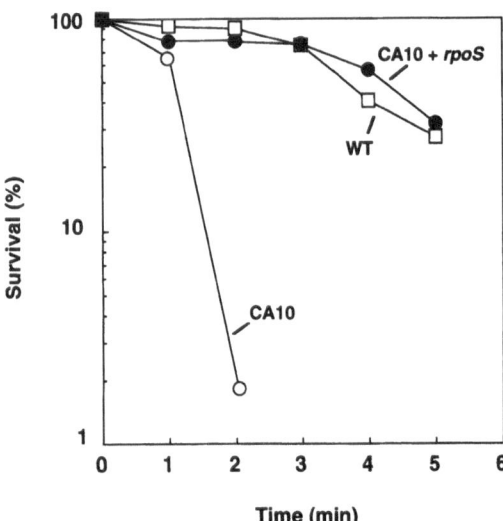

Fig. 2. Heat resistance. Wild type (WT), CA10 (*ppk ppx*) and CA10 pMM*katF3* (complemented with *katF* gene for *rpoS* expression) cells were grown overnight (ca. 20 h) at 37 °C in LB medium and then exposed to 55 °C. Their viability was determined by plating an LB agar. Percent survival is determined as viable cell number at each time point divided by viable cell number before exposure to heat

duction of oxygen radicals (Thor et al. 1982); cells lacking catalase, peroxidase, and superoxide dismutase show increased sensitivity (Greenberg and Demple 1988; Demple 1991). The *ppk* mutant in the stationary phase also proved to be highly sensitive to menadione toxicity and could be restored to resistance by complementation with the *ppk* gene.

3.3
Sensitivity to UV and Osmotic Challenge

Cells deficient in polyP were highly susceptible to UV as observed by loss of viability (T. Shiba, pers. comm.). Stationary phase cells show enhanced osmotic resistance compared with cultures in the exponential growth phase or preadapted to osmotic stress (Jenkins et al. 1990). After exposure of carbon-starved stationary phase cells to 2.5 M NaCl for 21.5 h, about 10 % of the wild type cells but only about 1 % of the *ppk* mutant cells survived. The sensitivity towards osmotic challenge, more pronounced in exponential cultures, showed greater losses of viability in the *ppk* mutant (Rao and Kornberg 1996). Accumulation of polyP in response to osmotic stress may be related to the lack of resistance of *ppk* mutants which fail to produce polyP. Mutants which fail to produce polyP have a decreased ability to survive in the stationary phase and are more susceptible to osmotic stress.

3.4
HPII Expression

Stationary phase cells exhibit strong *rpoS*-dependent resistance to H_2O_2. Catalase HPII, the *rpoS*-controlled *katE* product, is induced in stationary phase cells as a defense against H_2O_2 (Loewen et al. 1985). Direct assays revealed that the stationary phase cells possess a high level of HPII activity in contrast to the *ppk* mutant that showed only trace amounts of this enzyme activity. Complementation of the mutant with the *ppk* gene or an *rpoS* plasmid elevated synthesis of HPII to the wild type level.

3.5
Reduced Survival and Growth Defects

In *E. coli*, *ppk* and *ppx* genes are under the control of a common promoter (Akiyama et al. 1993). The inability to produce polyP upon deletion of the *ppk ppx* operon has produced several striking phenotypes. One is decreased long-term survival in the stationary phase of growth, and the appearance of morphological variants (Crooke et al. 1994; Fig. 3). Mutants that fail to express *ppk* not only failed to develop the characteristic stress-resistant properties of stationary phase cells but also lost their viability within a few days. After a few more days in stationary phase, the mutant cells were replaced by a small-colony variant that was genetically stable, more viable, and stress resistant; but neither PPK activity nor polyP was restored.

4
PolyP and Induction of *rpoS* Expression

Extra copies of *rpoS*, the gene that encodes the RNA polymerase sigma factor and regulates some 50 stationary phase genes, suppress the *ppk* feature

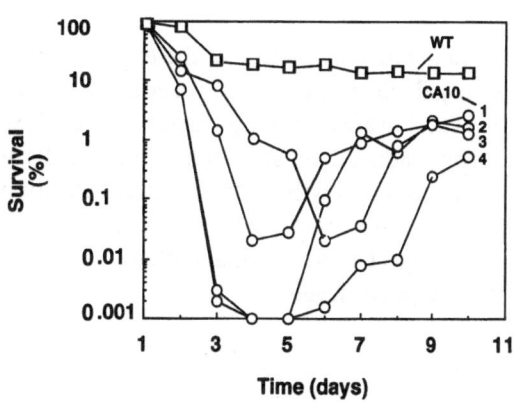

Fig. 3. Long-term survival. Wild type (WT) and CA10 (*ppk ppx*) cells were incubated in LB medium. Percent survival is expressed as viable cell numer at each time point divided by viable cell number of the same culture after 24 h from time of initial inoculation. Curves 1–4 represent four different CA10 cultures showing variations in survival and appearance of the small-colony variant

of heat sensitivity and lack of HPII. This result suggests that polyP is involved in the regulation of *rpoS* expression and function (Rao and Kornberg 1996). As another means to probe the functions of polyP in *rpoS* expression, polyP was reduced to undetectable levels by a plasmid-borne exopolyphosphatase (*PPX1*) (Shiba et al. 1997). In this polyP-less state, the high sensitivity to H_2O_2 could be ascribed to the lack of the *katE* gene product, HPII, and implicated polyP in the induction of the transcription not only of *katE*, but also of *rpoS* (see below).

4.1
Transcription of *katE* is not Induced in Cells Lacking PolyP

Sensitivity to hydrogen peroxide evinced by transformants that overproduce polyPase can be attributed to repressed levels of KatE, the stationary phase catalase (HPII). The dependence of *katE* on polyP, as measured by peroxide sensitivity, was 100-fold greater than that of the exponential phase catalase (HPI), product of the *katG* gene. The level of *katE* expression, as monitored by β-galactosidase activity with a *katE-lacZ* transcriptional fusion, was elevated by starvation. However, no significant induction of expression was observed in cells lacking polyP, achieved by polyPase overproduction (Shiba et al. 1997).

   That the low level of induction of *katE* transcription in a polyPase overproducer can be attributed to a decreased induction of *rpoS* expression was demonstrated in another way. When extra copies of *rpoS* gene on a multicopy plasmid were introduced into a polyPase overproducer, *katE* was induced as fully as in cells that harbored a plasmid for *rpoS* and also a control vector without the polyPase gene. However, no induction was observed in cells transformed with a plasmid bearing the polyPase gene and a control plasmid lacking *rpoS*.

4.2
Transcription of *rpoS* Gene

The dependence of polyP of the expression of *rpoS* was observed by measuring β-galactosidase activity produced in a λ lysogen of *E. coli* carrying an *rpoS-lacZ* transcriptional fusion gene. The activity of the enzyme was sharply reduced in stationary phase cells which overproduced polyPase. Inasmuch as growth rates of control and polyPase transformants were very much the same, implying no gross differences in metabolism, these observations suggest that stationary phase induction of *rpoS* expression is modulated by polyP levels.

4.3
Cellular Levels of RpoS in PolyP-Minus Cells

Levels of RpoS in cells harboring plasmids with either the polyPase gene or a corresponding control vector were measured during nitrogen starvation. The cellular content of RpoS increased with starvation time in control cells but failed to increase in cells that overproduced the recombinant polyPase (Shiba et al. 1997).

5
PolyP-Binding Proteins and Metal Complexes

PolyP exists in large crystalline masses (metachromatic granules) in many organisms and presumably also exists in smaller aggregates, as well as monodisperse molecules. In the course of accumulation of polyP in *E. coli*, either in stress responses or induced by *ppk* gene overexpression, the distribution of polyP among these physical states and the kinetics of exchange between these states have yet to be determined. To date, all measurements of polyP have estimated the total content of polyP in a cell.

5.1
PolyP-Binding Proteins

PolyP binds strongly to basic proteins (e.g., histones, basic domains of proteins) as do other polyanions (e.g., RNA, DNA, and lipid A). Using a nitrocellulose-binding assay of ^{32}P-polyP, an activity has been identified in *E. coli* lysates (Rashid and Kornberg, unpubl. observ.). This soluble activity is sensitive to heat and proteases, and is unattached to membranes and debris. It has been partially purified (ca. 100-fold) by ion exchange, gel filtration and polyP-affinity chromatography. The activity is polyP-specific and binds a very large amount of polyP, suggestive of nucleation of large aggregates.

5.2
Heavy Metals and Polyamines

Although Ca^{2+}, K^+, and Mg^{2+} are the metal ions most often associated with polyP, heavy metals have been found in the polyP granules of certain bacteria (Sicko-Goad and Lazinsky 1987; Scott and Palmer 1990). Rachlin et al. (1982) proposed that cells use polyP to detoxify heavy metals once they enter the cell. Accumulation of polyP in *Klebsiella aerogenes* and *E. coli* has been shown to impart tolerance towards cadmium (Aiking et al. 1984; Keasling and Hupf 1996). Polyamines contribute to the cationic pools in *E. coli* and might interact with polyP.

5.3
Metal–PolyP Complexes and Membrane Potential

PolyP, as a Mn^{2+} complex, has been shown to replace superoxide dismutase in *Lactobacillus plantarum* (Archibald and Fridovich 1982). Recently, Van Veen et al. (1994a) found that the P_i transport system in *Acinetobacter johnsonii* 210A transports metal–phosphate complexes into or out of the cell, depending upon the state of the cellular phosphate and energy metabolism. The proton motive force required for nutrient uptake from the medium is achieved by the excretion of a metal–phosphate complex in a polyP-accumulating cell (Van Veen et al. 1994b). PolyP-minus cells of *E. coli* could be used to probe the role of polyP in the generation of membrane potential by chelation of metal ions for nutrient uptake and for cell viability.

6
Summary

The molecular mechanisms responsible for polyP accumulation in *E. coli* remain largely obscure. Based on the available data, a tentative model is proposed (Fig. 1; Ault-Riché et al. 1998). Inhibition by (p)ppGpp of PPX interrupts the dynamic balance between the synthesis of polyP by PPK and its hydrolysis by PPX, accounting for polyP accumulation. However, mutants lacking PhoB, the response regulator of the Pho regulon, fail to accumulate polyP even in the face of high levels of (p)ppGpp. Clearly, PhoB is required in some undefined manner. With regard to osmotic stress, the pathway to polyP accumulation is also distinct from the one identified with the activation of *envZ* and the associated changes in membrane functions. A tentative scheme attempting to describe the metabolic turnover of polyP is given in Fig. 4.

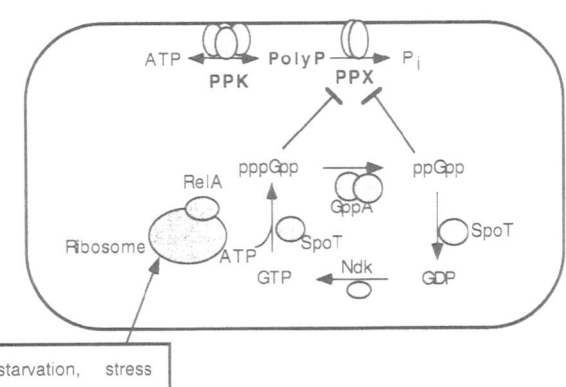

Fig. 4. Simplified scheme for polyphosphate metabolism in *E. coli*. Roles of RpoS and PhoB are not included. For details regarding pppGpp and ppGpp accumulation, see Cashel et al. (1996). *Ndk* Nucleoside diphosphate kinase

In adaptations to stress, cells must coordinate major changes in the rates of transcription, translation, and replication as well as make choices in the genes expressed (Kolter et al. 1993). PolyP could provide activated phosphates or coordinate an adaptive response by binding metals and/or specific proteins. Accumulation of polyP in *E. coli* and other organisms is commonly assumed to provide a reservoir of energy convertible to ATP. This seems implausible because of the turnover of ATP which consumes only a fraction of a second (Chapman and Atkinson 1977). Thus, other functions for polyP need to be considered, among them a regulatory role. PolyP, even at very low levels, is essential in *E. coli* for adaptations in stationary phase and for survival (Rao and Kornberg 1996).

As a polyanionic polymer, polyP has chemical similarities to DNA and RNA in interactions with basic domains of proteins. Further investigation of the cellular location of polyP, its state of metabolic availability and identification of its binding partners are needed. In view of the ubiquity of polyP in eukaryotic cells (including dynamic turnover in the nuclei of some mammalian cells), studies similar to those undertaken in *E. coli* may reveal comparable functions.

Acknowledgments. We are grateful to all our colleagues identified in the Acknowledgments in Chapter 1 for their contributions and to the National Institutes of Health (USA) for financial support.

References

Aiking H, Stijnman A, van Garderen C, Van Heerikhuizena H, van't Riet J (1984) Inorganic phosphate accumulation and cadmium detoxification in *Klebsiella aerogenes* NCTC 418 growing in continuous culture. Appl Environ Microbiol 47: 374–377

Akiyama M, Crooke E, Kornberg A (1993) An exopolyphosphatase of *Escherichia coli*. The enyzme and its *ppx* gene in a polyphosphate operon. J Biol Chem 268: 633–639

Archibald FS, Fridovich I (1982) Investigations of the state of the manganese in *Lactobacillus plantarum*. Arch Biochem Biophys 215: 589–596

Ault-Riché D, Fraley CD, Tzeng C-M, Kornberg A (1998) Novel assay reveals multiple pathways regulating stress-induced accumulations of inorganic polyphosphate in *Escherichia coli*. J Bacteriol 180: 1841–1847

Cashel M, Gentry DR, Hernandez VJ, Vinella D (1996) The stringent response. In: Neidhardt FC, Curtiss R III, Ingraham JL, Lin ECC, Low KB, Magasanik B, Reznikoff WS, Riley M, Shaechter M, Umbarger HE (eds) *Escherichia coli* and *Salmonella typhimurium*: cellular and molecular biology. American Society for Microbiology, Washington, DC, pp 1458–1496

Chapman AG, Atkinson DE (1977) Adenine nucleotide concentrations and turnover rates. Their correlation with biological activity in bacteria and yeast. Adv Microb Physiol 15: 253–306

Crooke E, Akiyama M, Rao NN, Kornberg A (1994) Genetically altered levels of inorganic polyphosphate in *Escherichia coli*. J Biol Chem 269: 6290–6295

Demple B (1991) Regulation of bacterial oxidative stress genes. Annu Rev Genet 25: 315–337

Greenberg JT, Demple B (1988) Overproduction of peroxide-scavenging enzymes in *Escherichia coli* suppresses spontaneous mutagenesis and sensitivity to redox-cycling agents in *oxy R⁻* mutants. EMBO J 7: 2611–2617

Hengge-Aronis R (1993) Survival of hunger and stress: the role of *rpoS* in early stationary phase gene regulation in *E. coli* Cell 72: 165-168

Irr JD (1972) Control of nucleotide metabolism and ribosomal ribonucleic acid synthesis during nitrogen starvation of *Escherichia coli*. J Bacteriol 110: 554-561

Jenkins DE, Chaisson SA, Matin A (1990) Starvation-induced cross protection against osmotic challenge in *Escherichia coli*. J Bacteriol 172: 2779-2781

Keasling JD, Hupf GA (1996) Genetic manipulation of polyphosphate metabolism affects cadmium tolerance in *Escherichia coli*. Appl Environ Microbiol 62: 743-746

Kolter R, Siegele DA, Tormo A (1993) The stationary phase of the bacterial life cycle. Annu Rev Microbiol 47: 855-874

Kuroda A, Murphy H, Cashel M, Kornberg A (1997) Guanosine tetra- and pentaphosphate promote accumulation of inorganic polyphosphate in *Escherichia coli*. J Biol Chem 272: 21240-21243

Lazzarini RA, Cashel M, Gallant J (1971) On the regulation of guanosine tetraphosphate levels in stringent and relaxed strains. J Biol Chem 246: 4381-4385

Loewen PC, Hengge-Aronis R (1994) The role of the sigma factor σ^S (KatF) in bacterial global regulation. Annu Rev Microbiol 48: 53-80

Loewen PC, Switala J, Triggs-Raine BL (1985) Catalases HPI and HPII in *Escherichia coli* are induced independently. Arch Biochem Biophys 243: 144-149

Rachlin JW, Jenson TE, Baxter M, Jani V (1982) Utilization of morphometric analysis in evaluating response of *Plectonema boryanum* (*Cyanophyceae*) to eight heavy metals. Arch Environ Contam Toxicol 11: 323-333

Rao NN, Kornberg A (1996) Inorganic polyphosphate supports resistance and survival of stationary-phase *Escherichia coli*. J Bacteriol 178: 1394-1400

Rao NN, Liu S, Kornberg A (1998) Inorganic polyphosphate in *Escherichia coli*: the phosphate regulon and the stringent response. J Bacteriol 180: 2186-2193

Scott JA, Palmer SJ (1990) Sites of cadmium uptake in bacteria used for biosorption. Appl Microbiol Biotechnol 33: 221-225

Shiba T, Tsutsumi K, Yano H, Ihara Y, Kameda A, Tanaka K, Takahashi H, Munekata M, Rao NN, Kornberg A (1997) Inorganic polyphosphate and the induction of *rpoS* expression. Proc Natl Acad Sci USA 94: 11210-11215

Shinagawa H, Makino K, Amemura M, Nakata A (1987) Structure and function of the regulatory genes for the phosphate regulon in *Escherichia coli*. In: Torriani-Gorini A, Rothman FG, Silver S, Wright A, Yagil E (eds) Phosphate metabolism and cellular regulation in microorganisms. American Society for Microbiology, Washington, DC, pp 20-25

Sicko-Goad L, Lazinskiy D (1986) Quantitative ultrastructural changes associated with lead-coupled luxury phosphate uptake and polyphosphate utilization. Arch Environ Contam Toxicol 15: 617-627

Spira B, Silberstein N, Yagil E (1995) Guanosine 3', 5'-bispyrophosphate (ppGpp) synthesis in cells of *Escherichia coli* starved for P_i. J Bacteriol 177: 4053-4058

Svitil AL, Cashel M, Zyskind JW (1993) Guanosine tetraphosphate inhibits protein synthesis in vivo. J Biol Chem 268: 2307-2311

Thor H, Smith MT, Hartzell P, Bellomo G, Jewell SA, Orrenius S (1982) The metabolism of menadione (2-methyl-1,4-naphthoquinone) by isolated hepatocytes. J Biol Chem 257: 12419-12425

Torriani-Gorini A (1994) Introduction: the Pho regulon of *Escherichia coli*. In: Torriani-Gorini A, Silver S, Yagil E (eds) Phosphate in microorganisms: cellular and molecular biology. American Society for Microbiology, Washington, DC, pp 1-4

Van Veen HW, Abee T, Kleefman AWF, Melgers B, Kortsee GJJ, Konings WN, Zehnder AJB (1994a) Energetics of alanine, lysine, and proline transport in cytoplasmic membranes of the polyphosphate-accumulating *Acinetobacter johnsonii* strain 210A. J Bacteriol 176: 2670-2676

Van Veen HW, Abee T, Periera H, Kortsee GJJ, Konings WN, Zehnder AJB (1994b) Generation of a proton motive force by the excretion of metal-phosphate in the polyphosphate-accumulating *Acinetobacter johnsonii* strain 210A. J Biol Chem 269: 29509–29514

Wanner BL (1994) Multiple controls of the *Escherichia coli* Pho regulon by the P_i sensor PhoR, the catabolite regulatory sensor CreC, and acetyl phosphate. In: Torriani-Gorini A, Silver S, Yagil E (eds) Phosphate in microorganisms. cellular and molecular biology. American Society for Microbiology, Washington, DC, pp 13–21

Wanner BL (1995) Signal transduction and cross regulation in the *Escherichia coli* phosphate regulon by PhoR, CreC, and acetyl phosphate. In: Hoch JA, Silhavy TJ (eds) Two-component signal transduction. ASM Press, Washington, DC, pp 203–221

Wanner BL (1996) Phosphorus assimilation and control of the phosphate regulon. In: Neidhardt FC, Curtiss R III, Ingraham JL, Lin ECC, Low KB, Magasanik B, Reznikoff WS, Riley M, Shaechter M, Umbarger HE (eds) *Escherichia coli* and *Salmonella typhimurium*: cellular and molecular biology. American Society for Microbiology, Washington, DC, pp 1357–1381

From Polyphosphates to Bisphosphonates and Their Role in Bone and Calcium Metabolism

H. Fleisch[1]

1
Introduction

This review will first describe the results obtained for the most part about 30–40 years ago on the effects and the possible role of pyrophosphate in bone and calcium metabolism. In a second part we shall describe how these results led to the development of a new class of compounds which are analogues of pyrophosphate, the bisphosphonates. The latter are today the main therapeutic drugs in the treatment of bone diseases characterized by increased osteolysis, such as Paget's disease, tumor bone disease, and osteoporosis. Lastly, the possibility of using and extrapolating the intracellular effects of these compounds on the possible action of pyrophosphate and polyphosphates in the cell will be discussed.

2
Pyrophosphate and Polyphosphate

Pyrophosphate was first described as present in bone about 40 years ago (Cartier 1957). It was originally thought to constitute a link between the mineral and the matrix. Later, it was found also in other calcified tissues such as dentine and enamel (Bisaz et al. 1968). A hypothesis of its role was developed from the finding that plasma (Fleisch and Neuman 1961) and urine (Fleisch and Bisaz 1962) contain an inhibitory activity against calcium phosphate crystal formation. Since part of this activity was destroyed by alkaline phosphatase (Fleisch and Neuman 1961), the inhibitor was likely to be some kind of a phosphate. We therefore studied whether various phosphates could have such an effect. Among the most likely candidates were the polyphosphates, since they were well-known inhibitors of calcium carbonate formation (Buehrer and Reitemeyer 1940) and therefore commonly used in industry to inhibit scaling, such as in washing powders (Rudy 1960). The hypothesis proved correct, since various polyphosphates, among other pyrophosphate, were shown to be very potent preventers of calcium phosphate

---

[1] Av. Désertes 5, 1009 Pully, Switzerland.

Progress in Molecular and Subcellular Biology, Vol. 23
H. C. Schröder, W. E. G. Müller (Eds.)
© Springer-Verlag Berlin Heidelberg 1999

crystal formation at micromolar doses (Fleisch and Neuman 1961; Fleisch and Bisaz 1962). Thus we looked for polyphosphates or pyrophosphate in biological fluids and indeed found the latter to be present first in urine (Fleisch and Bisaz 1962a), and later also in plasma (Fleisch and Bisaz 1962; Russell et al. 1972), saliva (Hausmann et al. 1970), and synovial fluid (Russell et al. 1970a), all locations where is had not been described before.

A further investigation of the physicochemical action of pyrophosphate showed that at biological concentrations it not only prevents the formation of calcium phosphate from solution, but also affects the formation of amorphous calcium phosphate, which is thought to be the first phase to be formed in calcification. Thus it activates the formation of the initial phase (Termine and Posner 1970), but inhibits the further transformation into its crystalline form (Fleisch et al. 1968a). Furthermore, it impairs the aggregation of crystals into larger clusters (Hansen et al. 1976) and inhibits formation and aggregation of another calcium salt, calcium oxalate (Fleisch and Bisaz 1964; Robertson et al. 1973). Of special interest was the finding that pyrophosphate also had an inhibiting effect on the dissolution of already formed apatite crystals (Fleisch et al. 1966a). All these effects are likely to be related to the high affinity of this compound to the surface of the mineral phase (Jung et al. 1973), where it blocks both formation and dissolution of the crystals.

These results prompted the question whether pyrophosphate would display these effects also in vivo. This proved to be the case. Thus, a first experiment showed that it prevents calcification of bone and cartilage in organ culture (Fleisch et al. 1966a). Then it was found to prevent calcification of arteries and kidneys (Schibler et al. 1968), as well as other tissues when induced by various means in vivo. All these effects were obtained only when pyrophosphate was given subcutaneously, not orally. In contrast, pyrophosphate did not impair mineralization of bone, and did not inhibit bone resorption in vivo. This lack of effect was explained by its rapid hydrolysis in vivo in the gastrointestinal tract as well as in the body after injection (Jung et al. 1970) and especially in bone (Jung et al. 1974).

Some results suggest that pyrophosphate can also promote mineralization in the early stages, for example in matrix vesicles in vitro (Anderson and Reynolds 1973, Hsu and Anderson 1984). It appears thus that pyrophosphate may have a dual effect: it first activates mineralization, possible through the formation of a nucleus of calcium pyrophosphate which is a very insoluble salt, and, at a later stage, inhibits it.

These various results suggest that pyrophosphate could play both a physiological and pathophysiological role in modulating mineralization and possibly demineralization (Fleisch et al. 1966b). It may protect soft tissues from calcification, and in bone itself it may influence the rate of mineralization as well as of demineralization. The regulation of the local concentration of pyrophosphate, and therefore of the mentioned rates, could be controlled by enzymes with a pyrophosphatase activity, such as alkaline phosphatase, which indeed has such an activity (Felix et al. 1974; Whyte 1994) and which has been for a long time associated spatially with calcification. Another such enzyme is nucleoside triphosphate pyrophosphatase which has been found in epiphyseal matrix vesicles (Caswell et al. 1987), as well as in human articular chondrocytes and bone osteoblast-like cells (Caswell and Russell 1988).

The best support for this theory was obtained in patients with hypophosphatasia, a disease characterized by a deficiency of alkaline phosphatase, in which there is a failure of normal mineralization of bone. In these patients pyrophosphate is increased in plasma (Russell et al. 1971) and in urine (Russell 1965). Pyrophosphate might also be of some importance as an inhibitor of stone formation, although this has not yet been proven. Its urinary excretion is lower in certain groups of patients, but not in others, and this hypothesis might offer in part an explanation for the sometimes beneficial effect of the oral administration of othophosphate in this disease. Indeed, oral othophosphate leads to an increase in urinary pyrophosphate (Fleisch et al. 1964).

The classical disease of pyrophosphate is chondrocalcinosis, also called pseudogout, a condition in which its calcium salt precipitates in the joint cavities, synovial membranes, articular cartilage and periarticular tissues and causes an arthropathy. Pyrophosphate concentration in the synovial fluid is strongly increased (Russell et al. 1970a; McCarty et al. 1971). Since it is not a bone disease, it will not be discussed here.

In view of these results, pyrophosphate was not a promising agent in disorders of calcium metabolism. However, its strong affinity to hydroxyapatite (Jung et al. 1973) and the fact that it passes rapidly from plasma to bone (Bisaz et al. 1978) have been exploited in clinical practice for scintigraphy by linking it to 99m-technetium. Furthermore, topical administration of pyrophosphate inhibits dental tartar formation (Mühlemann et al. 1970). This effect is currently made use of by adding pyrophosphate to toothpastes to inhibit the development of tartar.

An analogue of pyrophosphate which would be resistant to enzymatic breakdown and would have similar effects on crystal formation and dissolution might, however, overcome the drawbacks of pyrophosphate and the other polyphosphates. The bisphosphonates, which possess a P-C-P instead of P-O-P bond (Fig. 1), fulfilled this requirement. The second part of this review will be devoted to them.

Fig. 1. Chemical structure of pyrophosphate and bisphosphonates

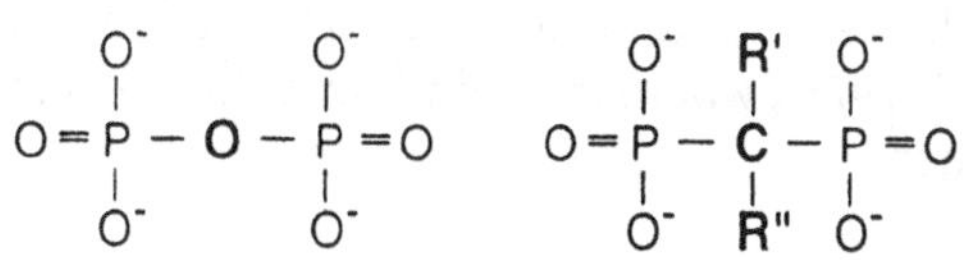

Pyrophosphate

Geminal bisphosphonate

3
Bisphosphonates

The bisphosphonats, erroneously called diphosphonates earlier, have been known to chemists since the middle of the nine teenth century, the fist synthesis dating back to 1865 in Germany (Menschutkin 1865). Their use was industrial, mainly in the textile, fertilizer and oil industries and, because of their property of inhibiting calcium carbonate precipitation, as preventers of scaling. Our knowledge of the biological characteristics of bisphosphonates dates back 30 years, the first report about them having been presented by the author's group in 1968 (Fleisch et al. 1968 b).

3.1
Chemistry

Bisphosphonates are compounds characterized by two C-P bonds (Fig. 2). If the two bonds are located on the same carbon atom, the compounds are called gemial bisphosphonates and are then stereometrically closely related to pyrophosphate. In practice they are simply called bisphosphonates. The P-C-P structure allows a great number of possible variations, mostly by changing the two lateral chains on the carbon. Many bisphosphonates have been investigated in humans with respect to their effects on bone, and six of them are commercially available today for treatment of bone disease.

Although the bisphosphonates shown in Fig. 2 exhibit certain similarities in their chemical, physicochemical, and biological characteristics, the potency of the various effects especially on bone resorption varies greatly from one compound to another, which implies that it is not a possible to extrapolate from the results of one compound to others with respect to their actions.

3.2
Physicochemical Effects

The bisphosphonates have very similar physicochemical effects on hydroxyapatite to pyrophosphate and polyphosphates. Thus, they are powerful inhibitors of the formation and aggregation of calcium phosphate crystals, even at very low concentrations (Francis et al. 1969; Fleisch et al. 1970). As pyrophosphate, they also inhibit the dissolution of these salts (Fleisch et al. 1969; Russell et al. 1970). These effects are related to the marked affinity of these compounds for the surface of solid phase calcium phosphate (Jung et al. 1973), where they bind onto the calcium in a bidentate or tridentate manner.

$$\begin{array}{ccc} & \text{NH}_2 & \\ & | & \\ \text{O}^- & (\text{CH}_2)_3 & \text{O}^- \\ | & | & | \\ \text{O}=\text{P} - & \text{C} - & \text{P}=\text{O} \\ | & | & | \\ \text{O}^- & \text{OH} & \text{O}^- \end{array}$$

(4-Amino-1-hydroxybutylidene)-
bis-phosphonate

alendronate*

Gentili; Merck Sharp & Dohme

$$\begin{array}{ccc} \text{O}^- & \text{Cl} & \text{O}^- \\ | & | & | \\ \text{O}=\text{P} - & \text{C} - & \text{P}=\text{O} \\ | & | & | \\ \text{O}^- & \text{Cl} & \text{O}^- \end{array}$$

(Dichloromethylene)-
bis-phosphonate

clodronate*

*Astra; Boehringer Mannheim;
Gentili; Leiras; Rhône-Poulenc Rorer*

$$\begin{array}{ccc} \text{O}^- & (\text{CH}_2)_2 & \text{O}^- \\ | & | & | \\ \text{O}=\text{P} - & \text{C} - & \text{P}=\text{O} \\ | & | & | \\ \text{O}^- & \text{OH} & \text{O}^- \end{array}$$

[1-Hydroxy-3-(1-pyrrolidinyl)-
propylidene]bis-phosphonate

EB-1053

Leo

$$\begin{array}{ccc} \text{O}^- & \text{CH}_3 & \text{O}^- \\ | & | & | \\ \text{O}=\text{P} - & \text{C} - & \text{P}=\text{O} \\ | & | & | \\ \text{O}^- & \text{OH} & \text{O}^- \end{array}$$

[1-Hydroxyethylidene)-
bis-phosphonate

etidronate*

Gentili; Procter & Gamble

$$\begin{array}{ccc} \text{O}^- & (\text{CH}_2)_2 & \text{O}^- \\ | & | & | \\ \text{O}=\text{P} - & \text{C} - & \text{P}=\text{O} \\ | & | & | \\ \text{O}^- & \text{OH} & \text{O}^- \end{array}$$

[1-Hydroxy-3-(methylpentylamino)
propylidene]bis-phosphonate

ibandronate*

Boehringer Mannheim

$$\begin{array}{ccc} \text{O}^- & \text{NH} & \text{O}^- \\ | & | & | \\ \text{O}=\text{P} - & \text{C} - & \text{P}=\text{O} \\ | & | & | \\ \text{O}^- & \text{H} & \text{O}^- \end{array}$$

[(Cycloheptylamino)-
methylene]bis-phosphonate

incadronate*

Yamanouchi

$$\begin{array}{ccc} & \text{NH}_2 & \\ & | & \\ \text{O}^- & (\text{CH}_2)_5 & \text{O}^- \\ | & | & | \\ \text{O}=\text{P} - & \text{C} - & \text{P}=\text{O} \\ | & | & | \\ \text{O}^- & \text{OH} & \text{O}^- \end{array}$$

(6-Amino-1-hydroxyhexy-lidine)bis-
phosphonate

neridronate

Gentili

$$\begin{array}{ccc} \text{H}_3\text{C} & \text{N} & \text{CH}_3 \\ & | & \\ \text{O}^- & (\text{CH}_2)_2 & \text{O}^- \\ | & | & | \\ \text{O}=\text{P} - & \text{C} - & \text{P}=\text{O} \\ | & | & | \\ \text{O}^- & \text{OH} & \text{O}^- \end{array}$$

[3-(Dimethylamino)-1-hydroxy-
propylidene]bis-phosphonate

olpadronate

Gador

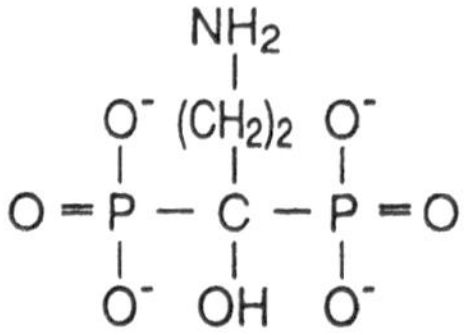

(3-Amino-1-hydroxypropylidene) bis-phosphonate
pamidronate*
Ciba-Geigy; Gador

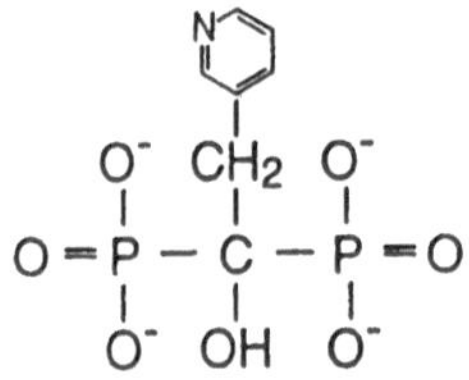

[1-Hydroxy-2-(3-pyridinyl)-ethylidene] bis-phosphonate
risedronate
Procter & Gamble

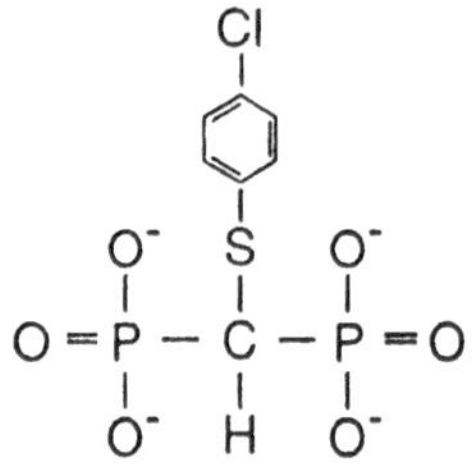

[[(4-Chlorophenyl)thio]-methylene] bis-phosphonate
tiludronate*
Sanofi

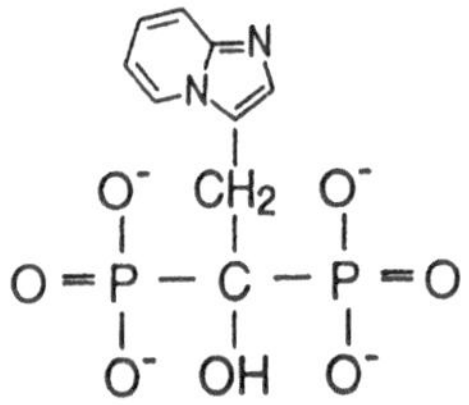

[1-Hydroxy-2-imidazo-(1,2-a) pyridin-3-ylethylidene]
bis-phosphonate
minodronate
Yamanouchi-Hoechst

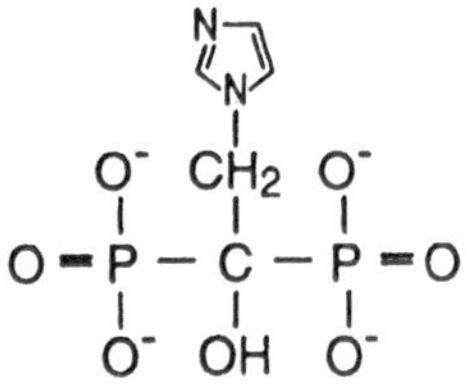

[1-Hydroxy-2-(1H-imidazole-1-yl)ethylidene] bis-phosphonate
zoledronate
Ciba-Geigy

Fig. 2. Chemical structure of the bisphosphonates investigated in man. (Fleisch 1997b)

3.3
Biological Effects

The bisphosphonates have two fundamental biological effects: inhibition of bone resorption and, when given at high doses, inhibition of calcification.

3.3.1
Inhibition of Bone Resorption

Bisphosphonates can be very powerful inhibitors of bone resorption (Fleisch et al. 1969; Russell et al. 1970b), their potency varying according to their structure. This was shown in vitro in cell and organ culture, as well as in vivo in both animals and humans. In animals the effect is present in normal animals, as well as under conditions where resorption is experimentally increased. Similarly, in humans, bone resorption is decreased in normal individuals as well as in patients afflicted with as series of conditions accompanied by increased bone resorption, such as Paget's disease, tumoral osteolysis, hyperparathyroidism, and osteoporosis (for review see Fleisch 1997b).

3.3.1.1
Effects on Organ and Cell Culture

Bisphosphonates block bone resorption induced by various means in organ culture (Russell et al. 1970; Reynolds et al. 1972). An inhibition can also be found when the effect of isolated osteoclasts is investigated on various mineralized matrices in vitro (Flanagan and Chambers 1989; Sato and Grasser 1990). In the presence of bisphosphonate the osteoclasts form fewer erosion cavities and these are of smaller size.

3.3.1.2
Effects in Vivo

Intact animals. In growing, intact rats, the bisphosphonates block the degradation of both bone and cartilage, thus arresting the remodeling of the metaphysis which becomes club-shaped and radiologically denser than normal (Schenk et al. 1973). This effect is often used as a model to study the potency of new compounds.

The inhibition of endogenous bone resorption has also been documented by ^{45}Ca kinetic studies and by markers of bone resorption (Gasser et al. 1972), and by morphology (Balena et al. 1993; Boyce et al. 1995). The decrease in resorption leads secondarily to a decrease in bone formation, resulting in diminished bone turnover.

The decrease in resorption is accompanied by an increase in calcium balance (Gasser et al. 1972) and in mineral content of bone. This is possible because of an increase in intestinal absorption of calcium (Gasser et al. 1972), consequent on an elevation of 1.25 $(OH)_2$ vitamin D. This increased balance is the reason for administering these compounds to humans suffering from osteoporosis.

Animals with Experimentally Increased Resorption. Bisphosphonates can also prevent experimentally induced increase in bone resorption. They impair, among other things resorption induced by agents such as parathyroid hormone, 1.25 $(OH)_2$ vitamin D, and retinoids, as well as in various models mimicking human diseases, among others osteoporosis and tumor bone disease.

Osteoporosis. Many osteoporosis models were investigated. They included sciatic nerve section, which was the first model investigated (Mühlbauer et al. 1971), spinal cord section, hypokinesis, ovariectomy, orchidectomy, heparin, lactation, low calcium diet, and corticosteroids. All bisphosphonates investigated have been effective, namely alendronate, clodronate, etidronate, ibandronate, incadronate, olpadronate, pamidronate, risedronate, tiludronate, and minodronate, and this in many different animals.

The inhibition of bone loss leads to less trabecular thinning, a decreased number of trabecular perforations, a decreased reduction in connectivity, and a smaller erosion of the cortex (Balena et al. 1993; Boyce et al. 1995), thus slowing down the decrease in bone strength. Indeed, if not given in excess, bisphosphonates improve biomechanical properties both in normal animals and in experimental models of osteoporosis (Balena et al. 1993). In humans an effect on both the mineral density and the incidence of vertebral and nonvertebral fractures has been documented recently (Liberman et al. 1995; Black et al. 1996).

The administration of bisphosphonates often leads not only to a halt in the loss of bone, but also to a positive calcium and bone balance both in animals (Gasser et al. 1972) and in humans. There are several explanations for this.

Tumor Bone Disease. Bisphosphonates partially or entirely correct the increase in bone resorption in experimental tumor bone disease. Thus, they prevent or correct the hypercalcemia induced in rats by subcutaneously implanted tumor cells (Martodam et al. 1983), slow down the bone resorption secondary to tumor invasion, and appear to lead to an actual decrease in tumor burden (Sasaki et al. 1995). This decrease is probably not due to a direct effect on the tumor cells, but may be caused by a diminished release of growth factors produced by the osteoclasts present in bone matrix, which may stimulate tumor cell growth during bone resorption (Mundy and Yoneda 1995). The effect of bisphosphonates on tumors is used in humans suffering from tumor bone disease. They correct hypercalcemia, reduce pain, prevent development of new osteolytic lesions as well as the occurrence of fractures and, consequently, improve the quality of life (Hortobagyi et al. 1996).

3.3.1.3
Structure–Activity Relationship

The activity of bisphosphonates on bone resorption varies greatly from compound to compound. Compounds have now been developed which are 5000–10 000 times more powerful in inhibiting bone resorption than etidronate, the first bisphosphonate investigated. The gradation of potency evaluated in the rat corresponds quite well with that found in humans.

However, up to now no clear-cut relationship between structure and activity could be demonstrated. The length of the aliphatic carbon is important since up to a certain length the activity increases, and then decreases later on (Shinoda et al. 1983). Adding a hydroxyl group to the carbon atom at position 1 increases potency (Van Beek et al. 1994), probably because it allows tridentate and therefore stronger binding to the apatite crystal. Derivatives with an amino group at the end of the side chain are very active. The first of these compounds to be described was pamidronate (Reitsma et al. 1980). Again the length of the side chain is relevant, the highest activity being found with a backbone of four carbons as present in alendronate (Schenk et al. 1986). A primary amine is not necessary for this activity since addition of certain groups to the amino nitrogen increases efficacy, as seen in the extremely active ibandronate (Mühlbauer et al. 1991). Cyclic geminal bisphosphonates are also very potent, especially those containing a nitrogen atom in the ring, such as risedronate. The most active compounds described so far, zoledronate (Green et al. 1994) and minodronate, belong to this class. This effect of nitrogen is very intriguing and not yet explained. A three-dimensional structural requirement appears to be involved. Indeed, stereo-isomers of the same chemical structure have shown tenfold differences in activity (Takeuchi et al. 1993). This opens the possibility of a binding to some kind of "receptor" or "active" site.

3.3.1.4
Mechanisms of Action

The inhibition of bone resorption is not explained, as initially thought, by the physicochemical effect on crystal dissolution observed in vitro, which is probably negligible, if at all effective, but largely, if not entirely, by cellular mechanisms. Indeed, the concentrations of bisphosphonates required to inhibit bone resorption with the newer, more potent compounds are so low that they are highly unlikely to have a significant impact on mineral dissolution. Moreover, structure–activity studies on a large array of compounds showed no correlation between the inhibition of mineral dissolution and the pharmacological activity on bone resorption in cell and organ culture in vitro (Sato and Grasser 1990) or in vivo (Shinoda et al. 1983).

There is now general agreement that the final target of bisphosphonate action is the osteoclast. A direct effect on the osteoclasts is supported by the finding that, under bisphosphonates, osteoclasts can show changes in morpholoy both in vitro (Sato and Grasser 1990) and in vivo (Schenk et al. 1973). These include changes in the cytoskeleton, especially loss of actin which is required for the formation of the sealing zone, in vinculin, and in the ruffled border (Murakami et al. 1995). A direct action on osteoclasts is also supported by the fact that, under certain conditions, bisphosphonates can enter cells (Felix et al. 1984), probably by fluid-phase pinocytosis or phagocytosis of complexes, particularly in cells of the macrophage lineage. The penetration into the osteoclasts is favored by the fact that the concentration of the bisphosphonate can attain very high values under these cells, partly because they deposit preferentially underneath them (Masarachia et al. 1996), and partly because those which are deposited on the surface of the bone are then released from the mineral when the osteoclasts dissolve the bone.

Four mechanisms appear to be involved in the inhibition of resorption: (1) inhibition of osteoclast recruitment; (2) inhibition of osteoclast adhesion; (3) shortening of osteoclast lifespan; (4) inhibition of osteoclast activity. The first three will lead to a decrease in the number of osteoclasts, which is observed in humans and often, although not always, in animals. All four effects could be due either to a direct action on the osteoclast or its precursors, or indirectly through action on cells which modulate the osteoclast.

1. *Inhibition of osteoclast recruitment*. Several bisphosphonates inhibit osteoclast differentiation in various culture systems of both cells (Hughes et al. 1989) and bones. Some experiments suggest that the effect occurs at the terminal step of the differentiation process.

2. *Inhibition of osteoclast adhesion*. The second possibility would be a decreased osteoclastic adhesion to the mineralized matrix. Whether this takes place is still debated since the results are ambiguous. One recent study reports such an effect (Colucci et al. 1995). In contrast, there is now excellent evidence that bisposphonates can inhibit in vitro the adhesion of other cells, such as tumor cells (Van der Plujm et al. 1996, Boissier et al. 1997).

3. *Shortening of osteoclast lifespan*. The third possibility is a shortening of the lifespan of the osteoclast. It has been proposed that this might be due to a toxic effect, but the results were obtained at very high concentrations. Recently it was found that bisphosphonates induce osteoclast programmed cell death (apoptosis), both in vitro and in vivo (Hughes et al. 1995). The ranking of effectiveness of various compounds was the same as seen in vivo. Whether this is a direct effect on osteoclasts, or an indirect one through the effect on other cells, is not known. A similar effect is seen in vitro on mouse macrophages (Rogers et al. 1996) and tumor cells (Shipman et al. 1997).

4. *Inhibition of osteoclast activity*. The last possibility is an inhibition of osteoclast activity after the bisphosphonate has been taken up by the osteclasts. Indeed, several facts suggest that the inhibition of recruitment is not the only mode of action of bisphosphonates in vivo. Thus, following bisphosphonate administration, the number of multinucleated osteoclasts on the bone surface often increases initially, despite a reduced bone resorption. However, the cells look inactive (Schenk et al. 1973) and show, as desribed above, cellular alterations. The cause for this initial increase is unknown. The decrease in activity can also be indirect through the liberation by the osteoblasts of substances leading to this effect (Owens et al. 1997).

In recent years our knowledge about the molecular and biochemical events leading to either osteoclast inactivation or diminished osteoclast formation by bisphosphonates has made substantial progress, although it is not yet fully elucidated. It has been known for a long time that bisphosphonates decrease acid production of various cells (Fast et al. 1978) as well as of calvaria. Recently, bisphosphonates have been shwon to reduce the amount of acid released through a sodium-independent mechanism by osteoclasts (Zimolo et al. 1995). Possibly part of this effect is due to the decrease of the proton transport by the vacuolar-type proton ATPase (David et al. 1996). However, until now no correlation between the effect in vitro on acid production and in vivo on bone resorption was evident. Actually some bisphosphonates, such as pamidronate, or long-chain bisphosphonates increase lactic acid production, possibly due to a toxic action.

Various bisphosphonates, especially clodronate, inhibit to some extent lysosomal enzymes in vitro, in cultured calvaria, or when given in vivo. Certain bisphosphonates, such as clodronate and etidronate, also inhibit prostaglandin synthesis by bone cells or calvaria, both in vitro and in vivo (Ohya et al. 1985). Since prostaglandins are involved in bone resorption, this inhibition may play a role in the resorption process.

In view of the homology between pyrophosphate and bisphosphonates, various enzymes involving pyrophosphate or ATP have been examined. While phosphatases and pyrophosphatases are influenced at relatively high concentrations only, or not influenced at all, some protein-tyrosine phosphatases, such as PTPσ, PTPε, and PTP-Meg1, are inhibited in vitro by micromolar conentrations (Endo et al. 1996; Schmidt et al. 1996, Opas et al. 1997). H_2O_2 and metal ions such as calcium are necessary for the inhibition, and it appears that the latter involves the oxidation of the active site cysteine of the PTPs (Opas et al. 1997; Skorey et al. 1997). These effects might be relevant since protein-tyrosine phosphorylation is important in the signal transduction pathways which control cell growth, differentiation, and activity. Unfortunately, the potency to inhibit the protein-tyrosine phosphatases of various bisphosphonates tested so far has no relationship to their pharmacological potency. This is also true for the osteoclast vacuolar H'-ATPase which is strongly inhibited by tiludronate, a weak inhibitor of bone resorption, but much less so by other more powerful bisphosphonates and by pyrophosphate (David et al. 1996).

Recently, an interesting new mechanism has been proposed for at least some of the bisphosphonates containing a nitrogen atom. It has been known for some time that bisphosphonates can inhibit squalene synthase and hence cholesterol biosynthesis (Ciosek et al. 1993). Some intermediates of the mevalonate patchway, such as farnesylpyrophosphate and geranylgeranylpyrophosphate, are also required for the prenylation of GTP-binding proteins, such as ras, rac, and rho which are involved in many cellular functions, among others proliferation, survival, membrane function, and cytoskeletal organization. It has now been shown that certain bisphosphonates, especially the nitrogen-containing compounds, also inhibit the prenylation of proteins (Luckman et al. 1998). This is, however, not the case for etidronate and clodronate. This effect might be due to the inhibition of several enzymes, among them farnesylpyrophosphate synthase, which has indeed been found to be inhibited by bisphosphonates. This mechanism might be the cause of apoptosis.

Another interesting observation is that some cells can metabolize primary etidronate and clodronate, but not the bisphosphonates containing a nitrogen atom, to a nonhydrolyzable toxic ATP analogue containing a β,γ-methylene group (AppCp nucleotides). It is thought that the bisphosphonates most likely replace pyrophosphate in a back reaction catalyzed by class 2 aminoacyl-tRNA synthetase enzymes (Frith et al. 1997). Thus, at least these compounds might act through this mechanism to induce apoptosis and necrotic cell death. The aminobisphosphonates may have a mechanism of action different from that of the bisphosphonates containing no nitrogen.

Unfortunately, no individual mechanism shows a good correlation with the potency in vivo when different bisphosphonates of various potencies are investigated. This suggests that, if any of the above mechanisms is relevant for bone resorption, it is not the only one and that the mechanism might be dissimilar for different bisphosphonates. There is good evidence that this is the case for etidronate and clodronate as compared to the more potent bisphosphonates.

Lastly, it is of interest that bisphosphonates inhibit the growth of amoebae of the slime mold *Dictyostelium discoideum*, and that there is a remarkable correlation between the order of potency of various bisphosphonates in this system and as inhibitors of bone resorption in vivo (Rogers et al. 1994, 1995). This might suggest that these compounds have similar molecular targets within the two systems.

3.3.1.4.1
Role of Osteoblasts

It appears more and more likely that the inhibitory effect is partly also mediated through other cells, especially cells of the osteoblast lineage. Indeed, it is now known that these cells do control the recruitment and activity of osteoclasts under physiological and pathological conditions through the production of as yet unkown substances.

It has recently been found that pretreating osteoblastic cell populations with bisphosphonates induced them to produce an inhibitory activity of osteoclastic resorption (Sahni et al. 1993; Yu et al. 1996). One part of this activity is labile to heat and proteinase and has a molecular weight around 3–4 kDa (Vitté et al. 1996). It acts not by decreasing the activity of osteoclasts but by decreasing their recruitment. Other cells such as fibroblasts do not seem to produce such inhibitors. In contrast, another part of the activity inhibits osteoclast activation (Owens et al. 1997).

3.3.1.4.2
Effect on Other Cells

Bisphosphonates appear to act also on macrophages. Such an effect might be the explanation for the acute phase response occurring in humans. Thus, some patients who receive an amino-bisphosphonate intravenously for the first time show a transient pyrexia of 1–2 °C, sometimes more, accompanied by flu-like symptoms (Adami et al. 1987). Recently, the pyrexia has been shown to be accompanied by an increase in circulating IL-6 and TNFα (Schweitzer et al. 1995; Sauty et al. 1996).

3.3.2
Inhibition of Calcification

3.3.2.1
Ectopic Mineralization and Ossification

Bisphosphonates can very efficiently inhibit ectopic calcification in vivo. Thus, they prevent, both when given parenterally and orally, experimentally induced calcification of many soft tissues (Fleisch et al. 1970). One of the bisphosphonates, etidronate, is used in humans to prevent ectopic calcification and ossification. Unfortunately, the results so far have not been as encouraging as expected. Topical administration can lead to a decreased formation of dental calculus. This effect is used to prevent tartar formation in humans by adding bisphosphonates to toothpastes.

3.3.2.2
Normal Mineralization

Unfortunately, when administered in doses approximating to those which inhibit soft tissue calcification, bisphosphonates can impair the mineralization of normal calcified tissues such as bone and cartilage (Schenk et al. 1973) as well as dentine, enamel, and cementum. An inhibition of the mineralization of bone and cartilage is seen also in humans when given in larger amounts, which has hampered the therapeutic use of these compounds in ectopic calcification.

While the different compounds vary greatly in their activity in bone resorption, they do not vary in the inhibition of mineralization. For most species the effective daily dose is in the order of 5–20 mg of compound phosphorus per kg, administered parenterally.

3.3.2.3
Mechanisms of Action

The mechanism of the inhibition of both normal and ectopic mineralization is most likely to be due, at least in part if not entirely, to a physicochemical mechanism. Indeed, there is a close relationship between the ability of an individual bisphosphonate to inhibit calcium phosphate formation in vitro and its effectiveness on calcification in vivo (Fleisch et al. 1970; Trechsel et al. 1977; Van Beek et al. 1994).

3.4
Pharmacokinetics

Bisphosphonates have a very low bioavailability, from a few percent for those given in larger amounts to below 1 % for the newer ones, which are given in low quantities. This is partly explained by their low lipophilicity which hampers transcellular transport, and their high negative charge which hampers paracellular transport. Furthermore, they are probably partly in an insoluble form in the gut, due to their chelation to calcium.

Once in the blood, bisphosphonates disappear very rapidly, mostly to bone. This is explained by their strong binding to the hydroxyapatite crystals (Jung et al. 1973). Consequently, soft tissues are exposed to these compounds for only short periods, explaining their bone-specific effects and their low toxicity.

It was generally thought that the bisphosphonates deposit in those locations within the bone where new bone is formed. More recently, however, they have also been found to deposit under the osteoclasts (Sato et al. 1991). The distribution of the amount deposited at bone formation and bone resorption sites depends upon the amount of bisphosphonate administered. When small amounts are given, they deposit mainly under the osteoclasts, while larger amounts go to both bone-forming and bone-resorbing sites (Mazarachia et al. 1996).

Once the bisphosphonates are buried in the skeleton, they are released for the most part only when the bone is destroyed in the course of its turnover. The skeletal half-life of various bisphosphonates is between 3 months and 1 year for mice and rats, but much longer for humans, sometimes over 10 years.

As could have been foreseen, the bisphosphonates are not metabolized in vivo and, up to now, all the bisphosphonates investigated were excreted unaltered.

4
Discussion

Initial work on the inhibitory activity of calcium phosphate precipitation in biological fluids led to the discovery of the yet unknown presence in these fluids of inorganic pyrophosphate, a strong inhibitor of both formation and dissolution of this calcium salt. The physiological role of pyrophosphate in the modulation of bone mineral formation and dissolution is not yet known. Work in this direction has practically stopped for the last two decades.

However, the work on pyrophosphate has led to the development of a new class of compounds that are close analogues of pyrophosphate. These compounds act on both mineralization and bone destruction. Although the mechanism of mineralization is, as foreseen, a physicochemical one, the mechanism of bone destruction acts in a completely different manner, namely through cellular mechanisms.

An interesting possibility emerges from this development. Since bisphosphonates have a structure similar to that of pyrophosphate, but which cannot be destroyed enzymatically, it is tempting to speculate that intracellular pyrophosphate and pholyphosphates may have some effects similar to those of bisphosphonates, and that the bisphosphonates could serve as a model for their study. This would open very interesting new perspectives for the role of polyphosphates in cell biology and might be worth investigating.

5
Further Information

More detailed information can be obtained in several recent reviews (Fleisch 1996, 1997a,b).

References

Adami S, Bhalla AK, Dorizzi R, Montesanti F, Rosini S, Salvagno G, Lo Cascio V (1987) The acute-phase response after bisphosphonate administration. Calcif Tissue Int 41: 326–331

Anderson HC, Reynolds JJ (1973) Pyrophosphate stimulation of calcium uptake into cultured embryonic bones. Fine structure of matrix vesicles and their role in calcification. Dev Biol 34: 211–227

Balena R, Toolan BC, Shea M, Markatos A, Myers ER, Lee SC, Opas EE, Seedor JG, Klein H., Frankenfield D, Quartuccio H, Fioravanti C, Clair J, Brown E, Hayes WC, Rodan GA (1993) The effects of 2-year treatment with the aminobisphosphonate alendronate on bone metabolism, bone histomorphometry, and bone strength in ovariectomized nonhuman primates. J Clin Invest 92: 2577–2586

Bisaz S, Russell RGG, Fleisch H (1968) Isolation of inorganic pyrophosphate from bovine and human teeth. Arch Oral Biol 13: 683–696

Bisaz S, Jung A, Fleisch H (1978) Uptake by bone of pyrophosphate, diphosphonates and their technetium derivatives. Clin Sci Mol Med 54: 265–272

Black DM, Cummings SR, Karpf DB, Cauley JA, Thompson DE, Nevitt MC, Bauer DC, Genant HK, Haskell WL, Marcus R, Ott SM, Torner JC, Quandt SA, Reiss TF, Ensrud KE (1996) Randomized trial of the effect of alendronate on the risk of fracture in women with existing vertebral fractures. Lancet 438: 1535–1541

Boissier S, Magnetto S, Frappart L, Cuzin B, Ebetino FH, Delmas PD, Clezardin P (1997) Bisphosphonates inhibit prostate and breast carcinoma cell adhesion to unmineralized and mineralized bone extracellular matrices. Cancer Res. 57: 3890–3894

Boyce RW, Paddock CL, Gleason JR, Sletsema WK, Eriksen EF (1995) The effects of risedronate on canine cancellous bone remodeling: three-dimensional kinetic reconstruction of the remodeling site. J Bone Miner Res 10: 211–221

Buehrer T, Reitemeyer R (1940) The inhibiting action of minute amounts of sodium hexametaphosphate on the precipitation of calcium carbonate from ammoniacal solutions. J Phys Chem 44: 552–574

Cartier P (1957) Les constituants minéraux des tissus cacifiés. V. Séparation et identification de pyrophosphates dans le tissu osseux. Bull Soc Chim Biol 39: 169–180

Caswell AM, Russell RG (1988) Evidence that ecto-nucleoside-triphosphate pyrophosphatase serves in the generation of extracellular inorganic pyrophosphate in human bone and articular cartilage. Biochim Biophys Acta 966: 310–317

Caswell AM, Ali SY, Russell RG (1987) Nucleoside triphosphate pyrophosphatase of rabbit matrix vesicles, a mechanism for the generation of inorganic pyrophosphate in epiphyseal cartilage. Biochim Biophys Acta 924: 276–283

Ciosek CP, Magnin DR, Harrity TW, Logan JVH, Dickson JK, Gordon EM, Hamilton KA, Jolibois KG, Kunselman LK, Lawrence RM, Mookhtiar KA, Rich LC, Slusarchyk DA, Sulsky RB, Biller SA (1993) Lipophilic 1,1-bisphosphonates are potent squalene synthase inhibitors and orally active cholesterol lowering agents in vivo. J Biol Chem 268: 24832–24837

Colucci S, Minielli V, Zambonin G, Grano M (1995) Etidronate inhibits osteoclast adhesion to bone surfaces but does not interfere with their specific recognition of single bone proteins. Ital J Miner Electrolyte Metab 9: 159–164

David P, Nguyen H, Barbier A, Baron R (1996) The bisphosphonate tiludronate is a potent inhibitor of the osteoclast vacuolar H+-ATPase. J Bone Miner Res 11: 1498–1507

Endo N, Rutledge SJ, Opas EE, Vogel R, Rodan GA, Schmidt A (1996) Human protein tyrosine phosphatase-σ: alternative splicing and inhibition by bisphosphonates. J Bone Miner Res 11: 535–543

Fast DK, Felix R, Dowse C, Neuman WF, Fleisch H (1978) The effects of diphosphonate on the growth and glycolysis of connective-tissue cells in culture. Biochem J 172: 97–107

Felix R, Fleisch H (1974) The pyrophosphatase and $(Ca^{2+}-Mg^{2+})$-ATPbase activity of purified calf bone alkaline phosphatase. Biochim Biophys Acta 350: 84–94

Felix R, Guenther HL, Fleich H (1984) The subcellular distribution of $[^{14}C]$ dichloromethylene-bisphosphonate and $[^{14}C]$ 1-hydroxyethylidene-1,1-bisphosphonate in cultured calvaria cells. Calcif Tissue Int 36: 108–113

Flanagan AM, Chambers TJ (1989) Dichloromethylene-bisphosphonate (Cl_2MBP) inhibits bone resorption through injury to osteoclasts that resorb Cl_2MBP-coated bone. Bone Miner 6: 33–43

Fleisch H (1996) Bisphosphonates: mechanisms of action and clinical use. In: Bilezikian JP, Raisz LG, Rodan GA (eds) Principles of bone biology. Academic Press, San Diego, pp 1037–1052

Fleisch H (1997a) Bisphosphonates: mechanisms of action. Endocr Rev 19(1): 80–100

Fleisch H (1997b) Bisphosphonates in bone disease. From the laboratory to the patient, 3rd edn. Parthenon, New York

Fleisch H, Bisaz S (1962a) Isolation from urine of pyrophosphate, a calcification inhibitor. Am J Physiol 203: 671–675

Fleisch H, Bisaz S (1962b) Mechanism of calcification: inhibitory role of pyrophosphate. Nature 1995: 911

Fleisch H, Bisaz S (1964) The inhibitory effect of pyrophosphate on calcium oxalate precipitation and its relation to urolithiasis. Experientia 20: 276

Fleisch H, Neuman WF (1961) Mechanism of calcification: role of collagen, polyphosphates and phosphatase. Am J Physiol 200: 1296–1300

Fleisch H, Bisaz S, Care AD (1964) Effect of orthophosphate on urinary pyrophosphate and the prevention of urolithiasis. Lancet 1: 1065–1067

Fleisch H, Maerki J, Russell RGG (1966a) Effect of pyrophosphate on dissolution of hydroxyapatite and its possible importance in calcium homeostasis. Proc Soc Exp Biol Med 122: 317–320

Fleisch H, Russell RGG, Straumann F (1966b) Effect of pyrophosphate on hydroxyapatite and its implications in calcium homeostasis. Nature 212: 901–903

Fleisch H, Straumann F, Schenk R, Bisaz S, Allgöwer M (1996c) Effect of condensed phosphates on calcification of chick embryo in tissue culture. Am J Physiol 211: 821–825

Fleisch H, Russell RGG, Bisaz S, Termine JD, Posner AS (1968a) Influence of pyrophosphate on the transformation of amorphous to crystalline calcium phosphate. Calcif Tissue Res 2: 49–59

Fleisch H, Russell RGG, Bisaz S, Casey PA, Mühlbauer RC (1968b) The influence of pyrophosphate analogues (diphosphonates) on the precipitation and dissolution of calcium phosphate in vitro and in vivo. Calcif Tissue Res 2: 10–10A

Fleisch H, Russell RGG, Bisaz S, Mühlbauer RC, Williams DA (1970) The inhibitory effect of phosphonates on the formation of calcium phosphate crystals in vitro and on aortic and kidney calcification in vivo. Eur J Clin Invest 1: 12–18

Fleisch H, Russell RGG, Francis MD (1969) Diphosphonates inhibit hydroxyapatite dissolution in vitro and bone resorption in tissue culture and in vivo. Science 165: 1262–1264

Francis MD, Russell RGG, Fleisch H (1969) Diphosphonates inhibit formation of calcium phoshate crystals in vitro and pathological calcification in vivo. Science 165: 1264–1266

Frith JC, Mönkkönen J, Blackburn GM, Russell RGG, Rogers MJ (1997) Clodronate and liposome-encapsulated clodronate are metabolized to a toxic ATP analog, adenosine 5'-(β-γ-dichloromethylene) triphosphate, by mammalian cells in vitro. J Bone Miner Res 12: 1358–1367

Gasser AB, Morgan DB, Fleisch HA, Richelle LJ (1972) The influence of two disphosphonates on calcium metabolism in the rat. Clin Sci 43: 31–45

Green JR, Müller K, Jaeggi KA (1994) Preclinical pharmacology of CGP 42'446, a new, potent, heterocyclic bisphophonate compound. J Bone Miner Res 9: 745–751

Hansen NM, Felix R, Bisaz S, Fleisch H (1976) Aggregation of hydroxyapatite crystals. Biochem Biophys. Acta 451: 549–559

Hausmann E, Bisaz S, Russell RGG, Fleisch H (1970) The concentration of inorganic pyrophosphate in human saliva and dental calculus. Arch Oral Biol 15: 1389–1392

Hortobagyi GN, Theriault RL, Porter L, Blayney D, Lipton A, Sinoff C, Wheeler H, Simeone JF, Seaman J, Knight RD, Heffernan M, Reitsma DJ (1996) Efficacy of pamidronate in reducing skeletal complications in patients with breast cancer and lytic bone metastases. N Engl J Med 335: 1785–1791

Hsu HH, Anderson HC (1984) The deposition of calcium, pyrophosphate and phosphate by matrix vesicles isolated from fetal bovine epiphyseal cartilage. Calcif Tissue Int 36: 615–621

Hughes DE, MacDonald BR, Russell RGG, Gowen M (1989) Inhibition of osteoclast-like cell formation by bisphosphonates in long-term cultures of human bone marrow. J Clin Invest 83: 1930–1935

Hughes DE, Wright KR, Uy HL, Sasaki A, Yoneda T, Roodman GD, Mundy GR, Boyce BF (1995) Bisphosphonates promote apoptosis in murine osteoclasts in vitro and in vivo. J Bone Miner Res 10: 1478–1487

Jung A, Russell RGG, Bisaz S, Morgan DB, Fleisch H (1970) Fate of intravenously injected pyrophosphate-32p in dogs. Am J Physiol 218: 1757–1764

Jung A, Bisaz S, Fleisch H (1973) The binding of pyrophosphate and two diphosphonates by hydroxyapatite crystals. Calcif Tissue Res 11: 269–280

Jung A, Bisaz S, Gebauer U, Russell RGG, Morgan DB, Fleisch H (1974) The uptake and metabolism of [32p]pyrophosphate by mouse calvaria in vitro. Biochem J 140: 175–183

Liberman UA, Weiss SR, Bröll J, Minne HW, Quan H, Bell NH, Rodriguez-Portales J, Downs RW Jr, Dequeker J, Favus M, Seeman E, Recker RR, Capizzi T, Santora AC II, Lombardi A, Shah RV, Hirsch LJ, Karpf DB (1995) effect of oral alendronate on bone mineral density and the incidence of fractures in postmenopausal osteoporosis. N Engl J Med 333: 1437–1443

Luckman SP, Hughes DE, Coxon FP, Russell RGG, Rogers MJ (1988) Nitrogen-containing bisphosphonates inhibit the mevalonate pathway and prevent post-translational prenylation of GTP-binding proteins, including Ras. J Bone Miner Res 13: 581–589

Martodam RR, Thornton KS, Sica DA, D'Souza SM, Flora L, Mundy GR (1983) The effects of dichloromethylene diphosphonate on hypercalcemia and other parameters of the humoral hypercalcemia of malignancy in the rat Leydig cell tumor. Calcif Tissue Int 35: 512–519

Masarachia P, Weinreb M, Balena R, Rodan GA (1996) Comparison of the distribution of ^{3}H-alendronate and ^{3}H-etidronate in rat and mouse bones. Bone 19: 281–290

McCarty DJ, Solomon SD, Warnock M, Paloyan (1971) Inorganic pyrophosphate concentrations in the synovial fluid of arthritic patients. J. Lab Clin Med 78: 216–229

Menschutkin N (1865) Über die Einwirkung des Chloracetyls auf phosphorige Säure. Ann Chem Pharm 133: 317–320

Mühlbauer RC, Russell RGG, Williams DA, Fleisch H (1971) The effects of diphosphonates, polyphosphates and calcitonin on "immobilisation osteoporosis" in rats. Eur J Clin Invest 1: 336: 344

Mühlbauer RC, Bauss F, Schenk R, Janner M, Bosies E, Strein K, Fleisch H (1991) BM 21.0955, a potent new bisphosphonate to inhibit bone resorption. J Bone Miner Res 6: 1003–1011

Mühlemann HR, Bowles D, Schatt A, Bernimoulin JP (1970) Effect of diphosphonate on human supragingival calculus. Helv Odontol Acta 14: 31–33

Mundy GR, Yoneda T (1995) Facilitation and suppression of bone metastasis. Clin Orthop 312: 34–44

Murakami H, Takahashi N, Sasaki T, Udagawa N, Tanaka S, Nakamura I, Zhang D, Barbier A, Suda T (1995) A possible mechanism of the specific action of bisphosphonate on osteoclasts: tiludronate perferentially affects polarized osteoclasts having ruffled borders. Bone 17: 137–144

Ohya K, Yamada S, Felix R, Fleisch H (1985) Effect of bisphosphonates on prostaglandin synthesis by rat bone cells and mouse calvaria in culture. Clin Sci 69: 403–411

Opas EE, Rutledge SJ, Golub E, Stern A, Zimolo Z, Rodan GA, Schmidt A (1997) Alendronate inhibition of protein-tyrosine-phosphatase-megl. Biochem Parmacol 54: 721–727

Owens JM, Fuller K, Chambers TJ (1997) Osteoclast activation: potent inhibition by the bisphosphonate alendronate through a nonresorptive mechanism. J Cell Physiol 172: 79–86

Reitsma PH, Bijvoet OLM, Verlinden-Ooms H, van der Wee-Pals LJA (1980) Kinetic studies of bone and mineral metabolism during treatment with (3-amino-1-hydroxy-propylidene)-1,1-bisphosphonate (APD) in rats. Calcif Tissue Int 32: 145–157

Reynolds JJ, Minkin C, Morgan DB, Spycher D, Fleisch H (1972) The effect of two diphosphonates on the resorption of mouse calvaria in vitro. Calcif Tissue Res 10: 302–313

Robertson WG, Peacock M, Nordin BEC (1973) Inhibitors of the growth and aggregation of calcium oxalate crystals in vitro. Clin Chim Acta 43: 31–37

Rogers MJ, Watts DJ, Russell RGG, Ji X, Xiong X, Blackburn GM, Bayless AW, Ebetino FH (1994) Inhibitory effects of bisphosphonates on growth of amoebae of the cellular slime mould *Dictyostelium discoideum*. J Bone Miner Res 10: 1029–1039

Rogers MJ, Xiong X, Brown RJ, Watts DJ, Russell RGG, Bayless AW, Ebetino FH (1995) Structure–activity relationships of new heterocycle-containing bisphosphonates as inhibitors of bone resorption and as inhibitors of growth of *Dictyostelium discoideum* amoebae. Mol Pharmacol 47: 398–402

Rogers MJ, Chilton KM, Coxon FP, Lawry J, Smith MO, Suri S, Russell RGG (1996) Bisphosphonates induce apoptosis in mouse macrophage-like cells in vitro by a nitric oxide-independent mechanism. J Bone Miner Res 11: 1482–1491

Rudy H (1960) Altes und Neues über kondensierte Phosphate. Benckiser, Ludwigshafen

Russell RGG (1965) Excretion of inorganic pyrophosphate in hypophosphatasia. Lancet 2: 462–464

Russell RGG, Bisaz S, Fleisch H, Currey HM, Rubinstein HM, Dietz A, Boussine I, Gabay R, Micheli A, Fallet G (1970a) Inorganic pyrophosphate in the plasma, urine and synovial fluid of patients with pyrophosphate arthropathy (chondrocalcinosis or pseudogout). Lancet 2: 899–902

Russell RGG, Mühlbauer RC, Bisaz S, Williams DA, Fleisch H (1970b) The influence of pyrophosphate, condensed phosphates, phosphonates and other phosphate compounds on the dissolution of hydroxyapatite in vitro and on bone resorption induced by parathyroid hormone in tissue culture and in thyroparathyroidectomised rats. Calcif Tissue Res 6: 183–196

Russell RGG, Bisaz S, Donath A, Morgan DB, Fleisch H (1971) Inorganic pyrophosphate in plasma in normal persons and in patients with hypophosphatasia, osteogenesis imperfecta and other disorders of bone. J Clin Invest 50: 961–969

Sahni M, Guenther HL, Fleisch H, Collin P, Martin TJ (1993) Bisphosphonates act on rat bone resorption through the mediation of osteoblasts. J Clin Invest 91: 2004–2011

Sasaki A, Boyce BF, Story B, Wright KR, Chapman M, Boyce R, Mundy GR, Yoneda T (1995) Bisphosphonate risedronate reduces metastatic human breast cancer burden in bone in nude mice. Cancer Res 55: 3551–3557

Sato M, Grasser W (1990) Effects of bisphosphonates on isolated rat osteoclasts as examined by reflected light microscopy. J Bone Miner Res 5: 31–40

Sato M, Grasser W, Endo N, Akins R, Simmons H, Thompson DD, Golub E, Rodan GA (1991) Bisphosphonate action. Alendronate localization in rat bone and effects on osteoclast ultrastructure. J Clin Invest 88: 2095–2105

Sauty A, Pecherstorfer M, Zimmer-Roth I, Fioroni P, Juillerat L, Markert M, Ludwig H, Leuenberger P, Burckhardt P, Thiébaud D (1996) Interleukin-6 and tumor necrosis factor α levels after bisphosphonate treatment in vitro and in patients with malignancy. Bone 18: 133–139

Schenk R, Merz WA, Mühlbauer R, Russell RGG, Fleisch H (1973) Effect of ethane-1-hydroxy-1,1-diphosphonate (EHDP) and dichloromethylene diphosphonate (Cl_2MDP) on the calcification and resorption of cartilage and bone in the tibial epiphysis and metaphysis of rats. Calcif Tissue Res 11: 196–214

Schenk R, Eggli P, Fleisch H, Rosini S (1986) Quantitative morphometric evaluation of the inhibitory activity of new aminobisphosphonates on bone resorption in the rat. Calcif Tissue Int 38: 342–439

Schibler D, Russell RGG, Fleisch H (1968) Inhibition by pyrophosphate and polyphosphate of aortic calcification induced by vitamin D_3 in rats. Clin Sci 35: 363–372

Schmidt A, Rutledge SJ, Endo N, Opas EE, Tanaka H, Wesolowski G, Leu CT, Huang Z, Ramachandaran C, Rodan SB, Rodan GA (1996) Protein-tyrosine phosphatase activity regulates osteoclast formation and function: inhibition by alendronate. Proc Natl Acad Sci USA 93: 3068–3073

Schweitzer DH, Oostendorp-van de Ruit M, van der Pluijm G, Löwik CWGM, Papapoulos SE (1995) Interleukin-6 and the acute phase response during treatment of patients with Paget's disease with the nitrogen-containing bisphosphonate dimethylaminohydroxypropylidene bisphosphonate. J Bone Miner Res 10: 956–962

Shinoda H, Adamek G, Felix R, Fleisch H, Schenk R, Hagan P (1983) Structure–activity relationships of various bisphosphonates. Calcif Tissue Int 35: 87–99

Shipman CM, Roges MJ, Apperley JF, Russell RG, Croucher PI (1997) Bisphosphonates induce apoptosis in human myeloma cell lines: a novel anti-tumour activity. Br J Haematol 98: 665–672

Skorey K, Ly HD, Kelly J, Hammond M, Ramachandran C, Huang Z, Gresser MJ, Wang Q (1997) How does alendronate inhibit protein-tyrosine phosphatases? J Biol Chem 272: 22472–22480

Takeuchi M, Sakamoto S, Yoshida M, Abe T, Isomura Y (1993) Studies on novel bone resorption inhibitors. I. Synthesis and pharmacological activities of aminomethylenebisphosphonate derivatives. Chem Pharm Bull Tokyo 41: 688–693

Termine JD, Posner AS (1970) Calcium phosphate formation in vitro. I. Factors affecting initial phase separation. Arch Biochem Biophys 140: 307–317

Trechsel U, Schenk R, Bonjour JP, Russell RGG, Fleisch H (1977) Relation between bone mineralization, Ca absorption, and plasma Ca in phosphonate-treated rats. Am J Physiol 232: E298–E305

Van Beek E, Hoekstra M, van de Ruit M, Löwik C, Papapoulos S (1994) Structural requirements for bisphosphonate actions in vitro. J Bone Miner Res 9: 1875–1882

Van der Pluijm G, Vloedgraven H, van Beek E, van der Wee-Pals L, Löwik C, Papapoulos S (1996) Bisphosphonates inhibit the adhesion of breast cancer cells to bone matrices in vitro. J Clin Invest 98: 698–705

Vitté C, Fleisch H, Guenther HL (1996) Bisphosphonates induce osteoblasts to secrete an inhibitor of osteoclast-mediated resorption. Endocrinology 137: 2324–2333

Whyte MP (1994) Hypophosphatasia and the role of alkaline phosphatase in skeletal mineralization. Endocr Rev 15: 439–461

Yu X, Scholler J, Foged NT (1996) Interaction between effects of parathyroid hormone and bisphosphonate on regulation of osteoclast activity by the osteoblast-like cell line UMR-106. Bone 19: 339: 345

Zimolo Z, Wesolowski G, Rodan GA (1995) Acid extrusion is induced by osteoclast attachment to bone: inhibition by alendronate and calcitonin. J Clin Invest 96: 2277–2283

Methods for Investigation of Inorganic Polyphosphates and Polyphosphate-Metabolizing Enzymes

B. Lorenz[1] and H. C. Schröder[2]

1
Introduction

One of the main reasons for the fact that inorganic polyphosphates were a neglected area of research over a long period of time was the lack of reliable experimental tools for the study of these polymers. In recent years, the range and specificity of the methods available have been revolutionized by the development of novel techniques especially in the field of molecular biology and their application to the investigation of polyphosphates. In particular, the availability of specific polyphosphate-dependent enzymes has been a considerable progress for the development of definite and exact methods. However, for certain questions, optimal methods are still missing.

The aim of this chapter is to summarize and to illustrate the most important and most widely used methods for investigation of polyphosphates, with their advantages and disadvantages. The efficiency and applicability of nuclear magnetic resonance spectroscopy for studying polyphosphates is described in Chapter 13. Furthermore, novel promising methods for the isolation and quantification of polyphosphate as well as for the measurement of polyphosphate-dependent enzyme activities by using the power of specific enzymes are explained and documented in Chapter 12.

2
Preparation of Inorganic Polyphosphates

2.1
Chemical Synthesis

Formally, inorganic polyphosphates are formed by condensation of orthophosphate under formation of phosphoanhydride bonds. Three different groups of basic structures can result: linear, branched and cyclic molecules. The latter are also called metaphosphates. When orthophosphoric acid or its salts are heated these substances form condensates by leaving of water.

[1] Institut für Biochemie, Universität, Leipziger Str. 44, 39120 Magdeburg, Germany.
[2] Institut für Physiologische Chemie, Universität, Duesbergweg 6, 55099 Mainz, Germany.

Progress in Molecular and Subcellular Biology, Vol. 23
H. C. Schröder, W. E. G. Müller (Eds.)
© Springer-Verlag Berlin Heidelberg 1999

Therefore the correct chemical nomenclature is "condensed phosphates". In this chapter only some general facts about the chemical synthesis of this group of molecules will be mentioned. Detailed descriptions of the synthesis, chemical structure, and properties of the obtained products are given by Thilo (1955, 1965) and Van Wazer (1958).

At temperatures above 160 °C diphosphates (pyrophosphates) are formed from monophosphates. When the primary product, monobasic phosphate, is heated to 200–300 °C a very high-molecular, crystalline compound [average chain length (n) > 1000] is formed, known as Madrell's salt. Another high molecular substance, the fibrous Kurrol's salt, with $n \approx 5000$ is formed when the melting mass of the Madrell's salt is heated up to 600 °C followed by a short incubation period at 500 °C. In contrast to all other salts of condensed phosphates with monovalent cations (with the exception of Ag^+) which are readily water-soluble, both these high molecular compounds are practically insoluble in water. A very small degree of solubility in water is also characteristic of phosphate salts of polyvalent cations (e.g. Ca^{2+}, Ba^{2+}, Fe^{3+}). By application of ion-exchange resins or by replacement of the corresponding cations by competitive ions (Li^+, NH_4^+) these substances can be dissolved in water (Van Wazer 1958; Griffin et al. 1965; Clark and Wood 1987). It should be noted that the terms "Kurrol's salt" and "Madrell's salt" sometimes are used differently in the literature.

The main compound of condensed phosphates is called Graham's salt. It is produced by heating of monobasic phosphates to temperatures higher than 620 °C (usually 700–1000 °C) followed by rapid cooling of the melting mass. A glass-like mass is formed, consisting of a heterogeneous mixture of linear polyphosphates with varying chain lengths, e.g. the commercially available, chemically synthesized polyphosphate with an average chain length of 15 (polyphosphate glass P15 from Sigma Co., St. Louis, USA) contains polyphosphate chains from 2 to about 140 phosphate residues and polyphosphate glass P35 from 2 to about 180 phosphate residues. The degree of condensation of the Graham's salt can reach more than 1000. The average chain length depends on the conditions during the synthesis (Thilo 1965). The higher the temperature and the lower the vapor pressure the higher is the average chain length. After a heating period of about 20 h an equilibrium distribution of the formed polyphosphate chains is reached. By mixing of different relative proportions of monobasic phosphates with dibasic phosphates the range of chain length of the formed product can be influenced. From the linear condensed phosphates only oligophosphates with a chain length of 2 to 5 can be synthesized as homogeneous, crystalline substances (Van Wazer 1958).

Besides the mainly formed linear molecules these chemical mixtures of polyphosphates also contain about 10 % of cyclic metaphosphates, predominantly trimetaphosphate (Thilo 1965). The degree of metaphosphate formation can be reduced by rapid cooling of the melting mass. The longer the process of cooling continues the higher is the amount of metaphosphates. An almost pure preparation of trimetaphosphate is obtained by very slowly cooling down of the melting mass as well as by heating of the monobasic phosphates to a temperature of only about 600 °C. The existence of cyclic condensed phosphates up to a condensation degree of 15 has been shown (Van Wazer 1958; Van Wazer and Karl-Kroupa 1958; Thilo and Schülke 1965), whereas the mono- and the dimetaphosphate probably do not exist (Van Wazer 1958; Thilo 1959, 1965).

In principle, it is possible that also the third hydroxy group of an intermediate phosphate residue might be linked by an acid anhydride bond. Actually, those branching points have been found to a small degree in chemically synthesized samples of Kurrol's and Graham's salts (Van Wazer and Holst 1950; Strauss and Treitler 1955a; Thilo and Sonntag 1957). They will be formed when the ratio between cations and phosphorus is lower than 1 and will occur at about 1 per 1000 residues (Strauss and Smith 1953; Strauss and Treitler 1955a,b; Thilo 1965). These molecules are called branched phosphates or ultraphosphates.

Branched phosphates have not been detected in biological material, whereas linear polyphosphates were found in almost all groups of organisms (for reviews, see Kulaev 1979; Kulaev and Vagabov 1983; Kornberg 1995). Cyclic compounds ($n = 3-8$) have been found in different algae (Niemeyer and Richter 1969; Niemeyer 1976) and higher plant species (Niemeyer 1975).

2.2
Radiolabeled Polyphosphates

For many investigations the use of radiolabeled polyphosphate is the only way to obtain results at all. By adding radioactive orthophosphate or monohydrogen phosphates to the primary products the corresponding radiolabeled polyphosphates can be synthesized. The most important advantage in using the radioactive isotope P-32 is the extreme sensitivity. The disadvantage is the risk of high radiation exposure during synthesis of labeled polyphosphate and high costs. Despite the fact that high specific radioactivity can be obtained with large amounts of P-32 (usually up to a few millicuries), the radiolabeled polyphosphates are efficient for 3 months at best due to the short half-life of P-32.

The substitution of P-32 by P-33, which is now commercially available, has emerged as an alternative to diminish considerably the problem of radiation exposure because the energy of β-rays emitted by the P-33 nuclide is only 1/7 that of P-32. Another advantage is the longer half-life, which is almost twice as long as that of P-32. On the other hand, the lower energy of P-33 causes a lower sensitivity of these assays and P-33 labeled compounds are considerably more expensive than their P-32 labeled counterparts to date.

2.3
Enzymatic Synthesis

More recently, a number of polyphosphate-synthesizing enzymes (polyphosphate kinases) have been purified to homogeneity from different microorganisms and some of the genes encoding these enzymes have been cloned, sequenced and expressed in *Escherichia coli* (Robinson et al. 1987; Ahn and Kornberg 1990; Akiyama et al. 1992; Kato et al. 1993; Tinsley et al. 1993; Tinsley and Gotschlich 1995; Tzeng and Kornberg 1998). The recombinant enzymes offer a superior experimental tool for the synthesis of polyphosphates, especially of radiolabeled substances under controlled conditions which are not so hazardous regarding radioactive exposure. The amounts of radioactivity necessary for the synthesis are considerably lower (10–50 μCi is sufficient). By using P-32 γ-ATP and the polyphosphate kinase, which catalyzes the processive synthesis of polyphosphates from ATP, a highly labeled polyphosphate molecule can be obtained. The use of polyphosphate kinase from *E. coli* leads to relatively homogeneous high-molecular polyphosphates of about 750 phosphate residues (Kornberg 1995). Fractions of smaller homogeneous polyphosphates can be obtained by partial alkaline or acidic hydrolysis, or by enzymatic degradation (Pepin et al. 1986; Robinson et al. 1987; Schuddemat et al. 1989; Pisoni and Lindley 1992; Kumble and Kornberg 1996), followed by a subsequent isolation of the fragments (see Sect. 5).

2.4
Synthesis by Living Cells or Organisms

The production of polyphosphates by living cells or organisms is a further method to synthesize polyphosphates. To obtain radiolabeled polymers radioactive orthophosphate is added to the cell culture medium. After cultivation the cells are harvested and the polyphosphates containing the incorporated radiophosphate are isolated. Both the yield and the size of the polyphosphates synthesized depend on several factors. Usually a higher degree and rate of incorporation can be reached when the cells are grown under starvation conditions, e.g. in phosphate-depleted medium or minimal phosphate medium, and then the (radioactive) phosphate is added (Liss and Langen 1962) or under nutritional limitations or stress conditions (Rao and Kornberg 1996; Rao et al. 1998; see also Chapt. 9). The isolation of polyphosphates from cells or tissues is relatively time-consuming. New extraction methods promise to allow a more simple and effective extraction (see Chap. 12). By choosing organisms, such as yeast, bacteria or algae, which are able to accumulate large amounts of polyphosphates, in vivo synthesis has been shown to be a usable alternative. Furthermore, this approach has often been used to study physiological changes in polyphosphate metabolism, e.g changes in concentration, size and distribution between different polyphosphate pools within the cells.

3
Isolation of Polyphosphates from Cells and Tissues

Inorganic polyphosphates have been reported to be nonrandomly distributed within the cell and may differently associate with several other compounds like proteins, nucleic acids, polysaccharides or polyhydroxybutyrate (Kulaev 1979; Kulaev and Vagabov 1983; Reusch and Sadoff 1988; Wood and Clark 1988). Furthermore, their solubility depends on the chain length as well as on the kind and concentration of chelating cations (Wood and Clark 1988). Moreover, polyphosphate turnover is very dynamic. After the breakdown of the physiological state of the cells by destroying their integrity, changes in size, in concentration, or rearrangements within different pools of polyphosphates may occur. Thus it is difficult to extract polyphosphates completely and to obtain them in their native state. To date, the only exact way to determine concentration, size and distribution of polyphosphates in vivo is nuclear magnetic resonance spectroscopy (see Chap. 13).

3.1
Isolation Procedures

A number of procedures for the isolation of polyphosphates from biological material has been developed. In the past most of these methods included drastic extraction steps like strong acid or alkaline solutions or higher temperature which led to a partial degradation of polyphosphate. MacFarlane (1936) first showed that in yeast two different polyphosphate fractions exist: an acid-soluble fraction, which contains the low molecular weight polyphosphates, and an acid-insoluble fraction, in which the polyphosphates with longer chain length are present. In the acid-soluble fraction orthophosphate, pyrophosphate, nucleosides, nucleotides and sugar phosphates are included, whereas the acid insoluble fraction contains the longer chain polyphosphates as well as nucleic acids, proteins and phospholipids. Since then the extraction of these two polyphosphate fractions as the first purification step has been used by a number of researchers (for reviews see Kulaev 1979; Kulaev and Vagabov 1983). In the 1950s and 1960s several different extraction procedures were tested, especially by Langen and coworkers (Langen 1958, 1961; Langen and Liss 1958), Kulaev and coworkers (Belozerski and Kulaev 1957; Kulaev et al. 1966; see also Kulaev 1979) and Harold (1963, 1966). They found varying results for different extraction procedures applied to the same organism. Moreover, it was shown that not every extraction procedure was suitable for all sources. For the extraction of polyphosphates from yeast Langen and Liss (1958) developed a five-step method which was widely used in the following years. They obtained four fractions which differed in size and metabolic state. The first step is the abovementioned extraction by acid (1 % trichloric acid) and the separation into an acid-soluble fraction and an acid-insoluble fraction. The following steps of this procedure are the extraction by highly concentrated salt solution (saturated sodium perchlorate solution), weak alkaline solution (NaOH, pH 9–10) and strong alkaline solution (1 N NaOH at 0 °C for separation of nucleic acids and subsequently 0.1 N NaOH at 20 °C).

Clark et al. (1986) showed that the methods of Langen and Liss (1958) as well as that of Harold (1963) led to a partial degradation of the isolated polyphosphates. Moreover, an incomplete isolation resulted when these methods were applied to *Propionibacterium shermanii*. An incomplete isolation after the Langen/Liss method was also found with other microorganisms (Kulaev and Vagabov 1983) and the fungus *Neurospora crassa* (Chernysheva et al. 1971). However, in contrast to this results, Langen and Liss found by applying their extraction procedure on yeast an almost complete extraction of total phosphate. Only irrelevant amounts remained in the final pellet after the last step of extraction. This shows again that the selection of the appropriate isolation method strongly depends on the source of the biological material.

The problem of degradation of polyphosphates can be avoided by using phenol/chloroform extraction as used for isolation of nucleic acids. This method was adapted for polyphosphates by Clark et al. (1986). Their studies showed that in contrast to the traditionally used procedure this extraction method caused no degradation of the polyphosphates. The evidence for this fact was given by addition of radioactive polyphosphate as a control to the cell suspensions before the extraction procedure was started. After completing the extraction the obtained fractions were checked by gel electrophoresis and no degradation of the radioactive polyphosphate could be observed. In the procedure of Clark et al. (1986) the proteins are cleaved by protease digestion and removed by phenol/chloroform extraction. The nucleic acids in the aqueous phase are degraded by nucleases and the polyphosphates are separated from the nucleotides by ethanol precipitation. In recent years this method has been the best known and most widely used procedure for isolation of polyphosphates. This basic procedure was also applied with small modifications to extraction of polyphosphates from a variety of eukaryotic cells (Schuddemat et al. 1989; Cowling and Birnboim 1994; Kumble and Kornberg 1995; Leitão et al. 1995; Lorenz et al. 1997a–c). The remaining amount of total phosphate which was not extracted by the Clark procedure was negligible, as shown for the different cell types. Moreover, the addition of the phosphatase inhibitor fluoride (Rao et al. 1985) and the presence of cold trichloric acid during the harvesting of cells (Clark et al. 1986) have been reported to prevent hydrolysis of polyphosphate by its enzymes.

Recently, two novel extraction methods have been developed which are described in detail in Chapter 12 (this Vol.). One follows a similar procedure for isolation of RNA by glass milk and was developed by Ault-Riché et al.

(1998). It extracts polyphosphates from *E. coli* almost completely and nearly all of the polyphosphate was found in a high molecular weight fraction ($n \approx 750-800$). By this method only polyphosphates with a chain length higher than 60 are completely isolated, whereas low molecular weight polyphosphates as well as nucleotides do not bind to the glass milk. In other organisms smaller physiological relevant fractions exist, which are not included in this extraction method. The second new method requires feeding of the cells in a medium containing P-32-orthophosphate. After the breakdown of the cells the polyphosphate is immobilized at filters and after different washing and incubation steps the polyphosphate is degraded by specific action of an exopolyphosphatase, and the remaining immobilized phosphate background as well as the released orthophosphate is measured.

4
Analysis of Polyphosphates and Measuring of Enzyme Activities

The accuracy of the analysis of polyphosphates as well as of related enzyme activities has been a problem for a long time. Most of the methods used are based on unspecific interactions of polyphosphates with indicator substances which usually benefit from electrostatic interactions between the polyphosphates and the indicator. The estimation of polyphosphates as inorganic orthophosphate after complete hydrolysis has also been applied. The accuracy of these basic methods depends on the absolute amounts of polyphosphates and on their relation to troublesome substances like other polyanions or competitive cations. These substances interfere with the electrostatic interactions. A high concentration of several phosphate labile substances in cells is also critical. For these reasons the error in the measurement of absolute values can be considerably large. In contrast to higher plants and animals, in most lower organisms the yield of polyphosphates is usually higher. This has a favorable effect on the relation to the amount of phosphate labile substances or other disturbing molecules. Therefore, the analysis of fractions of higher polyphosphate content can be reliable when these customary more or less unspecific methods are applied. However, the lower the polyphosphate content the more other compounds disturb the analysis and the higher is the error of the measurement. Thus, many earlier reports should be considered with care, especially when samples with only roughly purified polyphosphates or without effective prior separation have been measured. In contrast to the determination of absolute polyphosphate concentrations, experimental errors will be low when the change in the measurable signal over time, like enzymatic hydrolysis of polyphosphates, is analyzed. By using specific enzymes and modern separation techniques more exact methods are now available.

4.1
Basic Dyes

The use of basic dyes like toluidine blue is the oldest method for the detection of polyphosphates and has been applied for a long time. Polyphosphate shifts the absorption spectrum of toluidine blue towards shorter wavelengths. Therefore this phenomenon is called "metachromatic reaction". A weakly acidic solution of toluidine blue is blue-coloured and exhibits an absorption maximum at 630 nm. In the presence of polyphosphates the maximum is at 530 nm and the color of the solution turns to pink. The calculation of the absorbance ratio between 530 and 630 nm for a set of increasing polyphosphate concentrations allows the quantification of polyphosphates with the aid of a calibration curve. Although this method is simple, practicable and useful for some purposes, its weighty disadvantages result from the fact that (1) short-chain polyphosphates ($n < 10$) give only a weak or no metachromatic reaction and thus cannot be detected by this method, and (2) the interaction between the dye and polyphosphate is influenced by several compounds competing with the binding to polyphosphate, such as polyanions (e.g. nucleic acids or glucosaminoglycans), cations and proteins, and depends on several conditions, such as ionic strength and pH. Again, this may lead to an inaccurate quantification of polyphosphates.

The toluidine blue assay is a practicable method for determination of exo-polyphosphatase activity in samples from different origin (Lorenz 1997). This method can be easily carried out and evaluated. Due to the decreasing effect of magnesium ions, proteins or higher ionic strength on the metachromatic reaction the assay requires relatively high substrate concentrations. Thus, high enzymatic activity or long incubation times are necessary to obtain measurable changes in the metachromatic reaction.

4.2
Determination of Orthophosphate Using Ammonium Molybdate

For determination of polyphosphates these molecules are often hydrolyzed by acid and the released orthophosphate is then measured by the ammonium molybdate reaction – a very common and a long-known detection method for inorganic orthophosphate (Fiske and Subbarow 1925). The orthophosphate–molybdate complex formed after reduction can be measured colorimetrically. This method is sensitive, and the results can be influenced by many factors. So it is essential to clean all glassware very carefully with concentrated acid because of the high amounts of phosphate in commercial detergents. Serious pitfalls for this method arise from the fact that a large number of labile organic phosphate compounds are present in cell extracts and that chemically synthesized polyphosphates are contaminated with relatively high amounts of orthophosphate and pyrophosphate. These high concentrations of orthophosphate and pyrophosphate (which is degraded to orthophosphate by the action of widely distributed pyrophosphatase activities) can lead to such a high background that it will render the exact measurement almost impossible. This requires a separation of polyphosphates from orthophosphate, pyrophosphate as well as from labile organic phosphate compounds.

Polyphosphates belong to the acid labile phosphate-containing substances, i.e. the incubation of these molecules at pH 1 and 100 °C leads to a complete degradation within a period of 7 min, whereas nucleic acids and sugar phosphates are more stable. The amount of the latter substances, hydrolyzed under these conditions within the first 7 min, is nearly identical to that hydrolyzed within the period between 7 and 30 min. Therefore the amount of polyphosphates present in extracts from biological material was calculated by subtracting the amount of the orthophosphate released between 7 and 30 min (resembling the amount of orthophosphate cleaved off from more stable phosphate compounds) from the value measured between 0 and 7 min (the amount of orthophosphate resulting from the complete hydrolysis of acid labile phosphates and the quantity released from the more stable phosphate-containing substances) (Kulaev 1979; Kulaev and Vagabov 1983). In some cases the value 7 to 30 min is insignificantly small so that it can be ignored (Langen 1961). Another way of estimating polyphosphates is their selective precipitation with Ba^{2+} before the colorimetrical analysis with ammonium molybdate.

4.3
Radioassays

Radioactive substances have been used in a multitude of different cases: detection and measurement of polyphosphate-dependent enzyme activities, control of purification steps, investigation of reaction mechanisms and enzyme specificity, physiological studies like turnover measurement, changes in dependence of variations in the metabolism, environmental and stress conditions, separation of polyphosphates, etc. In the past the use of radioactive substrates was the only alternative to solve many experimental problems because of its unrivalled sensitivity. The use of radiolabeled substrates and subsequent measurement of radioactivity of the acid-soluble or -insoluble polyphosphate fraction has been the most appropriate method of detection of enzyme activities for a long time. To date, radioactive polyphosphate assays are still widely used for many different purposes and in combination with several other techniques.

4.4
Fura-2 Method

The novel fura-2 method has been developed recently (Lorenz et al. 1997c). It is an indirect system based on the Mn^{2+}-induced quenching of the fluorescence of the calcium indicator fura-2. Polyphosphates compete with the metal ion indicator fura-2 for the binding of the metal ions and cause a change in the fluorescence signal. This method has some advantages compared to conventional procedures for detection of polyphosphates. It can be applied for determination of pyrophosphate, tripolyphosphate and other short-chain polyphosphates not detectable by toluidine blue. Since the influence of orthophosphate on the quenching effect of Mn^{2+} on the fura-2 fluorescence is negligible at concentrations of up to 10 mM this method can be used for measurement of enzyme activities like pyrophosphatases and exopolyphosphatases catalyzing the cleavage of orthophosphate from polyphosphates. Thus, contamination of orthophosphate in samples is not as problematic as it is in the ammonium-molybdate assay. However, the main advantage of this method compared to the procedures described above is the possibility to monitor the degradation of pyrophosphate or polyphosphate continuously, as shown for the cleavage of pyrophosphate by the action of a pyrophosphatase (Fig. 1). The decrease in fluorescence reflects the enzymatic turnover of pyrophosphate. When the substrate has been degraded completely the fluorescence remains constant.

This novel method should be a useful tool for the measurement of pyrophosphate and polyphosphates in environmental or food samples. It should also be helpful in the determination of pyrophosphate content or pyrophosphatase activity in medical diagnosis. Besides the numerous benefits over other methods described above, this technique allows simultaneous processing of large quantities of samples (e.g. using an ELISA plate reader). Further-

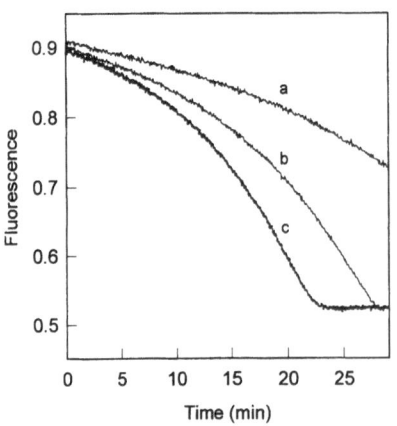

Fig. 1. Degradation of pyrophosphate continuously monitored by measuring fura-2 fluorescence at 360 nm. Assays were performed in a quartz cuvette, which contained 80 μM pyrophosphate, 80 μM $MgCl_2$, 0.5 μM fura-2, 0.5 μM $MnCl_2$ in a buffer of 30 mM Tris/HCl, pH 7.2, 30 mM $NaHCO_3$, 150 mM NaCl and 12 mM KCl. Reactions were started by addition of (a) 50, (b) 75 and (c) 100 ng pyrophosphatase (final volume of reaction mixture, 2 ml). Incubations were performed at 22 °C for 30 min

more, it is an alternative for the detection of polyphosphates after chromatographic separation (see Sect. 5.3).

4.5
Enzymatic Assays

The use of polyphosphate-dependent enzymes allows the most accurate estimation of polyphosphates. The most important advantage is the specificity of enzymatic reactions, i.e. only polyphosphates are able to react in the majority of cases, whereas those substances disturbing other procedures such as polyanions, nucleotides/nucleosides or sugar phosphates as well as other phosphate-containing substances or orthophosphate contamination usually are not registered by the enzymatic reactions.

Polyphosphate concentrations have been determined by application of a polyphosphate glucokinase from *P. shermanii* (Clark et al. 1986). The formed glucose-6-phosphate was converted by glucose-6-phosphate-dehydrogenase and the increasing concentration of NADPH was measured. As already mentioned, two novel methods using the enzymes polyphosphate kinase from *E. coli* and an exopolyphosphatase from *Saccharomyces cerevisiae* are described in detail in Chapter 12 (this Vol.)

Several indirect enzymatic assays have been used for measurement of polyphosphate-dependent enzyme activities:

- Polyphosphate kinase (polyphosphate synthesis): a photometric assay with pyruvate kinase reaction as ATP-regenerating system and lactate dehydrogenase as indicator reaction (measurement of NADH consumption; Robinson et al. 1986)
- Polyphosphate kinase (reverse reaction): photometric assay, in which the formed ATP was estimated by the combined action of hexokinase and glucose-6-phosphate dehydrogenase and measurement of NADPH formation (Robinson et al. 1987)
- Polyphosphate-AMP-phosphotransferase activity: measurement of ATP, ADP and AMP in combination with luciferase and luciferase/adenylate kinase, respectively (Van Groenestijn et al. 1987).

Moreover, enzymatic reactions were measured using radioactive substances in combination with different separation techniques such as thin-layer chromatography or gel electrophoresis.

One novel method to exploit the specificity of enzymes for the determination of polyphosphates as well as for measuring exopolyphosphatase activities should be introduced here for the first time. Figure 2 shows the indicator system used. This is a commercially available system, which can be used in principle as a detection method for all kinds of reactions in which orthophosphate is released or consumed. This assay has been applied for the measurement of different enzyme activities: ATPase, pyrophosphatase as well as several phosphatases and kinases (Haugland 1996).

We applied this method for the determination of exopolyphosphatase activities as well as for the determination of polyphosphates. The degradation of tetrapolyphosphate by an exopolyphosphatase from *S. cerevisiae* (a homogeneous enzyme preparation kindly provided by Arthur Kornberg) is shown in Fig. 3. As pyrophosphate is known to be the final product of this reaction (Wurst and Kornberg 1994) half the tetrapolyphosphate is converted to orthophosphate after the action of exopolyphosphatase. An additional treatment with pyrophosphatase then leads to a complete degradation of the remaining pyrophosphate. Taking the end point absorbance after both enzymatic reactions, a linear dependence between the tetrapolyphosphate concentration and the resulting absorbance is obtained. The experimentally obtained amounts of tetrapolyphosphate determined after complete enzymatic hydrolysis and by comparison with a calibration curve of orthophosphate excellently agree with the theoretical values (Table 1). Thus, by combination of both enzymes the concentration of low molecular weight polyphosphates can be estimated. For longer polyphosphate chains ($n \geq 20$) only the exopolyphosphatase activity is necessary as a prerequisite, because the remaining pyrophosphate is negligible. This method is also applicable to the estimation of exopolyphosphatase activities from extracts or partially purified fractions. For fractions devoid of unspecific phosphatase activities both the initial velocity and the endpoint of the reaction can be used to measure the turnover rate. In contrast, for samples containing unspecific phosphatase activity only the initial velocity can be used for the evaluation because

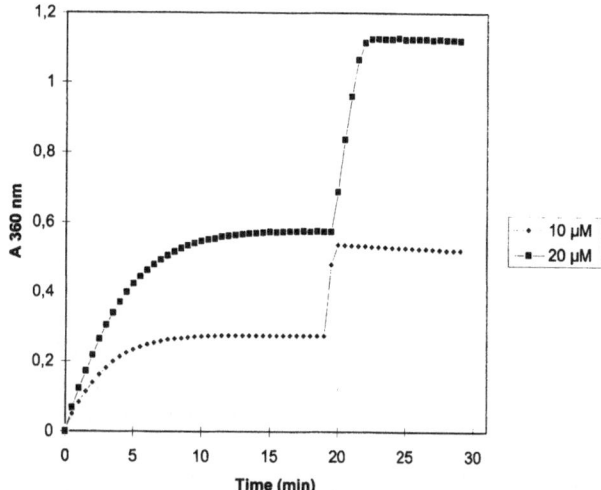

Fig. 3. Measurement of degradation of tetrapolyphosphate. The reaction was done in an assay volume of 300 µl consisting of 200 µM 2-amino-6-mercapto-7-methylpurine ribonucleoside, 10 or 20 µM tetrapolyphosphate, and 0.5 units purine nucleoside phosphorylase in a buffer of 20 mM Tris/HCl, pH 7.5, 1 mM MgCl$_2$. Phosphate contamination was removed by preincubation at 22 °C for 10 min. This background was corrected, the reaction was started by addition of exo-polyphosphatase, and absorbance at 360 nm was recorded. After termination of the reaction pyrophosphatase from yeast (Boehringer Mannheim) was added to cleave pyrophosphate and complete the degradation

Table 1. Determination of tetrapolyphosphate (polyP$_4$). The experiment was carried out as described in Fig. 3. Reaction of the assay with indicated amounts of tetrapolyphosphate was started by addition of exopolyphosphatase. After termination of the reaction, pyrophosphatase was added. Differences in absorption at 360 nm were estimated and amount of released ortho-phosphate was calculated by comparison with values of a calibration curve of orthophosphate. Results are means (± SD) of four independent experiments

nmol polyP$_4$/assay	nmol P$_i$ (theoretical value after exopoly-phosphatase reaction)	nmol P$_i$ (experimental value after exo-polyphosphatase reaction)	nmol P$_i$ (theoretical value after pyrophos-phatase reaction)	nmol P$_i$ (experimental value after pyro-phosphatase reaction)
1.5	3	3.35 ± 0.5	6	6.3 ± 0.6
3	6	6.16 ± 0.7	12	12.1 ± 1.1
6	12	12.06 ± 1.5	24	23.4 ± 2.6

unspecific phosphatases will cleave off orthophosphate from the end product of the indicator reaction (ribose-1-phosphate). Thus, a reaction cycle is created which leads to a complete consumption of the MESG substrate independent of the present exopolyphosphatase activity.

4.6
Detection of Polyphosphates in Whole Cells and Tissues

The staining of cells or tissues using basic dyes such as toluidine blue, neutral red or methylene blue has been used for the discovery of polyphosphate-containing structures in many different species (for review see Kulaev 1979). However, polyphosphates cannot be clearly distinguished from other polyanions in this way. Nucleic acids or proteins are often associated with polyphosphates in complex structures – so-called volutin granula. Although this cytochemical detection of polyphosphates in most cases results from the staining of polyphosphates it was shown that some positive stained structures did not contain polyphosphates (Martinez 1963). Thus, results obtained by the use of the staining method have to be considered with caution.

As shown by Allan and Miller (1980) 4'6'-diamino-2-phenylindole (DAPI) – a prominent staining reagent for nucleic acids – can be applied for detection of polyphosphates, too. In contrast to nucleic acids and other polyanions polyphosphates are able to shift the spectrum of the dye in a specific way. Nuclear magnetic resonance spectroscopy is another method used for detection of polyphosphates in living cells (see Chap. 13). Furthermore, by application of electron microscopy in combination with X-ray dispersion analysis polyphosphate-containing structures can be seen as electron-dense areas (Coleman et al. 1972; Doonan et al. 1979; Baxter and Jensen 1980). This method also allows us to determine the nature of the cations interacting with the polyphosphates in the granula.

5
Separation of Polyphosphates and Determination of Their Chain Lengths

As mentioned above, chemically synthesized condensed phosphates as well as those produced by cells are heterogeneous mixtures of molecules with varying chain length. However, for a number of experiments it is necessary to separate the mixtures into more defined fractions or to remove orthophosphate, pyrophosphate or the low molecular weight polyphosphates from the samples. In addition, separation methods are necessary for the isolation and identification of polyphosphates from cells and tissues, for the estimation of their concentration, for the determination of enzymatic activities, or for the investigation of changes in polyphosphate metabolism. The knowledge of the chain length or the average chain length of polyphosphate samples is important for many purposes, e.g. for determination of K_m, enzyme specificity, turnover of polyphosphates, and metabolic changes under varying conditions.

5.1
Thin-Layer Chromatography

Thin-layer chromatography (TLC) was one of the first methods applied for separation and estimation of the chain length of condensed phosphates. Molecules up to a chain length of eight can be separated by paper or thin-layer chromatography. Although their separation on paper is most effective (Grunze and Thilo 1955) thin-layer plates or folia (e.g. cellulose, polyethylenimine cellulose) were used in most cases. A two-dimensional chromatography allows the separation of linear and cyclic compounds up to a condensation grade of about eight. The first run is done with an acid solvent whereas in the second a basic solvent is used as the mobile phase. In the first dimension the linear compounds migrate faster, whereas the corresponding cyclic substances in the second dimension are more mobile (see Chap. 5; Fig. 1).

The applicability of the thin-layer chromatography is considerably restricted because of its limitation for shorter polyphosphate chains. Nevertheless, it is a practicable and fast method for detection of polyphosphate-dependent enzymes, predominantly by using radiolabeled substrates (Robinson et al. 1984; Pick et al. 1990; Keasling et al. 1993; Booth and Guidotti 1995; Lorenz et al. 1997b).

5.2
Ion-Exchange Chromatography

Anion-exchange chromatography has been reported to be successful in separating linear and cyclic condensed phosphates. Several examples are described where anion-exchange resins were used, especially polystyrene-based resins such as "Dowex". This technique allows a quantitative separation of linear polyphosphates of up to a chain length of eight and a clear distinction of up to 12 residues (Matsuhashi 1963). An effective separation of higher molecular polyphosphates has not been described with such common materials and usual chromatography.

5.3
High-Performance Liquid Chromatography

Conventional anion-exchange chromatography is relatively laborious and time-consuming. Moreover, this method is limited to the separation of condensed phosphates with low molecular weights. The use of high-performance liquid chromatography (HPLC) offers a much more effective and more sensitive procedure for separation and determination of polyphosphates. The separation of linear polyphosphates and metaphosphates by HPLC is a relatively new application and was first described only

15–20 years ago (Yamaguchi et al. 1979; Baba et al. 1984; Biggs et al. 1984; Brazell et al. 1984; Chester and Smith 1984). The separation quality has been improved within the last few years (Baba et al. 1985a–c) so that this technique can also be used for determination of enzymatic degradation of pyrophosphate and tripolyphosphate (Yoza et al. 1985, 1997), for the estimation of their concentration in food (Matsunaga et al. 1988, 1990) or environmental samples (Muessik-Zefika et al. 1994). Linear polyphosphates up to a chain length of about 35 can be exactly separated (Baba et al. 1985a–c). The separation is usually performed by application of ion-exchange matrices. In addition, the separation by ion pair chromatography on a reversed phase column (Biggs et al. 1984) and the fractionation of higher polyphosphates on a size exclusion column (Muessik-Zefika et al. 1994) have been successful. The detection of condensed phosphates following chromatographic separation is not easy, since they cannot be monitored directly in the UV or visible light range because the molecules are lacking chromophore groups. Most commonly used for detection is a highly sophisticated automatic flow injection system with post-column acid hydrolysis of polyphosphate at temperatures of about 140 °C followed by a detection of the formed orthophosphate by ammonium molybdate reaction. Another possibility is the use of a post-column reaction with Fe^{3+} and salicylic acid under UV detection. Other detection methods include the monitoring by special technologies such as plasma atom emission spectroscopy (Biggs et al. 1984), phosphorus selective dual flame photometry (Chester and Smith 1984), or the application of suppressor conductivity detectors (Stover et al. 1994). The sensitivity of these techniques is high; the detection limit is in the nanogram range: plasma atom emission spectroscopy (200 ng phosphorus), dual flame photometry (100 ng for orthophosphate as sodium salt, slightly higher for pyrophosphate and tripolyphosphate), flow injection system with ammonium molybdate reaction (about 100 ng for orthophosphate as sodium salt) and even about 1 ng using suppressor conductivity detector (for oligophosphates).

Obviously, the greatest handicap of the use of HPLC for separations of condensed phosphates is that it requires special equipment such as a column oven for optimal separation as well as special detection systems which are usually not part of the standard laboratory equipment. A new alternative in detection of polyphosphates by HPLC is offered by the indirect methods using 2-(4-pyridylazo)resorcinol and yttrium ions developed originally for detection of inositol polyphosphates by Mayr (1988) and detection in the visible light range. The application of the method with the fluorescent dye fura-2 and manganese ions (see Sect. 4.4) and detection by a fluorescence detector are also possible. In both cases polyphosphates compete with the cation-sensitive dye for the binding of the cation thereby causing a change in the resulting signal.

5.4
Gel Electrophoresis

Today the most effective, powerful and widely used method for separation of polyphosphates is gel electrophoresis. Following the method for separation of nucleic acids by polyacrylamide gel electrophoresis (PAGE) with high urea concentrations published by Maxam and Gilbert (1977), Robinson et al. (1984) adapted this procedure for polyphosphates. The addition of urea to the gel is required for a clear separation of different polyphosphates into distinct bands. The detection of polyphosphates is achieved by staining with toluidine blue or by autoradiography if radioactive polyphosphate is used. As stated above, the detection of orthophosphate or pyrophosphate by staining with toluidine blue is not possible. Tripoly- and tetrapolyphosphate can be made visible with this dye, although they diffuse rapidly out of the gel. Polyphosphates with higher chain lengths are tightly enough fixed inside the gel matrix. Only a few micrograms of nonradioactive polyphosphate mixtures are necessary for this method. The limit of detection for a single band is even lower. The electrophoresis in high percentage polyacrylamide gels (15–20%) allows us to resolve polyphosphates up to a chain length of about 100 into distinct bands. The separation of polyphosphates with higher chain lengths than 100 can be reached by using low percentage polyacrylamide gels for polyphosphates up to a chain length of about 800 or agarose gels for long-chain polyphosphates ($n \approx 500–1700$; Clark and Wood 1987). These authors showed that the migration of the polyphosphate is a function of the molecular mass and the pore size. A plot of the logarithms of the chain length of polyphosphates versus the R_f for the same concentrated polyacrylamide gel is linear over a limited range. By using different concentrated gels or by using different homogeneous molecular weight markers the chain length of polyphosphates up to about 1700 can be estimated. Furthermore, the combination of this method with the glucokinase reaction is very accurate in the determination of chain lengths up to 450 phosphate residues (Pepin et al. 1986).

Because of its efficiency, power and capacity PAGE was used for investigation of many questions: e.g. determination of enzyme activity, examination of reaction mechanism and properties of enzymes, determination of polyphosphates isolated from cells and tissues, evidence of different polyphosphate pools in cells, their dynamic changes, and changes of enzyme activity and polyphosphate content under different conditions, e.g., energy source, development and ageing, stress or apoptosis (Clark et al. 1986; Pepin et al. 1986; Robinson et al. 1987; Schuddemat et al. 1989; Lorenz et al. 1994, 1995, 1997b; Leitão et al. 1995). Moreover, PAGE has also been applied for isolation of polyphosphates of limited size (Pepin and Wood 1986; Lorenz et al. 1994).

The application of a commercially available system for preparative gel electrophoresis called "Prep cell" from BioRad (Hercules, USA) allows a very effective separation of polyphosphates (Lorenz, unpubl. results). This system has an additional cycle for collecting the separated substances, when the molecules leave the gel. The enormous advantage of this system is that it allows a simple collection of polyphosphate fractions or a further online analysis of the polyphosphate sample.

5.5
Gel Filtration

Gel filtration for separation or isolation of polyphosphates has also been applied. As a matter of principle all common materials are suitable. The pore size of the material should be chosen according to the molecular range. The quality of the separation depends on the heterogeneity of the sample and on the length of the column. For a number of other methods, e.g. for polyacrylamide gel electrophoresis, toluidine blue assay and fura-2 assay, a low ion concentration in the samples is necessary. Usually applied procedures for sample concentration such as precipitation with ethanol or aceton and lyophilization do not lower the ionic strength; in contrast, the ion concentration increases. Therefore the separation of polyphosphates by gel filtration with water as eluent has been tested (Lorenz 1997). Since polyphosphates are negatively charged molecules, they should not bind to commonly used gel filtration media like Sephadex (Pharmacia) or similar matrices when a low ionic strength is used. It was found that relatively high amounts of polyphosphates can be loaded, e.g., on a 90×1 cm column up to 10 mg at a maximal concentration of 20 mg/ml. Higher concentrated polyphosphate solutions are too viscous and lead to inefficient separation. The quality of the separation of a polyphosphate mixture of an average chain length of 15 is shown in Fig. 4. The separated fractions were analyzed by polyacrylamide gel electrophoresis. As shown, homogeneous fractions of polyphosphates with a chain length varying in the range of about 30–40 were obtained after gel filtration. In comparison, the chain length of the original sample ranges from 1 to about 140.

In summary, gel filtration of polyphosphates with pure water as eluent is a practicable and effective method for separation of polyphosphates with concomitant removal of salt and is profitable for further investigations after concentration of polyphosphate fractions, e.g. in enzyme assays. Gel filtration is also a simple and effective method for separation of orthophosphate, pyrophosphate and oligophosphates from polyphosphate mixtures. A separation on Sephadex G-25, G-15 or G-10 (Pharmacia) or similar materials, e.g. columns designed for desalting of proteins or nucleic acids, is suitable.

Fig. 4. Analysis of chain length of polyphosphate fractions obtained after gel filtration on Sephadex G-50 (90 × 1 cm): 0.5 ml of a 20 mg/ ml solution of polyphosphate glass with an average chain length of 15 (Sigma) was loaded onto the column and the sample was eluted with distilled water. Aliquots (5 μl) from each 450-μl fraction collected were analyzed by gel electrophoresis in a 20 % poly-acrylamide/7-M urea gel. Estimation of chain length was done by comparing the migration of each fraction with defined markers

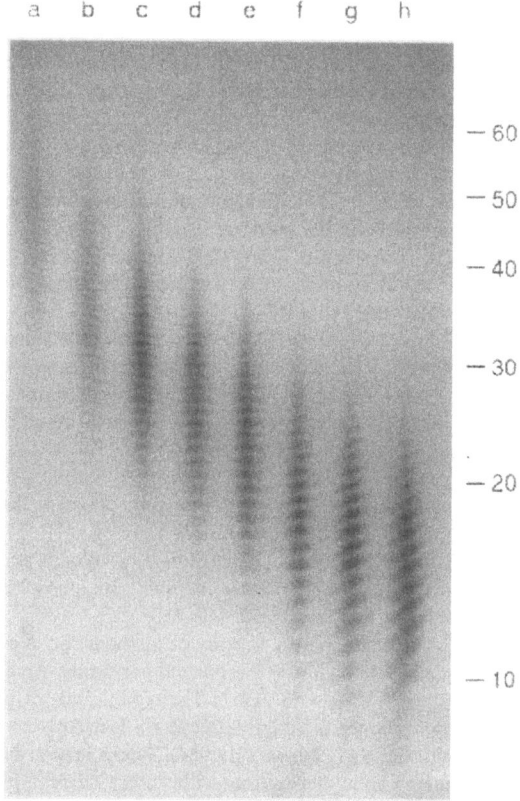

5.6
Other Methods for Determination of Chain Lengths

The end-point titration of the linear polyphosphates is based on the fact that intermediate hydroxy groups of the polyphosphate chain are strongly acidic in contrast to the terminal groups. By estimation of the consumption of bases referring to content of phosphates the chain length can be calculated (Van Wazer 1958; Liss and Langen 1960).

Beyond this analysis of viscosity has been used for determination of poly-phosphates. The basis of the method is that viscosity increases with increasing chain length. This method allows only an approximate estimation. Moreover, this method has been used for calculation of endopolyphosphatase activity (Ingelman and Malmgren 1948; Malmgren 1952). Further methods for determination of chain length are nuclear magnetic resonance spectroscopy (Chap. 13) or the measuring of polyphosphate–protein interaction by a nephelometer (Lyons and Siebenthal 1966). All these methods require relatively large amounts of substances. This strongly limits the applicability of these procedures. To date, they are not applied very often.

References

Ahn K, Kornberg A (1990) Polyphosphate kinase from *Escherichia coli*. Purification and demonstration of a phosphoenzyme intermediate. J Biol Chem 265: 11734–11739

Akiyama M, Crooke E, Kornberg A (1992) The polyphosphate kinase gene of *Escherichia coli*. Isolation and sequence of the *ppk* gene and membrane location of the protein. J Biol Chem 267: 22556–22561

Allan RA, Miller JJ (1984) Influence of S-adenosylmethionine on DAPI-induced fluorescence of polyphosphate in the yeast vacuole. Can J Microbiol 26: 912–920

Ault-Riché D, Fraley CD, Tzeng CM, Kornberg A (1998) Novel assay reveals multiple pathways regulating stress-induced accumulations of inorganic polyphosphate in *Escherichia coli*. J Bacteriol 180: 1841–1847

Baba Y, Yoza N, Ohashi S (1984) Simultanous determination of phosphate and phosphonate by flow injection analysis with a parallel detection system. J Chromatogr 295: 153–160

Baba Y, Yoza N, Ohashi S (1985a) Simultanous determination of phosphate and phosphonate by flow injection analysis and high-performance liquid chromatography with a series detection system. J. Chromatogr 318: 319–324

Baba Y, Yoza N, Ohashi S (1985b) Effect of column temperature on high-performance liquid chromatographic behaviour of inorganic polyphosphates. I. Isocratic ion-exchange chromatography. J Chromatogr 348: 27–37

Baba Y, Yoza N, Ohashi S (1985c) Effect of column temperature on high-performance liquid chromatographic behaviour of inorganic polyphosphates. II. Gradient ion-exchange chromatography. J Chromatogr 350: 119–125

Baxter M, Jensen T (1980) A study of methods for in situ X-ray energy dispersive analysis of polyphosphate bodies in *Plectonema boryanum*. Arch Microbiol 126: 213–215

Belozerski AN, Kulaev IS (1957) The significance of polyphosphates in the development of *Aspergillus niger*. Biokhimiya 22: 29 (in Russian)

Biggs WR, Gano JT, Brown RJ (1984) Determination of polyphosphate distribution by liquid chromatographic separation with direct current plasma-atomic emission spectrometric detection. Anal Chem 56: 2653–2657

Booth JW, Guidotti G (1995) An alleged yeast polyphosphate kinase is actually diadenosine-5',5'''-P^1,P^4-tetraphosphate α,β-phosphorylase. J Biol Chem 270: 19377–19382

Brazell RS, Holmberg RW, Moneyhun JH (1984) Application of high-performance liquid chromatography flow injection analysis for the determination of polyphosphoric acids in phosphorus smokes. J Chromatogr 290: 163–172

Chernysheva EK, Kritskii MS, Kulaev IS (1971) The determination of the degree of polymerization of different inorganic polyphosphate fractions from mycelia of *Neurospora crassa*. Biokhimiya 36: 138–142 (in Russian)

Chester TL, Smith CA (1984) Determination of inorganic phosphates in detergents by high-performance liquid chromatography on PRP-1 with phosphorus-selective detection. J Chromatogr 287: 447–451

Clark JE, Wood HG (1987) Preparation of standards and determination of sizes of long-chain polyphosphates by gel electrophoresis. Anal Biochem 161: 280–290

Clark JE, Beegen H, Wood HG (1986) Isolation of intact chains of polyphosphate from "*Propionibacterium shermanii*" grown on glucose or lactate. J Bacteriol 168: 1212–1219

Coleman JR, Nilsson JR, Warner RR, Batt P (1972) Qualitative and quantitative electron probe analysis of cytoplasmatic granules in *Tetrahymena pyriformis*. Exp Cell Res 74: 207–219

Cowling RT, Birnboim HC (1994) Incorporation of [^{32}P]orthophosphate into inorganic polyphosphates by human granulocytes and other human cell types. J Biol Chem 269: 9480–9485

Doonan BB, Crang RE, Jensen TE, Baxter M (1979) In situ X-ray energy dispersive microanalysis of polyphosphate bodies in *Aureobasidium pullulans*. J Ultrastruct Res 69: 232–238

Fiske CH, Subbarow Y (1925) The colorimetric determination of phosphorus. J Biol Chem 66: 375–400

Griffin JB, Davidian NM, Penniall R (1965) Studies of phosphorus metabolism by isolated nuclei. VII. Identification of polyphosphate as a product. J Biol Chem 240: 4427–4434

Grunze H, Thilo E (1955) Die Papierchromatographie der kondensierten Phosphate, 2nd edn. Akademie-Verlag, Berlin

Harold FM (1963) Inorganic polyphosphate of high molecular weight from Aerobacter aerogenes. J Bacteriol 86: 885–887

Harold FM (1966) Inorganic polyphosphates in biology: structure, metabolism and functions. Bacteriol Rev 30: 772–794

Haugland RP (1996) Handbook of fluorescent probes and research chemicals. 6th edn. Molecular Probes, Eugene, p 221

Ingelman B, Malmgren H (1948) Enzymatic breakdown of polymetaphosphate. Acta Chem Scand 2: 365–380

Kato J, Yamamoto T, Yamada K, Ohtake H (1993) Cloning, sequence and characterization of the polyphosphate kinase-encoding gene (ppk) of Klebsiella aerogenes. Gene 137: 237–242

Keasling JD, Bertsch L, Kornberg A (1993) Guanosine pentaphosphate phosphoydrolase of Escherichia coli is a long-chain exopolyphosphatase. Proc Natl Acad Sci USA 90: 7029–7033

Kornberg A (1995) Inorganic polyphosphate: toward making a forgotten polymer unforgettable. J Bacteriol 177: 491–496

Kulaev IS (1979) The biochemistry of inorganic polyphosphates. Wiley, New York

Kulaev IS, Vagabov VM (1983) Polyphosphate metabolism in microorganism. Adv Microbiol 24: 83–171

Kulaev IS, Krasheninnikov IA, Kokurina NK (1966) The localization of inorganic polyphosphates and nucleotides in mycelia of Neurospora crassa. Biokhimiya 31: 850 (in Russian)

Kumble KD, Kornberg A (1995) Inorganic polyphosphate in mammalian cells and tissues. J Biol Chem 270: 5818–5822

Kumble KD, Kornberg A (1996) Endopolyphosphatases for long chain inorganic polyphosphate in yeast and mammals. J Biol Chem 271: 27146–27151

Langen P (1958) Über die Polyphosphate in Ostsee-Algen. Acta Biol Med Ger 1: 368–372

Langen P (1961) Über Unterschiede im Gehalt an labilem Phosphat zwischen Rot- und Braunalgen. Pubbl Staz Zool Napoli 32: 130–133

Langen P, Liss E (1958) Über Bildung und Umsatz der Polyphosphate der Hefe. Biochem Z 330: 455–466

Leitão JM, Lorenz B, Bachinski N, Wilhelm C, Müller WEG, Schröder HC (1995) Osmotic-stress-induced synthesis and degradation of inorganic polyphosphates in the alga Phaeodactylum tricornutum. Mar Ecol Progr Ser 121: 279–288

Liss E, Langen P (1960) Über ein hochmolekulares Polyphosphat der Hefe. Biochem Z 333: 193–201

Liss E, Langen P (1962) Versuche zur Phosphat-Überkompensation in Hefezellen nach Phosphatverarmung. Arch Microbiol 41: 383–392

Lorenz B (1997) Optimisation of methods for determination of polyphosphate and polyphosphate-hydrolyzing enzyme activities in environmental samples. In: Müller WEG (ed) Modern aspects in monitoring of environmental pollution in the sea. Akademie gemeinnütziger Wissenschaften zu Erfurt, Mathematisch-Naturwissenschaftliche Klasse, Sitzungsberichte 8, pp 238–250

Lorenz B, Müller WEG, Kulaev IS, Schröder HC (1994) Purification and characterization of an exopolyphosphatase from Saccharomyces cerevisiae. J Biol Chem 269: 22198–22204

Lorenz B, Batel R, Bachinski N, Müller WEG, Schröder HC (1995) Purification and characterization of two exopolyphosphatases from the marine sponge Tethya lyncurium. Biochim Biophys Acta 1245: 17–28

Lorenz B, Leuck J, Köhl D, Müller WEG, Schröder HC (1997a) Anti-HIV-1 activity of inorganic polyphosphates. J Acquir Immune Defic Syndr Hum Retrovirol 14: 110–118

Lorenz B, Münkner J, Oliveira MP, Kuusksalu A, Leitão JM, Müller WEG, Schröder HC (1997b) Changes in metabolism of inorganic polyphosphate in rat tissues and human cells during development and apoptosis. Biochim Biophys Acta 1335: 51–60

Lorenz B, Münkner J, Oliveira MP, Leitão JM, Müller WEG, Schröder HC (1997c) A novel method for determination of inorganic polyphosphates using the fluorescent dye fura-2. Anal Biochem 246: 176–184

Lyons JW, Siebenthal CD (1996) On the binding of condensed phosphates by proteins. Biochim Biophys Acta 120: 174–176

MacFarlane M (1936) Phosphorylation in living cells. Biochem J 30: 1369

Malmgren H (1952) Enzymatic breakdown of polymetaphosphate. V. Purification and inhibition of the enzyme. Acta Scand 6: 16–26

Martinez RJ (1963) On the nature of the granules of the genus *Spirillum*. Arch Microbiol 44: 334–343

Matsuhashi M (1963) Die Trennung von Polyphosphaten durch Ionenaustauschchromatographie. Anwendung auf die Hefe-Polyphosphate. Hoppe Seilers Z Physiol Chem 333: 28–34

Matsunaga A, Yamamoto A, Mizukami E, Hayakawa K, Miyazaki M (1988) Determination of polyphosphates by high-performance liquid chromatography. Application to soft drinks. Eisei Kagaku 34: 70–74

Matsunaga A, Yamamoto A, Mizukami E, Kawasaki K, Oozumi T (1990) Determination of polyphosphates in foods by high-performance liquid chromatography. Nippon Shokuhin Kogyo Gakkaishi 37: 20–25

Maxam AM, Gilbert W (1977) A new method for sequencing DNA. Proc Natl Acad Sci USA 74: 560–564

Mayr GW (1988) A novel metal-dye detection system permits picomolar-range h.p.l.c. analysis of inositol polyphosphates from non-radioactively labeled cell or tissue specimens. Biochem J 254: 585–591

Muessik-Zefika M, Kornmueller A, Merkelbach B, Jekel M (1994) Isolation and analysis of intact polyphosphate chains from activated sludges associated with biological phosphate removal. Water Res 28: 1725–1733

Niemeyer R (1975) Poly- and Metaphosphate in höheren Pflanzen (*Lemnaceae*). Planta 122: 303–305

Niemeyer R (1976) Cyclic condensed metaphosphates and linear polyphosphates in brown and red algae. Arch Microbiol 108: 243–247

Niemeyer R, Richter G (1969) Schnellmarkierte Poly- und Metaphosphate bei der Blaualge *Anacystis nidulans*. Arch Mikrobiol 69: 54–59

Pick U, Bental M, Chitlaru E, Weiss M (1990) Polyphosphate-hydrolysis - a protective mechanism against alkaline stress? FEBS Lett 274: 15–18

Pepin CA, Wood HG (1986) Polyphosphate glucokinase from *Propionibacterium shermanii*. Kinetics and demonstration that the mechanism involves both processive and nonprocessive type reactions. J Biol Chem 261: 4476–4480

Pepin CA, Wood HG, Robinson NA (1986) Determination of the size of polyphosphates with polyphosphate glucokinase. Biochem Int 12: 111–123

Pisoni RL, Lindley ER (1992) Incorporation of [^{32}P]orthophosphate into long chains of inorganic polyphosphate within lysosomes of human fibroblasts. J Biol Chem 267: 3626–3631

Rao NN, Kornberg A (1996) Inorganic polyphosphate supports resistance and survival of stationary-phase *Escherichia coli*. J Bacteriol 178: 1394–1400

Rao NN, Roberts MF, Torriani A (1985) Amount and chain length of polyphosphates in *Escherichia coli* depend on cell growth conditions. J Bacteriol 162: 242–247

Rao NN, Liu S, Kornberg A (1998) Inorganic polyphosphate in *Escherichia coli*: the phosphate regulon and the stringent response. J Bacteriol 180: 2186–2193

Reusch RN, Sadoff HL (1988) Putative structure and functions of a poly-β-hydroxybutyrate/calcium polyphosphate channel in bacterial plasma membranes. Proc Natl Acad Sci USA 85: 4176–4180

Robinson NA, Wood HG (1986) Polyphosphate kinase from *Propionibacterium shermanii*. Demonstration that the synthesis and utilization of polyphosphate is by a processive mechanism. J Biol Chem 261: 4481–4485

Robinson NA, Goss NH, Wood HG (1984) Polyphosphate kinase from *Propionibacterium shermanii*: formation of an enzymatically active insoluble complex with basic proteins and characterization of synthesized polyphosphate. Biochem Int 8: 757–769

Robinson NA, Clark JE, Wood HG (1987) Polyphosphate kinase from *Propionibacterium shermanii*. Demonstration that polyphosphates are primers and determination of the size of the synthesized polyphosphate. J Biol Chem 262: 5216–5222

Schuddemat J, de Boo R, van Leeuwen CCM, van der Broek PJA, van Steveninck J (1989) Polyphosphate synthesis in yeast. Biochim Biophys Acta 1010: 191–198

Stover FS, Bulmahn JA, Gard JK (1994) Polyphosphate separations and chain length characterization using minibore ion chromatography with conductivity detection. J Chromatogr 668: 89–95

Strauss UP, Smith EH (1953) Polyphosphates as polyelectrolytes. II. Viscosity of aqueous solutions of Graham's salts. J Am Chem Soc 75: 3935

Strauss UP, Treitler TL (1995a) Chain branching in glassy polyphosphates: dependence on the Na/P ratio and rate of degradation at 25 °C. J Am Chem Soc 77: 1473

Strauss UP, Treitler TL (1955b) Degradation of polyphosphates in solution. I. Kinetics and mechanism of the hydrolysis at branching points in polyphosphate chains. J Am Chem Soc 78: 3553

Thilo E (1955) Die kondensierten Phosphate. Angew Chem 67: 141–149

Thilo E (1959) Die kondensierten Phosphate. Naturwissenschaften 46: 367–374

Thilo E (1965) Zur Strukturchemie der kondensierten anorganischen Phosphate. Angew Chemie 77: 1056–1066

Thilo E, Schülke U (1965) Über Metaphosphate die mehr als vier Phosphoratome in Ringanionen enthalten. Z Anorg Allg Chem 341: 293

Thilo E, Sonntag A (1957) Zur Chemie der kondensierten Phosphate und Arsenate. XIX. Über die Bildung und Eigenschaften vernetzter Phosphate des Natriums. Z Anorg Allg Chem 291: 186

Tinsley CR, Gotschlich EC (1995) Cloning and characterization of the meningococcal polyphosphate kinase gene: production of polyphosphate synthesis mutants. Infect Immun 63: 1624–1630

Tinsley CR, Manjula BN, Gotschlich EC (1993) Purification and characterization of polyphosphate kinase from *Neisseria meningitidis*. Infect Immun 61: 3703–3710

Tzeng CM, Kornberg A (1998) Polyphosphate kinase is highly conserved in many bacterial pathogens. Mol Microbiol 29: 381–382

Van Groenestijn JW, Deinema MH, Zehnder AJB (1987) ATP-production from polyphosphate in *Acinetobacter* strain 210A. Arch Microbiol 148: 14–19

Van Wazer JR (1958) Phosphorus and its compounds. Interscience, New York

Van Wazer JR, Holst K (1950) Structure and properties of the condensed phosphates. I. Some general considerations about phosphoric acids. J Am Chem Soc 72: 639–644

Van Wazer JR, Karl-Kroupa EJ (1956) Existence of ring phosphates higher than tetrametaphosphate. J Am Chem Soc 78: 1172

Wood HG, Clark JE (1988) Biological aspects of inorganic polyphosphates. Annu Rev Biochem 57: 235–260

Wurst H, Kornberg A (1994) A soluble exopolyphosphatase of *Saccharomyces cerevisiae*. J Biol Chem 269: 10996–11001

Yamaguchi H, Nakamura T, Hirai Y, Ohashi S (1979) High-performance liquid chromatographic separation of linear and cyclic condensed phosphates. J Chromatogr 172: 131–140

Yoza N, Hirano H, Baba Y, Ohashi S (1985) Characterization of enzymatic hydrolysis of inorganic polyphosphates by flow injection analysis and high-performance liquid chromatography. J Chromatogr 325: 385–393

Yoza N, Onoue S, Kuwahara Y (1997) Catalytic ability of alkaline phosphatase to promote P-O-P bond hydrolysis of inorganic diphosphate and triphosphate. Chem Lett 1997: 491–492

Definitive Enzymatic Assays in Polyphosphate Analysis

D. Ault-Riché and A. Kornberg[1]

1
Introduction

Inorganic polyphosphate (polyP), an anhydride-linked linear polymer of orthophosphate, is present in a wide variety of organisms from bacteria to mammals (Kulaev 1979; Kulaev and Vegabov 1983; Wood and Clark 1988). Despite the prominence of polyP (as in the vacuolar deposits in yeast that may represent 10 to 20 % of cellular dry weight), this molecule has received scant attention. Studies by Kulaev, Wood and a few others (Kulaev et al. 1987; Wood and Clark 1988; Kornberg 1995) disclosed the ubiquity of polyP and identified a few related enzyme activities. Yet, polyP has remained a largely forgotten polymer. Among the reasons is the lack of evidence describing its metabolic role, due in measure to the inadequacy of specific quantitative methods for polyP analysis.

PolyP was first seen as metachromatic granules in microorganisms in the form of particles stained pink by basic blue dyes, and was culled "volutin" early in this century (Meyer 1904). Historically, the polyP particle was recognized as a diagnostic feature of medically important bacteria, such as *Corynebacterium diphtheriae*. For some time they were mistaken for nuclcic acids. Early analytic methods were based on the ability of polyP, like other polyanions such as nucleic acids or heparin, to shift the absorption of a bound basic dye, such as toluidine blue, to a shorter wavelength (630 to 530 nm). With the advent of electron microscopy, these particles were seen to be highly refractive and appeared to volatilize while viewed under the electron beam; they were then identified as polyP (Wiame 1947). In addition to the staining and appearance of the granular accumulations observed in the light and electron microscopes, NMR analysis has been used to identify polyP in intact cells. However, NMR detection requires high concentrations and fails to measure the polyP in aggregates and in metal complexes.

Additional methods for identifying polyP have rested on crude and cumbersome separations in cell extracts from the known phosphate-containing polymers, followed by determination of the acid-lability characteristic of

[1] Department of Biochemistry, Stanford University School of Medicine, Stanford, California 94305-5307, USA.

Progress in Molecular and Subcellular Biology, Vol. 23
H. C. Schröder, W. E. G. Müller (Eds.)
© Springer-Verlag Berlin Heidelberg 1999

phosphoanhydride bonds (i.e., conversion to P_i in 7 min at 100 °C in 1 M HCl). These assay methods were not sufficiently quantitative to be conclusive, especially for low concentrations of polyP. In addition, the requirement for extensive extractions allowed only a few samples to be analyzed at a time. In sum, these assay methods are nonspecific, insensitive and unwieldy.

2
Enzymes Used in Assays of PolyP

PolyP-specific enzymes are unique and valuable reagents for specific, sensitive and facile assays (Ahn and Kornberg 1990; Akiyama et al. 1993; Wurst and Kornberg 1994). The availability of purified polyP-specific enzymes allowed us to synthesize ^{32}P-labeled polyP for use as a substrate. We have also developed both ^{32}P-based and non-radioactive enzyme-based methods to rapidly assay polyP and have adapted the non-radioactive method to a high throughput (HTP) format. Beyond that, the purified enzyme has opened the route of reverse genetics and, thereby, the means to identify the encoding gene, knock it out or to overexpress it. By manipulating expression of the gene and the cellular levels of its product, phenotypes have been created which provide clues to metabolic functions (see Chap. 9).

2.1
PolyP Kinase (PPK)

PolyP kinase (PPK) catalyzes the readily reversible transfer of the terminal (γ-) phosphate of ATP to polyP (Kornberg et al. 1956; Kornberg 1957; Ahn and Kornberg 1990) as follows:

$$\gamma\text{-}[^{32}\text{P}]n\text{ATP} \longleftrightarrow \text{poly}^{32}\text{P}_n + n\text{ADP}$$

The enzyme, purified to homogeneity from *Escherichia coli*, is a tetramer of 80-kDa subunits. It is responsible for the processive synthesis of long polyP chains (ca. 750 residues); labeling with ^{32}P in vitro provides such chains for use as substrates and standards. Addition of a primer in the synthetic reaction is not required, nor do ATP or P_i serve in this role. The reaction, in the presence of ADP, strongly favors the depolymerization reaction. The K_m for polyP$_{750}$ (as polymer) is 60 nM and that for ADP is 360 µM, whereas the K_m for ATP is 2 mM. ADP as a substrate for the depolymerization reaction inhibits the polyP synthesis reaction; the K_i value of ADP is 90 µM. With ADP in ten fold excess, PPK converts nearly all of polyP to ATP, identified by the use of either [^{14}C]ADP or [^{32}P]polyP as substrates.

PPK is autophosphorylated under the conditions of polyP synthesis and an N-P-linked phosphoenzyme was identified as the intermediate (Ahn and Kornberg 1990). Tryptic digests of [^{32}P]PPK contain a predominant ^{32}P-labeled peptide that includes His-441 (H441). Of the 16 histidine residues in

E. coli PPK, 4 are conserved among several bacterial species (Kumble et al. 1996). Mutagenesis of these four histidines shows that two (H441 and H460) when mutated to glutamine or alanine fail to be phosphorylated, show no enzymatic activities, and fail to support polyP accumulation in cells bearing these mutant enzymes.

The *E. coli* gene (*ppk*) encoding PPK has been cloned, sequenced (Akiyama et al. 1992) and overexpressed (about 100-fold in *E. coli*) (Crooke et al. 1994). The gene possesses an open reading frame for 687 amino acids (mass of 80 278 daltons). PPK has attachment to the cell outer membrane; the purified soluble PPK reassociates with cell membrane fractions.

Generation of a wide variety of nucleoside (and deoxynucleoside) triphosphates (NTPs) from their cognate nucleoside diphosphates (NDPs) is of critical importance in virtually every aspect of cellular life. Their function is fulfilled largely by the ubiquitous and potent nucleoside diphosphate kinase (NDK), most commonly using ATP as the donor. An additional and possibly auxiliary NDK-like activity is the capacity of PPK to use polyP as the donor in place of ATP, thereby converting GDP and other NDPs to NTPs (Kuroda and Kornberg 1997). This reaction was observed with the PPK activity in crude membrane fractions from *E. coli* and *Pseudomonas aeruginosa* (*P. aeruginosa*) as well as with the purified PPK from *E. coli*; the activity was absent from the membrane fractions obtained from *E. coli* mutants lacking the *ppk* gene. The enzyme also transferred a pyrophosphate group to GDP to form the linear guanosine 5' tetraphosphate (Kuroda and Kornberg 1997).

2.2
Exopolyphosphatase (PPX)

Exopolyphosphatase (PPX) catalyzes hydrolysis of the terminal residues of polyP to P_i processively and nearly completely (Akiyama et al. 1993) as follows:

$$polyP_n \longrightarrow n\text{-}2P_i + PP_i$$
$$polyP_n \longrightarrow n\text{-}3P_i + PPP_i$$

In *E. coli*, PPX is encoded by a gene in the *ppk* operon located immediately downstream of *ppk*; *E. coli ppx* contains 513 amino acids (58 133 daltons). PPX, purified to homogeneity, is judged to be a dimer. The K_m for $polyP_{750}$ (as polymer) is 9 nM. Substrates of bacterial alkaline phosphatase (glucose-1-phosphate, glucose-6-phosphate, ATP or ADP) are ineffective inhibitors of PPX. PPX does have a preference for longer chain sizes inasmuch as short chain $polyP_{3,4,15}$ compete poorly with $polyP_{750}$. PPX is inhibited by high levels of guanosine tetraphosphate (ppGpp) and pentaphosphate (pppGpp) with K_i values of 10 and 200 µM, respectively (Kuroda et al. 1997). High levels of ppGpp and pppGpp are generated in response to various types of physiological stress, such as amino acid starvation in *E. coli*, and lead to massive accumulations of polyP. Inasmuch as PPK and PPX levels fluctuate only slightly, the polyP accumulation can be attributed to inhibition by pppGpp and/or ppGpp of the hydrolytic breakdown of polyP by PPX, thereby blocking the dynamic turnover of polyP (see Chap. 9).

Another PPX (scPPX1), isolated from *Saccharomyces cerevisiae*, is a more powerful enzyme and analytic reagent than the *E. coli* PPX, releasing 30 000 P_i residues per min per enzyme molecule at 37 °C (Wurst and Kornberg 1994). It is 40 times more active than *E. coli* PPX, and can act on a far broader size range among polyP chains (i.e., 3 to 1000 residues), with a preference for those of about 250 residues (K_m values for $polyP_{10, 25, 250}$ and $_{750}$ are 3900, 160, 4 and 50 nM, respectively). Cloning the gene for scPPX1 (Wurst et al. 1995) enabled the enzyme to be knocked out in *S. cerevisiae* and overproduced in *E. coli*. Disruption of the gene did not affect the growth rate of *S. cerevisiae*. The scPPX1 is cytosolic and can be differentiated from other PPXs located in the yeast vacuole where the vast accumulations of polyP are located. Overexpression of this potent PPX can deplete *E. coli* of its polyP content (Shiba et al. 1997).

2.3
Other Enzymes

Other enzymes which could be used as reagents for analysis of polyP are the glucokinase which attacks the terminal residues of the polyP chain with glucose (Hsieh et al. 1993) and a phosphotransferase which attacks the termini with AMP (Bonting et al. 1991; Fig. 1). These enzymes have not yet been cloned and definitive analytic assays using these enzymes have not been developed.

In contrast to PPX purified from *E. coli* and *S. cerevisiae*, an endopolyphosphatase (PPN) which acts on long chain polyP was purified to apparent homogeneity from *S. cerevisiae* (Kumble and Kornberg 1996). PPN is a dimer of 35-kDa subunits. Distributive action on long chain polyP ($\sim$750 P_i residues) produces polyP of shorter chains to a limit of about 60 P_i residues, as well as a more abundant release of tripolyP. An analysis of the reaction products suggests a non-processive mechanism in which three PPP_i mole-

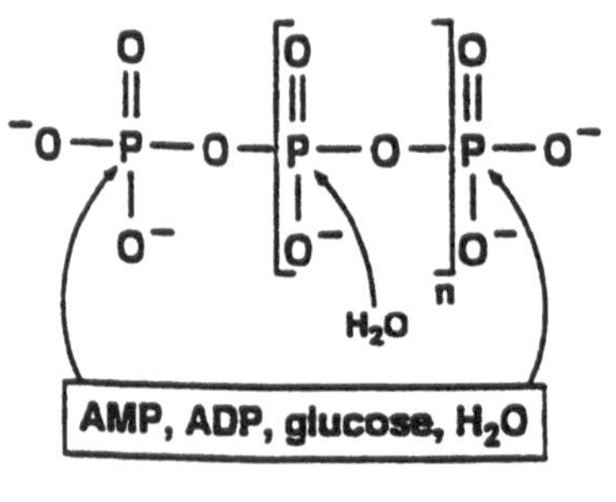

Fig. 1. PolyP metabolism. PolyP chains are attacked at their termini by AMP, ADP, glucose or H_2O, catalyzed, respectively, by polyP-AMP phosphotransferase, PPK, polyP glucokinase and PPX

cules are cleaved from the chain ends for every internal chain cleavage until a length of 60 residues is reached. PPN activities were identified in a wide variety of sources including archaea and human cells. PPN activity has been partially purified from rat and bovine brain where its abundance is ten fold higher than in other mammalian tissues. Inclusion of this enzyme in analytic methods could be useful by creating additional polyP ends available to PPX for attack.

Several lines of evidence suggest that polyP is formed in a much different way in eukaryotes than in prokaryotes. Pulse-chase studies of PC-12 cells (mammalian adrenal pheochromocytoma cells) showed that polyP is rapidly turned over (the complete pool of polyP is turned over in 60 min) and that the external $^{32}P_i$ appeared to be incorporated into polyP by a route which avoided the internal pools of P_i or ATP (the specific radioactivity of the polyP synthesized from external P_i corresponded to that of the P_i in the medium and not to the specific radioactivity of the P_i or ATP in the cells) (Kumble and Kornberg 1995). Further, a strain of yeast deficient in a vacuolar ATPase used in establishing the proton motive force across the vacuolar membrane was found to lack polyP (Wurst, unpubl.). This could be a direct or indirect effect of the mutation.

3
Purification and Assay of Recombinant PPK

As a reagent, PPK is used in both the preparative synthesis of labeled polyP and in analytic assays for estimating polyP in biological samples (see Sect. 6). Methods are described here for purification of recombinant, overexpressed E. coli PPK and histidine-tagged E. coli PPK. The recombinant native PPK has an activity 100-fold greater than the histidine-tagged PPK. However, the histidine-tagged PPK has the advantage of a simpler purification. Both enzymes have been used successfully.

Recombinant PPK is obtained by growth of strain W3110 containing the plasmid pBC10 (Akiyama et al. 1992). This high-copy number plasmid contains both the *ppk* and *ppx* full length genes. Cells are grown aerobically in Luria-Bertani (LB) broth in the presence of ampicillin to an OD_{600} of about 1.5. The cells are lysed by exposure to lysozyme followed by brief heat (37 °C) shock. Low-speed centrifugation produces a lysate cleared of debris. Nucleic acids are sheared by brief sonication and hydolyzed with DNase and RNase. Enzymes are released from the membranes by the addition of KCl and sodium bicarbonate followed by a high speed centrifugation. After exchanging the buffer by dialysis, the enzyme is precipitated by ammonium sulfate and then redialyzed to remove the salt. The dialyzed extract is applied to an SP-Sepharose FF column and the PPK eluted with a linear gradient of KCl. Pooled active fractions are dialyzed, then applied to a Mono S HR column, and PPK is eluted with a linear KCl gradient. This method typically gives a 50- to 60-fold purification with a 20 % yield to give 6×10^8 units per mg protein: 1 unit equals 1 pmol ATP converted/min. The enzyme is divided into small samples and stored at $-80\,°C$ in 15 % glycerol.

PPK activity is determined in a reaction containing γ-[^{32}P]ATP and an ATP regeneration system and 10^4 units of PPK (Ahn and Kornberg 1990; Akiyama et al. 1992). The ATP regeneration system removes ADP inhibition during the course of the reaction. Conversion of ATP to polyP is monitored by ascending thin layer chromatography (TLC) with polyethyleneimine (PEI)-cellulose in a solvent of formic acid and lithium chloride; polyP remains at the origin while ATP, ADP and P_i differentially migrate with the solvent.

The method described above for PPK activity works well with purified or partially purified samples of polyP. However, it does not consistently measure PPK activity in crude lysates due to inconsistent resolution of the reaction products in the TLC. Instead, conversion of γ-[^{32}P]ATP into polyP can be followed by capture of the polyP as an acid precipitate on glass filters (Ahn and Kornberg 1990). The filters are washed and the bound radioactivity determined by liquid scintillation counting.

Histidine-tagged PPK is obtained by growth of strain M15 containing the plasmid pGexPPK (Kumble et al. 1996). Cells are grown in LB broth containing antibiotics to an OD_{600} of 0.8 and then induced to produce PPK by the addition of isopropyl-1-thio-β-D-galactopyranoside (IPTG) to 50 μM and continued for 2.5 h, resulting in a six- to ten-fold overproduction. The cells are lysed, cleared, and the membranes solubilized as described above. The supernatant is desalted using a Sephadex G-25 column and the histidine-tagged PPK bound in batch to Ni-NTA agarose in the presense of 200 mM imidazole to minimize non-specific protein binding to the Ni-NTA resin. After binding, the resin is washed with a buffer lacking imidazole and the PPK eluted by chelation of the Ni with 10 mM EDTA. Addition of 15 mM $MgCl_2$ to the eluate restores PPK acitvity. This method results in a ten-fold purification with a 45 % yield. The typical specific activity is 10^7 units per mg which is three-fold lower than specific activities obtained for the wild-type PPK. The lower specific activity is probably due to the presence of the histidine tag.

4
Purification and Assay of Recombinant PPX

The *S. cerevisiae* PPX (scPPX1) is a powerful reagent (Wurst and Kornberg 1994). A histidine tag has been added to the carboxy terminal end of the gene to simplify its purification. Recombinant PPX is obtained by growth of strain BL21 (DE3) containing the plasmid pTrcPPX1. Cells are grown in LB broth in the presence of ampicillin to an OD_{540} of 0.5, induced by the addition of 0.5 mM IPTG, and continued for 6 h. After harvesting, the cells are lysed with lysozyme and sonication and a cleared supernatant is prepared as described for PPK above. The supernatant is loaded onto an Ni-NTA Sepharose column and the PPX is eluted with a linear gradient of imidazole (0–150 mM). Active fractions are pooled and loaded onto a hydroxyapatite column and eluted with a potassium phosphate buffer. A typical 20-fold purification with a 50% yield gives 2–5×10^9 units of activity (one unit is 1 pmol P_i released/min) with a specific activity of 10–20×10^7 units per mg protein. The enzyme is divided into small samples and stored at $-80\,°C$ in 15% glycerol.

PPX activity is determined in a reaction containing $1\,\mu g$ PPX and $[^{32}P]polyP_{750}$ (Akiyama et al. 1993). The reaction is warmed to $37\,°C$ and the polyP is added last to the reaction to prevent precipitation of the polyP in a complex with magnesium. The reaction is terminated by spotting onto the PEI-cellulose TLC plate. PolyP and P_i are separated by ascending TLC in a solvent of formic acid and lithium chloride (polyP remains at the origin while P_i migrates with the solvent). As with methods for measuring PPK activity in crude lysates, polyP degradation by PPX can be monitored by trapping the remaining $[^{32}P]polyP$ as an acid precipitate on glass GF/C filters during the course of the reaction (Akiyama et al. 1993).

5
Rapid Preparation of ^{32}P-PolyP

The key to developing better assays was the ability to use PPK to synthesize monodisperse ^{32}P-labeled polyP for use as a substrate. A rapid method for preparative-scale purification of $polyP_{750}$ (700–800 residues long) entails sedimentation through CsCl and precipitation with isopropanol (Ault-Riché et al. 1998). In this method, a reaction containing γ-$[^{32}P]ATP$, PPK and an ATP-regeneration system converts ATP to polyP, monitored by ascending TLC as described in Section 4. Yields of 50% of the $[^{32}P]ATP$ are typical.

The polyP reaction mixture is loaded over a cushion of 2.5 M CsCl and sedimented. Salt is removed from the sediment by precipitation with isopropanol followed by washes with 70% ethanol. Recoveries are greater than 90%. The polyP product consists of chains of approximately 750 P_i residues and consistently has a purity of 99% (as determined by its susceptibility to PPX hydrolysis). The entire procedure can be completed in about 2 h.

Limited acid hydrolysis of the purified polyP can be used to generate polyP ladders of various chain lengths (Ault-Riché et al. 1998). $PolyP_{750}$, 2 mM, is warmed to $70\,°C$ and added to pre-warmed 20 mM HCl and maintained at $70\,°C$ for between 20 and 90 s, after which the reaction is stopped by transfer to an ice bath and the pH neutralized by the addition of a buffer. Electrophoresis of the polyP through a polyacrylamide-urea gel can be used to determine polyP chain sizes.

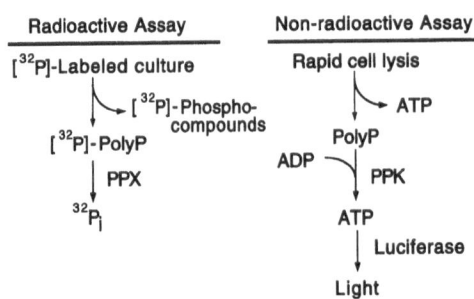

Fig. 2. Strategies for enzyme-based assays of polyP in biological samples

6
Estimating PolyP in Biological Samples

Two methods for estimating polyP in biological samples are presented in Fig. 2. The first requires prior labeling of the culture with $^{32}P_i$ followed by adsorption of the labeled polyP onto ion exchange filters and then release of $^{32}P_i$ from the filters by the action of PPX. The second involves the rapid isolation of polyP using powdered glass or glass filters followed by its conversion to ATP by the action of PPK and estimation of the generated ATP using the luciferase system and a luminometer. The radioactive method is limited in several ways: by its requirement for prior labeling of the sample with $^{32}P_i$, by poor and inconsistent polyP extractions from cultures grown in LB broth, and by the need to avoid high phosphate-buffered media (100 mM), inasmuch as $[^{32}P]P_i$ is added at a level of 20 µCi/ml.

6.1
Radioactive Assay of PolyP with PPX

The method of radioactive assay of polyP with PPX was optimized for use with defined MOPS-buffered minimal medium (Rao et al. 1998) containing glucose (4 mg/ml) as a carbon source and K_2HPO_4 as the P_i source. $[^{32}P]P_i$ is added to 20 µCi/ml of the medium. PolyP is extracted by lysis of the cells in a solution containing formic acid, urea, SDS, EDTA and carrier polyP (type P_{65}). The suspension is sonicated and the polyP enriched by binding to DE81 ion-exchange filter disks. The disks are washed and the polyP eluted with KCl. ATP and other radiolabeled material are removed from the eluate by treatment with Norit charcoal. The DE81 adsorption of polyP and the removal of ^{32}P contaminants by Norit each produce at least a ten-fold enrichment of polyP. The polyP is concentrated by re-adsorption to DE81 disks. After washes, the disks are treated directly with PPX. Decrease in ^{32}P on the filters or increase in ^{32}P released from the filters corresponds to polyP.

6.2
Non-radioactive Assay of PolyP with PPK

Because of the limitations in the radioactive assay described above, a non-radioactive method for estimating polyP using PPK was developed (Ault-Riché et al. 1998). PolyP is first extracted by lysis of cells in 4 M guanidinium isothiocynate and then bound to powdered glass (Glassmilk; Bio101) by addition of ethanol; ATP fails to bind to Glassmilk under these buffer conditions. The detergent SDS is included during the binding step to reduce protein binding to the glass which facilitates quantitative release of the polyP in the elution step. PolyP is eluted from the Glassmilk with hot water or a low ionic strength buffer. PolyP is quantitatively converted to ATP by PPK in the presence of a ten-fold excess of ADP. The ATP generated is then quantitated using the luciferase system and a luminometer. Analysis of E. coli extracts gives a linear correlation in six determinations with a correlation coefficient of 1.00. The signal is completely removed by digestion with PPX prior to analysis with PPK. The assays are accurate to low levels (i.e., 100 pmol P_i residues per mg protein), require very little source material (i.e., 0.5 ml E. coli culture of OD_{600} 0.5), and permit 30 or more samples to be processed in a few hours. The strong lysis buffer used to extract polyP should make adaptation of this assay for use with other types of cells relatively easy. We have used the assay with P. aeruginosa, Acinetobacter johnsonii, S. cerevisiae, Methanococcus thermolithotrophicus (a methanogenic Archea) and Chlamydomonas reinhardi (a unicellular green alga).

6.3
High Throughput (HTP) Assay of PolyP

The non-radioactive method was adapted to a high throughput (HTP) format using 96-well GF/C filter plates (Polyfiltronics) instead of powdered glass (Ault-Riché, unpubl.). Small cultures (0.8 ml) are grown in 96-well Bioblocks (Rainin) and the growth monitored with a 96-well spectrophotometer (BioRad). Wells in the blocks have a volume of 3 ml and are square to enhance aeration. Doubling times of E. coli grown in these blocks are similar to the doubling times of 1-ml cultures grown in 10-ml culture tubes on a roller drum. After pelleting, the cells are lysed using the GITC/SDS/ethanol buffer described above and the polyP extracted by filtration through GF/C filter plates using a vacuum manifold (Qiagen). After washing and elution from the 96-well plates, the polyP is estimated as described earlier using PPK and the luciferase system; the light is quantitated using a Topcount 96-well luminometer (Packard). Reagent handling is simplified with the use of multichannel and repeating pipettors and robots such that over a thousand samples can be analyzed in a day.

The HTP assay has been successfully used in a mutagenesis study of PPK (Ault-Riché and Tzeng, unpubl.). Overexpression of PPK in *E. coli* results in elevated cellular polyP levels (>300 nmol/mg protein) while overexpression of an inactive mutant PPK fails to generate elevated polyP levels (<100 pmol/mg protein) (Kumble et al. 1996). Oligonucleotide primer pairs (provided by the Stanford Genome Center, Palo Alto, California) were designed to create separate mutations in 43 conserved amino acids in *ppk*. The plasmids were transformed into super-competent XL1-blue *E. coli* (Stratagene, La Jolla, California) and 12 clones from each mutation (516 clones in all) were arrayed into 96-well blocks, grown, induced to overexpress the mutant PPK, and assayed for polyP accumulation using the HTP assay.

The assay identified 11 mutations which led to a failure to accumulate polyP in vivo. The mutant enzymes were purified and the effects of the mutations were confirmed by in vitro activity assays. Several of the mutations had differential effects on the forward (polyP formation) and reverse (ADP kinase) activities, consistent with observation that the two activities in the wild type kinase could be differentiated with regard to temperature, pH and denaturant lability (Tzeng, unpubl.).

The HTP assay can be used as a tool in several areas of investigation including: further PPK site-directed mutagenesis, cloning genes involved in accumulations of polyP, purification of polyP granules and polyP complexes, and in creating additional analytic tools.

7
Future Directions in Assay Development

The availability of purified polyP-specific enzymes has led to the development of rapid, sensitive and definitive assays. Several further improvements in methods are apparent. PolyP exists in non-uniform distributions of chain lengths which vary between the types of cells (bacteria to mammalian cells). The possible meaning of these distributions remains a mystery. Experiments analyzing these distributions require labeling cells with ^{32}P, enrichment for polyP, and size analysis using polyacrylamide gel electrophoresis. This approach is extremely laborious and the requirement for prior labeling limits the types of samples which can be analyzed. Rapid methods which can discriminate and quantitate the distribution of chain sizes in non-labeled samples are needed. High performance liquid chromatography (HPLC) coupled with innovative detection methods may provide such a method.

A major impediment to investigating the biological functions of polyP has been the lack of a specific stain for polyP. Basic dyes can be used to visualize large concentrations of pure polyP, but they are non-specific, readily binding other polymers such as DNA and RNA. A specific stain for polyP would allow us to visualize where and how polyP is distributed in the cell and would allow greater description of the complexes that it forms (e.g., immunofluorescence). It would also enable a variety of cloning strategies which are dependent on discriminating polyP-producing and non-producing clones on a plate (e.g., colony lift hybridization screens). Such a stain would allow us to address a much wider range of questions regarding the role of polyP in development and aging.

Looking for PPK and PPX mutants which maintain the ability to specifically and tightly bind polyP but lack catalytic activity could provide a possible route towards developing a specific stain. Rapid in vivo screening of PPK and PPX mutants has already been demonstrated using the HTP assay. *E. coli* cells overexpressing an inactive PPK fail to accumulate polyP. Cells overexpressing an active PPK and inactive PPX accumulate 100-fold higher polyP levels than cells expressing both an active PPK and PPX. Inactive mutants identified with the HTP assay can then be screened for polyP binding activity. The rapid development of facile, easy-to-use methods for polyP analysis has created the opportunity to make this forgotten polymer memorable.

Acknowledgements. We express our gratitude to the colleagues identified in the acknowledgements in Chapter 1 and to the National Institutes of Health for research support.

References

Ahn K, Kornberg A (1990) Polyphosphate kinase from *Escherichia coli*. Purification and demonstration of a phosphoenzyme intermediate. J Biol Chem 265: 11734–11739

Akiyama M, Crooke E, Kornberg A (1992) The polyphosphate kinase gene of *Escherichia coli*: isolation and sequence of the *ppk* gene and membrane location of the protein. J Biol Chem 267: 22556–22561

Akiyama M, Crooke E, Kornberg A (1993) An exopolyphosphatase of *Escherichia coli*. J Biol Chem 268: 633–639

Ault-Riché D, Fraley CD, Tzeng CM, Kornberg A (1998) Novel assay reveals multiple pathways regulating stress-induced accumulations of inorganic polyphosphate in *Escherichia coli*. J Bacteriol 180: 1841–1847

Bonting CFC, Kortstee GJJ, Zehnder AJB (1991) Properties of polyphosphate: AMP phosphotransferase of *Acinetobacter* strain 210A, J Bacteriol 173: 6484–6488

Crooke E, Akiyama M, Rao NN, Kornberg A (1994) Genetically altered levels of inorganic polyphosphate in *Escherichia coli*. J Biol Chem 269: 6290–6295

Hsieh PC, Shenoy BC, Jentoft JE, Phillips NFB (1993) Purification of polyphosphate and ATP glucose phosphotransferase from *Mycobacterium tuberculosis* H37Ra: evidence that poly(P) and ATP glucokinase activities are catalyzed by the same enzyme. Protein Exp Purif 4: 76–84

Kornberg SR (1957) Adenosine triphosphate synthesis from polyphosphate by an enzyme from *Escherichia coli*. Biochim Biophys Acta 26: 294–300

Kornberg A (1995) Inorganic polyphosphate: toward making a forgotten polymer unforgettable. J Bacteriol 177: 491–496

Kornberg A, Kornberg SR, Simms ES (1956) Metaphosphate synthesis by an enzyme from *Escherichia coli*. Biochim Biophys Acta 20: 215–227

Kulaev IS (1979) The biochemistry of inorganic polyphosphates. Wiley, New York

Kulaev IS, Vegabov VM (1983) Polyphosphate metabolism in microorganisms. Adv Microbiol 15: 731–738

Kulaev IS, Vegabov VM, Shabalin YA (1987) New data on biosynthesis of polyphosphates in yeast. In: Torriani-Gorini A, Rothman FG, Silver A, Wright A, Yagil E (eds) Phosphate metabolism and cellular regulation in microorganisms. American Society for Microbiology, Washington, DC, pp 233–238

Kumble KD, Kornberg A (1995) Inorganic polyphosphate in mammalian cells and tissues. J Biol Chem 270: 5818–5822

Kumble KD, Kornberg A (1996) Endopolyphosphatases for long-chain inorganic polyphosphate in yeast and mammals. J Biol Chem 271: 27146–27151

Kumble KD, Ahn K, Kornberg A (1996) Phosphohistidyl active sites in polyphosphate kinase of *Escherichia coli*. Proc Natl Acad Sci USA 93: 14391–14395

Kuroda A, Kornberg A (1997) Polyphosphate kinase as a nucleoside diphosphate kinase in *Escherichia coli* and *Pseudomonas aeruginosa*. Proc Natl Acad Sci USA 94: 439–442

Kuroda A, Murphy H, Cashel M, Kornberg A (1997) Guanosine tetra- and pentaphosphate promote accumulation of inorganic polyphosphate in *Escherichia coli*. J Biol Chem 272: 21240–21243

Meyer A (1904) Orientierende Untersuchungen über Verbreitung, Morphologie, und Chemie des Volutins. Bot Z 62: 113–152

Rao NN, Liu S, Kornberg A (1998) Inorganic polyphosphate in *Escherichia coli*. I. The phosphate regulon and the stringent response. J Bacteriol 180: 2186–2193

Shiba T, Tsutsumi K, Yano H, Ihara Y, Kameda A, Tanaka K, Takahashi H, Munekata M, Rao NN, Kornberg A (1997) Inorganic polyphosphate and the induction of rpoS expression. Proc Natl Acad Sci USA 94: 11210–11215

Wiame JM (1947) Yeast metaphosphate. Fed Proc 6: 302

Wood HG, Clark JE (1988) Biological aspects of inorganic polyphosphates. Annu Rev Biochem 57: 235–260

Wurst H, Kornberg A (1994) A soluble exopolyphosphatase of *Saccharomyces cerevisiae*: purification and characterization. J Biol Chem 269: 10996–11001

Wurst H, Shiba T, Kornberg A (1995) The gene for a major exopolyphosphatase of *Saccharomyces cerevisiae*. J Bacteriol 177: 898–906

Study of Polyphosphate Metabolism in Intact Cells by 31-P Nuclear Magnetic Resonance Spectroscopy

Kuang Yu Chen[1]

1
Introduction

Inorganic polyphosphates (polyP) are naturally occurring linear polymers of orthophosphate that are found in microorganisms, lower eukaryotes such as yeast, and animals (Harold 1966, Kulaev and Vagabov 1983; Wood and Clark 1988). In certain cases, polyP can accumulate to more than 10 % of total dry mass. The ubiquitous presence of polyP suggests that they may have important physiological functions. For example, polyP has been proposed to function as a high-energy reserve or a phosphate reserve and may play an important role in regulating the levels of ATP (Harold 1966). Due to their polyanionic nature, polyP can also serve as counter ions for cationic species such as Mg^{2+}, Mn^{2+}, basic amino acids, and polyamines. In this regard, they may function in counteracting the osmotic pressure exerted by basic amino acids and various cations accumulated in fungal vacuoles. In addition, the presence of polyP in nuclei and membrane in certain organisms would suggest that polyP may have other unknown functions. It has been difficult to study the metabolism, regulation, and function of polyP for the following reasons: (1) polyP does not possess chromophores in its chemical structures; (2) polyP cannot easily be derivatized with specific chromophore or fluorescent probe; and (3) polyP may exist as a mixture of polymers with varying chain length, ranging from 3 to 1000 residues. Several analytical methods have been developed to study polyP, including enzymatic assay, HPLC and electrophoresis (reviewed by Wood and Clark 1988). Among them, in vivo phosphorus-31 nuclear magnetic resonance (^{31}P-NMR) remains, arguably, the unique one, being the least disruptive and quantitative (Roberts 1987).

In vivo NMR is a rapid and non-invasive technique suitable for studying metabolic processes in intact cells, tissues, and organelles. High resolution ^{31}P-NMR has provided valuable information on the identification and quantitation of phosphorus metabolites, on the intracellular pH and compartmentation, and on the kinetics and pathways of biochemical reactions. The

[1] Department of Chemistry and The Cancer Institute of New Jersey, Rutgers – The State University of New Jersey, Piscataway, New Jersey 08854-8081, USA.

Progress in Molecular and Subcellular Biology, Vol. 23
H. C. Schröder, W. E. G. Müller (Eds.)
© Springer-Verlag Berlin Heidelberg 1999

use of phosphorus NMR to study polyP began almost a quarter of a century ago. The subject was reviewed more than a decade ago (Roberts 1987).Thus, this chapter will emphasize more on the work appearing in recent literature.

2
Phosphorus NMR

2.1
Basic Principles of NMR

Atomic nuclei with spin quantum number other than zero behave like a microscale magnetic bar. In the presence of external magnetic field, the nuclei with spin number of 1/2 can assume two orientations, each with associated quantum number of +1/2 or -1/2. The energy difference between these two states is proportional to the external field B_0 as shown by the following equation:

$$\Delta E = (\gamma/2\pi)hB_0,$$

where γ is gyromagnetic ratio of the nucleus, h is the Planck constant, and B_0 is the external field. Transistion can be induced by applying an oscillating magnetic field B_1, in the plane perendicular to the direction of B_0, with a frequency, υ_0, where υ_0 satisfies the following equation:

$$\upsilon_0 = (\gamma/2\pi)B_0.$$

Due to the nature of the chemical environment of the nuclei, the true effective field, B_{eff}, that the nucleus can experience may differ somewhat from B_0 as expressed with equation:

$$B_{eff} = B_0 (1-\sigma),$$

where σ, shielding constant, is the small secondary field generated due to the electronic structures surrounding the nucleus.

The nuclear magnetic resonance (NMR) spectrum of a sample in a magnetic field can be generated by either sweeping the applied irradiation or by sweeping the external magnetic field (continuous-wave mode approach). The NMR spectrum thus generated appears as an array of resonance signals characteristic of the sample and reveals the chemical nature of the compounds containing the particular nuclei and the abundance of the nuclei.

This continuous-wave mode detection method has been replaced by the Fourier-transform (FT) NMR since 1966. In FT NMR, the radio frequency field is applied as pulses that cover sufficient range to excite all nuclei in the sample. The signal received after each pulse contains many frequency components and is called free induction decay (FID). This time-dependent signal is then converted to all its frequency-dependent components via a Fourier transform as shown by the equation:

$$F(\omega) = \int\limits_{-\infty}^{+\infty} f(t)e^{iwt}dt,$$

where ω is frequency and equals $2\pi\upsilon$ and i is the imaginary. The introduction of pulse-radio frequency, Fourier transform and improvement of superconducting magnets have greatly enhanced the sensitivity and power of NMR spectroscopy. There are three major types of application of NMR in biological systems. First, it can be used to study the structure of biological molecules such as proteins and nucleic acid in solution. It can also be used to examine the metabolism and native environment of biological molecules in the biological samples (e.g. intact cells). Finally, with additional magnetic field gradients, it can be used to generate images of biological samples.

2.2
Phosphorus NMR

The characteristics of ^{31}P-NMR are: (1) the ^{31}P nucleus is 100% naturally abundant, eliminating the need for enrichment; (2) the ^{31}P nucleus has a nuclear spin of 1/2 with a magnetic moment of 1.1305, and no nuclear quadrupole moment, thus eliminating the more complicated signal quadrapolar splitting: (3) the ^{31}P nucleus has moderate relaxation times, allowing relatively rapid signal averaging; (4) the ^{31}P-NMR chemical shifts of phosphorus compounds span over 600 ppm. The use of the multinuclear FT NMR Spectrometer (80–800 MHz, proton frequency) has also greatly alleviated the problem associated with low sensitivity of the ^{31}P nucleus that is only about 1/15 of that of ^{1}H at the same field. Phosphorus compounds of biological interest include phosphates, phosphonates, and various easters of phosphates and phosphonates. The chemical shift of ^{31}P in these biological compounds can span over a 30-ppm range, making ^{31}P-NMR an attractive tool for examining phosphorus metabolites in microorganisms, plants, and animal tissues. In addition, ^{31}P-NMR does not suffer from the problem of solvent suppression since no water signal will appear in the ^{31}P resonance region. Figure 1 shows schematically the chemical shifts of common biological compounds containing phosphorus. As can be seen, the phosphorus compounds can be grouped according to their chemical shifts. The simplicity of the ^{31}P spectrum, usually containing 8 to 12 resonances, is due to the fact that narrow signals are generated only from relatively mobile compounds. The insoluble or highly immobilized species such as membrane-bound phospholipids usually give very broad signals that are either NMR invisible or appear as broad components underlying the narrow metabolite signals.

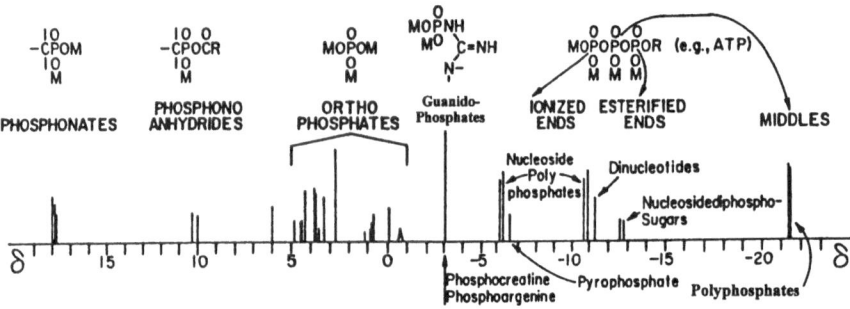

Fig. 1. Chemical shifts of biological phosphorus compounds at pH 10.0. (Adapted from Van Wazer and Ditchfield 1987)

3
Polyphosphates

3.1
Inorganic and Organic

A phosphate group can be defined as a molecular structure in which each phosphorus atom is surrounded by an approximately tetrahedral array of four oxygen atoms. Phosphates with one or more P-O-P linkages are formed by sharing of an oxygen between two different phosphate groups. Straight P-O-P chains are thermodynamically more preferred to branched chains so that linear and simple cyclic molecules are the common forms of phosphate molecules. Inorganic polyphosphate (polyP) refers to linear P-O-P chains containing more than three PO_4 units. In living systems, linear inorganic polyP can be as long as 1000 PO_4 units (Wood and Clark 1988). In general, polyP with a chain length of less than ~20 units is considered short-chain polyP whereas polyP with a chain length greater than 50 units is termed long-chain polyP. In vivo NMR only detects polyP, both short-chain and long-chain, that are sufficiently mobile. The term polyphosphate has also been used to refer to biological compounds such as diadenosine polyphosphates or inositol polyphosphates. This chapter is only concerned with phosphorus NMR study of linear inorganic polyphosphates.

3.2
Metabolism, Regulation, and Function

Most living microorganisms, from bacteria to yeast and algae, contain polyP to as much as 10~20 % of their dry weight (Kulaev and Vagabov 1983). The regulation and the function of polyP are of considerable interest in view of their high content in these microorganisms. PolyP is also present in higher eukaryotes, albeit at lower concentrations (Kumble and Kornberg 1995).

Several enzymes specifically involved in polyP metabolism have been purified or identified (Wood and Clark 1988). Some of these enzymes, including
polyphosphate kinase and exopolyphosphatase from *E. coli* and yeast, have
been cloned (Akiyama et al. 1993; Wurst et al. 1995), making it possible to
employ genetic and molecular biological approach to study the polyP regulation.

The ubiquitous occurrence of polyP, the wide range of polyP content in
different organisms, and the subcellular distribution of polyP all suggest that
polyP may have diverse functions depending on the cell types, organisms,
and environments. So far polyP has been reported to function as (1) an
energy storage source (Kulaev 1979), (2) a phosphate reserve (Kulaev 1979),
(3) a substrate for glucokinase (Phillips et al. 1993), (4) a substrate for adenylate kinase (Bonting et al. 1991), (5) a buffer against pH stress (Pick et al.
1990) (6) a counterion to neutralize cationic species in vacuole (Cramer and
Davis 1984), (7) a component of specific membrane channel for DNA entry
(Reusch and Sadoff 1988), and (8) a regulator in response to environmental
stress (Yang et al. 1993). It can be anticipated that, with the advancement of
polyP research, additional functions will be discovered.

4
Use of Phosphorus NMR to Study Polyphosphate in Intact Cells

4.1
NMR Apparatus and Sample Preparations

Figure 2 (above) shows the schematic block diagram of the pulse-Fourier-
transform NMR spectrometer which uses high-power radio frequency
pulses and Fourier transformation to achieve high sensitivity. In brief, very
strong and homogeneous magnetic fields (2.0–18.9 Telsa) are produced by
superconducting magnets where the sample will reside. A short, high-
powered radio frequency pulse, lasting for microseconds, will be applied to
the sample. The pulse is equivalent to a full range of frequencies for the particular nucleus examined. The response of the sample to the pulse is called
time-domain signal, or free induction decay (FID), which is stored in the
computer. Thousands of these accumulated time-domain signals are added
and mathematically treated by Fourier transform which converts amplitude
vs time signals into amplitude vs frequency signals. Figure 2 (below) illus-
trates that the Fourier transform of the FID gives NMR spectrum as a function of frequency.

Intact viable cells can be presented to NMR instrument in three types: (1)
cells in concentrated suspension are transferred into NMR tube (Salhany et
al. 1975); (2) cells or tissues, imbedded in agarose gel or beads, are trans-

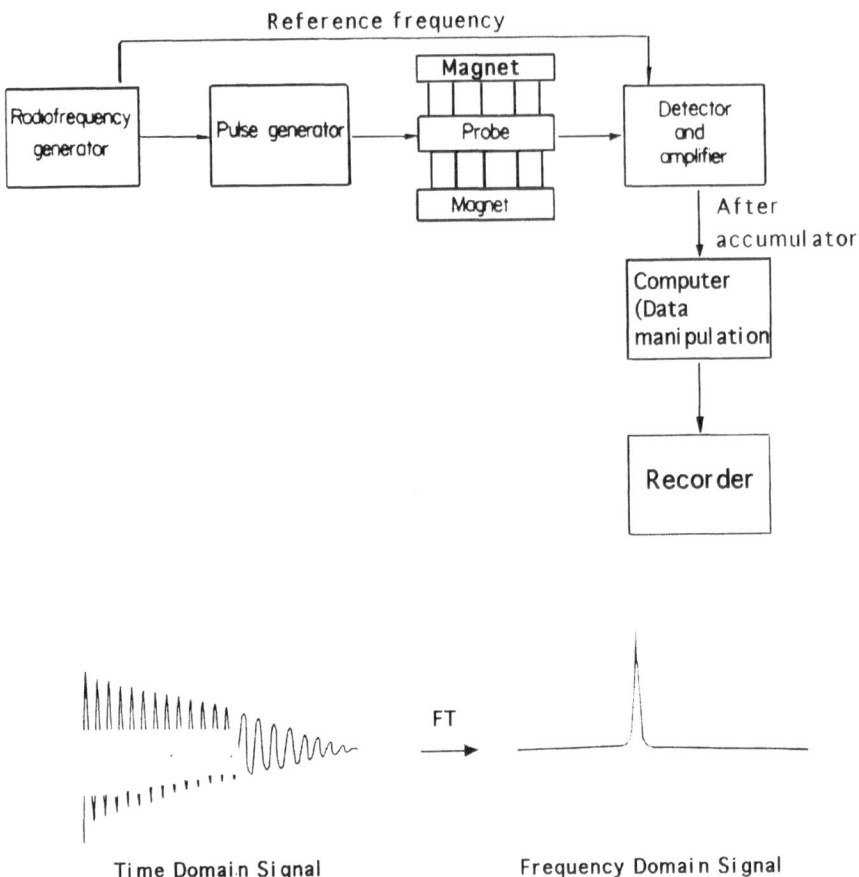

Fig. 2. *Above* Fourier-transform NMR spectroscopy. *Below* A time-domain FID is converted by Fourier transform (*FT*) to a frequency-domain signal

ferred into NMR tube that is equipped with perfusion device (e.g. Bental et al. 1990); and (3) cells are maintained in a cultivator which has radio frequency coil inserted (e.g. Mechan et al. 1992). There has been no rigorous and systematic study to compare these methods in order to determine whether the conditions of sample preparation may affect the polyP metabolism as monitored by NMR. Nevertheless, magnets with bore size up to 40 cm are available at 4.7 Telsa (200 MHz NMR frequency for ^{1}H). The increase in bore size would allow the use of larger sample size in an environment (e.g. cultivator) that bears more resemblance to the physiological conditions, not mentioning the possibility of generating NMR spectrum directly from live animals or plants.

4.2
Specific Information Gained on PolyP from Phosphorus NMR

PolyP generally gives three resonance peaks, terminal P (PP1) at about −7 ppm, penultimate P (PP2–PP3) at about −21.7 ppm, and internal P (PP4) at about −22.5 ppm. The chemical shift of PP1 may overlap with β-phosphorus of nucleotide disphosphate and γ-phosphorus of nucleotide triphosphate whereas PP2–3 may overlap with β-phosphorus of nucleotide triphosphate. The internal P of polyP, with a chemical shift around −21 to −24 ppm can be easily identified since only β-phosphorus of nucleotide triphosphate may show resonance close to polyP. In contrast, the terminal or penultimate P resonance occurs in regions where other resonance peaks cluster. PolyP that can be detected by NMR (i.e. is NMR visible) represents the more mobile fraction of total polyP. Thus, a lack of NMR-visible polyP signal does not necessarily indicate an absence of polyP in the sample. Figure 3 shows a typical ^{31}P-NMR spectrum for *Neurospora crassa*. Peak 1 represents internal P of polyP. Peak 7 represents vacuolar phosphate and peak 8 represents cytosolic phosphate. Since the chemical shift of orthophosphate peak is the weighted average of shifts of monobasic and dibasic phosphates, it can be used to directly estimate the precise pH of intracellular compartments.

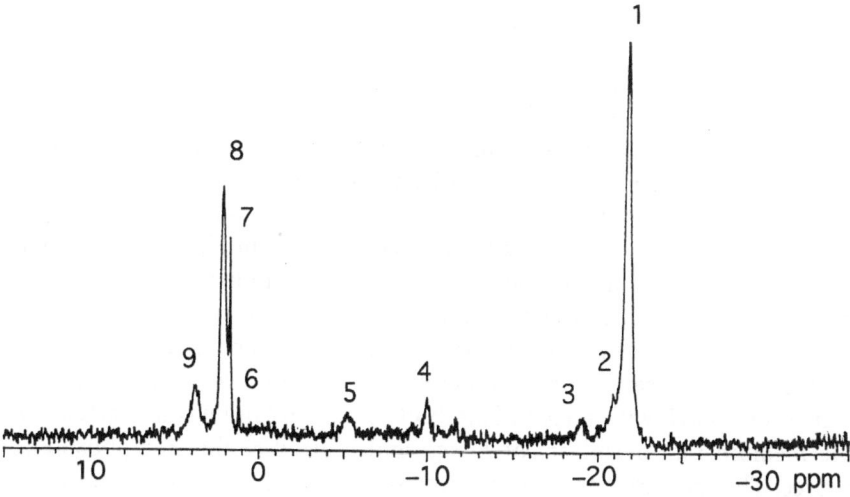

Fig. 3. Phosphorus NMR spectrum of wild-type *N. crassa* in logarithmic phase. Resonance peaks are assigned as follow: *1* polyP (inner phosphate); *2* polyP (penultimate phosphate); *3* β-phosphorus of nucleotide triphosphates; *4* diphosphates and sugar dinucleotides; *5* terminal residue of nucleotide triphosphates; *6* glycerophosphocholine or α-glycerol phosphoryl ethanolamine; *7* vacuolar orthophosphate; *8* cytosolic orthophosphate; *9* sugar monophosphates. (reprinted from Yang et al. 1993 with permission)

The ability to identify polyP signal and other phosphorus metabolites in the same spectrum makes the in vivo ^{31}P-NMR a powerful tool to monitor the fluctuation of polyP under various experimental conditions. The following parameters related to polyP metabolism and regulation in various organisms have been examined using phosphorus NMR technique: (1) compartmentation; (2) effects of growth conditions (e.g. energized state vs de-energized state, low phosphate vs high phosphate, logarithmic stage vs stationary phase; and (3) effects of environmental stress (e.g. alkalization, osmotic stress, heat, metal ions). Some of these studies are highlighted and discussed below.

4.3
Eubacteria

4.3.1
Escherichia coli

The genes involved in polyP metabolism in *Escherichia coli*, polyP kinase (*ppk*) and polyphosphatase (*ppx*), have been cloned (Akiyama et al. 1992, 1993). Both genes have been placed behind inducible promoters and could be overexpressed, thus allowing tight regulation of polyP content in *E. coli* (Van Dien et al. 1997). For example, when *ppk* is induced early in growth, overproducing PPK could lead to an accumulation of large amounts of polyP of up to 80 μmol in phosphate monomer units per g of dry cell weight. In contrast, the induction of the *ppx* gene subsequently could cause polyP degradation with a half-life of 210 min. By controlling the concentration of inducer in the medium, the steady-state polyP level can be precisely controlled. The availability of these molecular tools and genetic information should make *E. coli* an ideal system to study polyP regulation and function by the combined use of NMR and molecular biology. Sharfstein and Keasling (1994) have compared the difference in polyP biosynthesis between wild type *E. coli* and *ppx* defect mutants. As can be seen in Fig. 4 the accumulation of polyP occurs in wild type *E. coli*, but not in mutant *E. coli* that has a defect in the *ppx* gene after phosphate-shift. In contrast, polyP is constitutively present in the mutants due to the lack of PPX activity.

Phosphate is clearly stored as polyP in *E. coli*. However, under phosphate starvation, polyP with a chain length of up to 100 can be transported from the medium into periplasm of *E. coli* and utilized as the sole phosphorus source. This process has been demonstrated with the use of ^{31}P-NMR (Rao and Torriani 1988). Furthermore, with the use of NMR, the same group was able to show that porins PhoE and OmpF facilitate a higher permeability for poly-P100 than porin OmpC does.

It has been shown that polyP may be involved in DNA uptake in competent *E. coli*. Reusch and Sadoff (1988) reported that polyP forms a complex

Fig. 4. PolyP in *E. coli* W3110 (*top*) and polyP in *E. coli* CA38/ pBC29 (*bottom*) during phosphate shift. *A* Phosphate-starved; *B* 30 min after phosphate shift; *C* 60 min after shift; *D* 95 min after shift; *E* 130 min after shift; *G* 255 min after shift. (Adapted from Sharfstein and Keasling 1994)

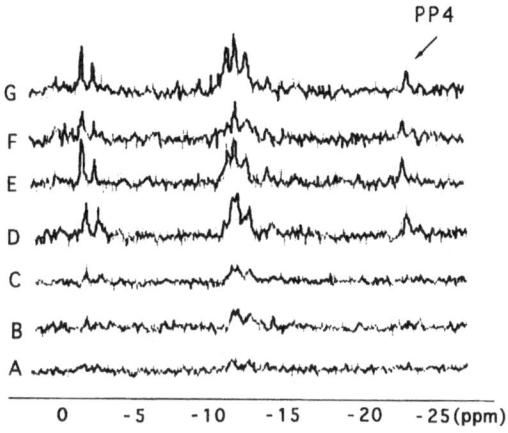

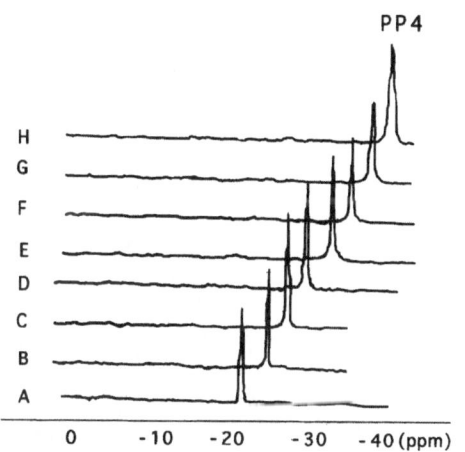

with polyhydroxybutyrate (PHB), and Ca^{2+}, and such complex may act as the membrane component responsible for DNA entry in *E. coli*. Such polyP-containing membrane-bound complex may also behave as voltage-activated calcium channels (Huang and Reusch 1996). However, this interesting aspect of polyP function may be difficult to study by NMR since polyP in the complex is not NMR visible.

4.3.2
Propionibacterium acnes

The cells of light-sensitive skin bacterium *Propionibacterium acnes* have been examined by ^{31}P-NMR. The spectra show a large accumulation of polyP when grown on Eagles medium. Addition of glucose to the cell suspension gives rise to a change in the pH gradient across the cell membrane and a decrease in the polyP peak. A lethal dose of broad-band near-ultraviolet light (corresponding to a 10% survival in a survival test) increases the amount of polyP visible in the NMR spectra. The increase is likely due to an equilibrium between long-chain invisible polyP in granules and short-chain polyP in cytoplasm (Kjeldstad and Johnsson 1987). When the cells were exposed to temperatures from 15 to 45°C, the amount of polyP increases with increasing temperature (see Fig. 5). Similar to UV irradiation, this increase is due to equilibrium between NMR-invisible long-chain polyP in granules and free short-chain polyP in cytoplasm. In both cases, there are no UV- or temperature-induced changes in the other phosphorous components seen in the spectra, except a decrease in ATP for higher temperatures (Kjeldstad et al. 1989). The physiological significance of the observation is not clear. However, it is of interest to note that the short-chain polyphosphates such as tetra- or penta-polyphosphate have been proposed to serve as alarmones for the production of heat-response proteins (Lee et al. 1983).

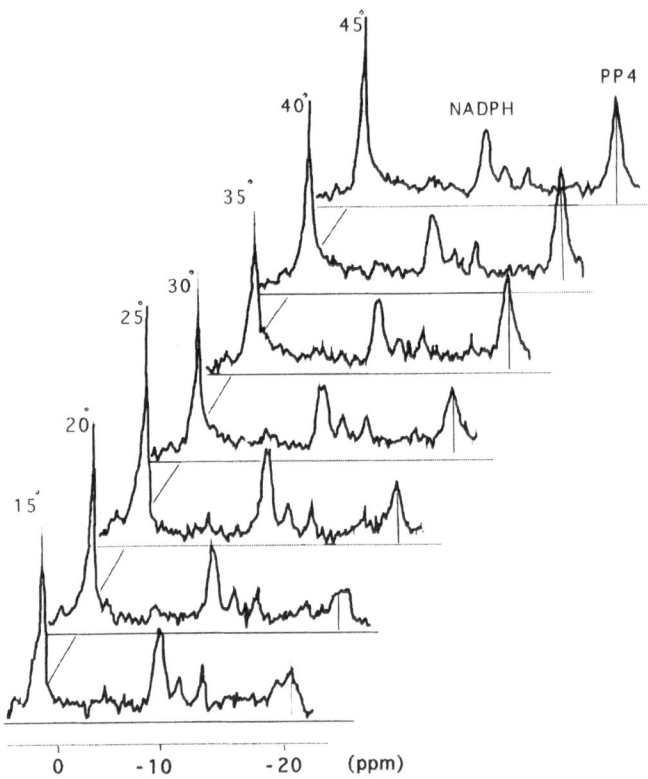

Fig. 5. Effect of temperature on polyP in the bacterium *P. acnes*. Cells were suspended in buffer with 30–40% cell volume for 15 min at indicated temperatures (°C) and then spectra were recorded at 22°C. PP4 is the resonance peak of polyP. (Adapted from Kjieldstad et al. 1989)

4.3.3
Acinetobacter johnsonii

Activated sludge in wastewater treatment plants is enriched with polyP-accumulating bacteria such as *Acinetobacter*. The strictly aerobic, polyP-accumulating *Acinetobacter johnsonii* strain 210A degrades its polyP when oxidative phosphorylation is impaired. The end products of this degradation, divalent metal ions and inorganic phosphate, are excreted as a neutral metal-phosphate via the electrogenic symport system of the organism. In vivo ^{31}P-NMR studies of polyP degradation in anaerobic cell suspensions reveal the presence of a considerable outwardly directed phosphate gradient across the cytoplasmic membrane corresponding to a gradient of at least 100 mV. Thus, energy recycling by metal phosphate efflux contributes significantly to the overall production of metabolic energy in *A. johnsonii* 210A (van Veen et al. 1994).

4.4
Lower Eukaryotes

4.4.1
Yeast

Salhany et al. (1975) first applied high-resolution ^{31}P-NMR to study phosphorus metabolites in intact yeast cells. Since then, yeast has become a popular system for intact cell NMR research (Gillies et al. 1981; Grimmecke et al. 1981; Reidl et al. 1989). Figure 6 shows that the added orthophosphate is rapidly accumulated by the cells and stored mainly in a stable pool of polyP with an average chain length of about 200 units during phosphate-shift experiments (Bourne 1990). The average chain length is estimated from the ratio of the area of internal P peak to that of the terminal P peak. Figure 7 shows the ^{31}P-NMR spectra of yeast-form cells at early stationary phase of growth, as well as germ tubes and hyphae. The intensity of most signals, as measured relative to that of Pi, is clearly modulated both at the different phases of growth and during yeast-to-mycelium conversion. In particular, the intensity of the polyP signal is high in exponentially growing cells, then progressively declines in the stationary phase, is very low in germ tubes and, finally, becomes undetectable in hyphae (Cassone et al. 1983). However, Greenfield et al. (1987) did not observe significant difference in total polyP between log phase and stationary phase yeast cells.

A recent ^{31}P-NMR study by Shirahama et al. (1996) shows that under phosphate starvation, wild-type yeast cells continue to grow for two to three generations, implying that wild-type cells can use polyP to sustain the growth. In contrast, delta slp1 cells, which are defective in the vacuolar compartment and lack polyP, cease their growth immediately under phos-

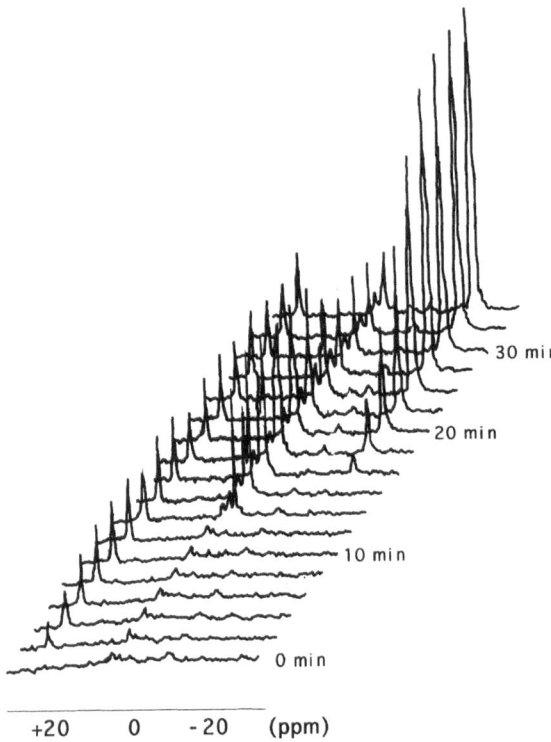

Fig. 6. Phosphate-shift experiments with yeast *C. utilis*. At time 2 min methyl-phosphonate (200 μmol) was added to the culture. At time 15 min and 38 min ortho-phosphate (800 μmol) was added to the culture. (reprinted from Bourne 1990 with permission)

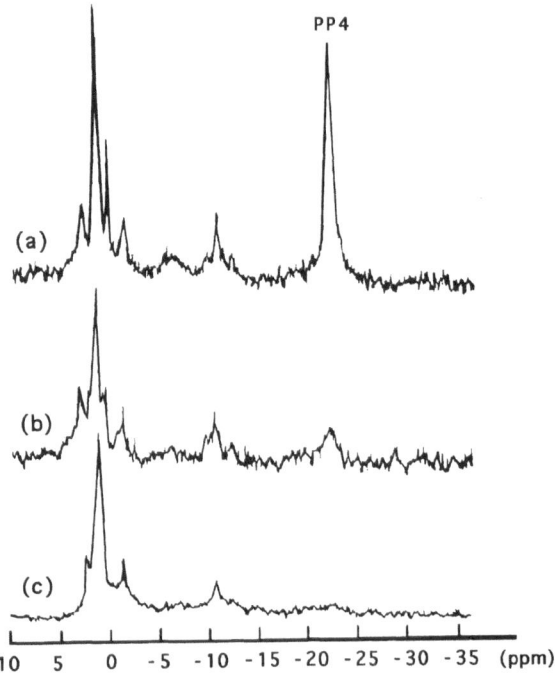

Fig. 7. Phosphorus NMR spectra of yeast *C. albicans* at different stages of growth. **a** Early stationary phase. **b** Germ tubes. **c** Hyphae. (Adapted from Cassone et al. 1983)

phate starvation. This study provided strong evidence that vacuolar polyP represent an active pool for phosphate and is mobilized to cytosol during phosphate starvation and sustained cell growth for a couple of rounds of cell cycle.

The proposed pH buffering and phosphogenic functions of polyP are investigated by subjecting chemostat-cultivated *Saccharomyces cerevisiae* to alkalization by NaOH addition and anaerobiosis (Castro et al. 1995). The subsequent changes in intracellular phosphate-containing species were observed by ^{31}P-NMR. The alkalization increases cytosolic pH and causes rapid polyP degradation to short chains. The pH changes and extent of polyP degradation depend inversely on initial polyP content. In contrast, anaerobiosis results in the complete hydrolysis of polyP to orthophosphates as opposed to short-chain polyP. The bulk of NMR-visible polyP (vacuolar) degradation to short polymers conceivably contributes to neutralizing added alkalinity. During anaerobiosis, however, polyP degradation may serve other functions, such as phosphorylation potential regulation.

Loureiro-Dias and Santos (1989) have shown that 2-deoxyglucose also causes a significant decrease in the polyP level and a downshift of cytosolic pH by 0.4 pH units as monitored by ^{31}P-NMR. Beauvoit et al. (1991) reported that a rapid hydrolysis of polyP occurs in yeast cells in the presence of either uncoupler CCCP or a vacuolar membrane ATPase specific inhibitor, bafilomycin A1, as monitored by ^{31}P-NMR. These studies support the notion that polyP is coupled to the energy state of the organism. The vacuolar polyP content appears to depend on two factors: vacuolar pH, strictly linked to the vacuolar ATPase activity, and inorganic phosphate concentration. Thus, polyP is totally absent in a null vacuolar ATPase activity mutant as indicated by ^{31}P-NMR. However, it is not clear whether these mutants may contain polyP in the NMR-invisible form (c.g. membrane or DNA bound). These ATPase mutants will be useful for probing polyP functions.

Possible involvement of polyP in the cellular repair of ionization damage has been studied with ^{31}P-NMR. Using a novel NMR spectroscopy probe which incorporated a bioreactor, a radiation source, and a radio frequency detection circuit tunable between 100 and 300 MHz for in vivo NMR spectroscopy of ^{23}Na, ^{13}C, and ^{31}P at 11.7 Tesla, it has been shown that a rapid decrease in adenosine triphosphate (ATP) and polyP occurs at the onset of irradiation (8 Gy/h) followed by a slow recovery of polyP (Magness and McFarland 1997). Similar study has been made by Holahan et al. (1988). They employed ^{31}P-NMR spectroscopy to examine alterations in phosphate pools during cellular recovery from radiation damage in yeast cells. They suggest that the polyP is hydrolyzed as a source of phosphates for repair of radiation damage.

The thermotolerant yeast *Hansenula polymorpha* is able to grow on vanadate concentrations (>96 mM) that are toxic to other organisms. *Hansenula polymorpha* cells growing on a vanadate-containing medium undergo a sig-

(A) Control

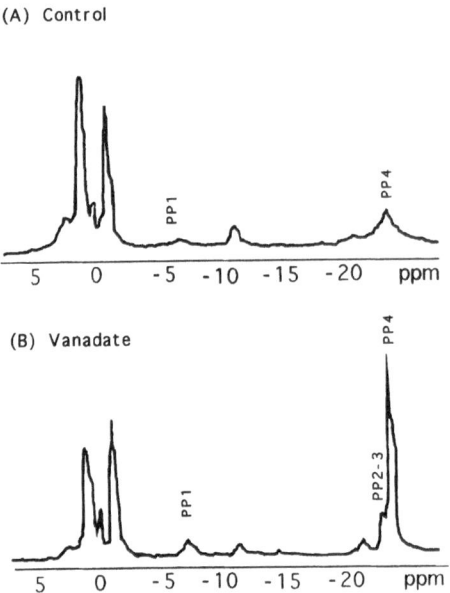

PP1

PP4

5   0   -5   -10 -15 -20   ppm

Fig. 8. Effect of vanadium on polyP in
H. polymorpha cells. A Control cells.
B Cells cultured in presence of 50 mM
sodium orthovanadate. (reprinted from
Mannazzu et al. 1997 with permission)

(B) Vanadate

PP4

PP1

PP2-3

5   0   -5   -10 -15 -20   ppm

nificant increase in cell vacuoles and a thickening of the cell wall. The presence of small cytoplasmic vesicles and an increase in cristae at the level of the plasma membrane were also observed. These ultrastructural modifications were accompanied by a large increase in the intracellular polyP level, as detected by in vivo ^{31}P-NMR (Fig. 8). The increases in vacuoles and polyP in vacuoles suggest that these changes may be involved in vanadium detoxification (Mannazzu et al. 1997). However, more quantitative analysis is needed to assess the role of polyP in metal detoxification.

4.4.2
Neurospora crassa

Greenfield et al. (1988) used ^{31}P-NMR to study the effect of insulin on phosphorus metabolites in wall-less *Neurospora crassa* mutant. The spectra show millimolar levels of intracellular inorganic phosphate (Pi), phosphodiesters, and diphosphates including sugar diphosphates and polyP. Although insulin affects the growth rate of *N. crassa* mutants, it apparently has no effect on polyP level.

Yang et al. (1993) used high-resolution ^{31}P-NMR to investigate the effects of growth stage and environmental osmolarity on changes of polyP metabolism and intracellular pH in wild type *Neurospora crassa* cells. They found that both polyP and cytosolic pH are growth-dependent. As shown in Fig. 9, the ratio of polyP to orthophosphate in vacuoles increases from 2.4 to 13.5 in *N. crassa* as cells grew from early log phase to stationary phase. This

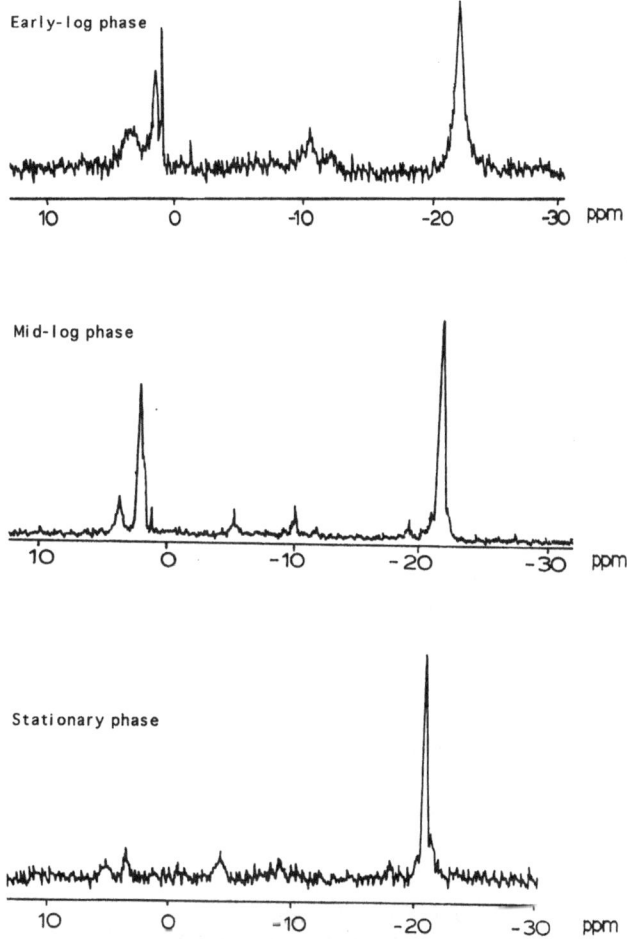

Fig. 9. Phosphorus NMR spectra of *N. crassa* at different stages of growth

observation is consistent with the notion that polyP serves as a phosphate reserve when growth rate slows down. From the same ^{31}P-NMR spectrum, it can be shown that cytosolic pH increases from 6.91 to 7.25 and vacuolar pH from 6.49 to 6.84 in *N. crassa* from early log phase to stationary phase. They also reported that hypoosmotic shock of *N. crassa* produces growth-dependent changes including: (1) a rapid hydrolysis of polyP with a concomitant increase in the concentration of the cytoplasmic phosphate; (2) an increase in cytoplasmic pH: and (3) an increase in vacuolar pH. Early log phase cells produced the most dramatic response (Fig. 10) whereas the stationary phase cells appeared to be recalcitrant to the osmotic stress. Thus, 95 and 60 % of polyphosphate in the early log phase and mid-log phase cells, respectively, disappeared in response to hypoosmotic shock, but little or no hydrolysis of polyP occurrend in stationary cells. The cytosolic pH and the

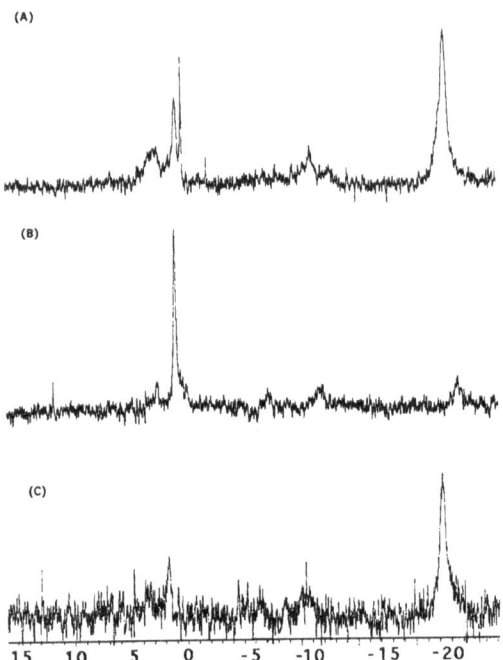

Fig. 10. Effect of hypoosmotic stress on polyP in *N. crassa* cells at early logarithmic phase of growth. **A** control. **B** Hypoosmotically treated with water for 2 h. **C** Hypoosmotically treated for 2 h and then re-incubated in Vogel's medium for 30 min. (reprinted from Yang et al. 1993 with permission)

vacuolar pH increased in response to hypoosmotic shock by 0.4 and 0.53 unit, respectively, in early log phase cells, and by 0.22 and 0.27 unit, respectively, in the mid-log phase cells. In contrast, hypoosmotic shock of the stationary phase cells did not cause any change in intracellular pH. The osmotic stress-induced polyphosphate hydrolysis and pH changes in early log and mid-log phase cells are reversible, suggesting that these changes are related to environment osmolarity (Yang et al. 1993).

4.4.3
Physarum polycephalum

^{31}P-NMR spectroscopic analysis of the polyP pool in cellular and nuclear extracts of *Physarum polycephalum* demonstrates that plasmodia and cysts contain inorganic polyP with an average chain length of about 100 phosphates. However, long-chain polyP is degraded to short-chain polyP with an average chain length of about 10 phosphates only during sporulation. Since the degradation of polyP occurs even in the presence of a sufficiently large pool of inorganic phosphate, Pilatus et al. (1989) concluded that the degradation of polyP serves in supplying energy for biosynthesis during sporulation rather than in increasing the availability of phosphate. Their results, based on NMR analysis, show that 25 % of total polyP resides in nuclei. The data as presented, however, do not clearly indicate whether the degradation occurs in the nuclei or in cytoplasm. Additional study is also needed to confirm the conversion of polyP to ATP through the action of polyP:ADP phosphotransferase activity.

4.4.4
Aspergillus terreus

Lyngstad and Grasdalen (1993) have used a 10-mm-diameter airlift bioreactor for in vivo NMR studies of *Aspergillus terreus*, another mycelial/pellet forming organism, grown in suspension. Signals were observed for intra- and extracellular orthophosphate, glycerol-3-phosphorylethanolamine, glycerol-3-phosphorylcholine, sugar phosphates and polyP. They also showed that polyP signals disappeared when the respiratory gas was exchanged for pure N_2 and the intracellular pH was estimated at 6.2 from the spectra.

4.4.5
Algae

Microorganisms including unicellular algae can be trapped inside agarose beads (3 % w/v) at a concentration of around 7×10^8 cells per ml beads (Bental et al. 1990). During in vivo NMR experiments, the cells in the beads can be perfused continuously with the desired medium at a rate of 1 ml/min. They have used this set-up to examine the effect of osmotic shock on polyP in algae. In addition, by perfusing phosphate-depleted algal cells trapped inside agarose beads with orthophosphate-containing medium, the authors were able to follow the process of polyP synthesis in whole, living cells (Bental et al. 1991). The results suggest that, in *Dunaliella*, low molecular weight, probably cyclic, polyP intermediates are synthesized from Pi, and are then condensed to high molecular weight polymers. Studies of the intracellular organization of the polyP by electron microscopy and NMR techniques indicate that most of these polymers are stored in the cell in a soluble form, and not in solid-like structures (Bental et al. 1991).

Pick et al. (1990) carried out in vivo ^{31}P-NMR studies in the unicellular-alga *Dunaliella salina*, and demonstrated that the cytoplasmic alkalization as induced by ammonium ions (20 mM) induces hydrolysis of polyP, which is correlated kinetically with the recovery of cytoplasmic pH. Analysis of acid extracts of the cells indicates that long-chain polyP is hydrolyzed mainly to tripolyphosphate. The results suggest that the hydrolysis of polyP provides a pH-stat mechanism to counterbalance alkaline stress (Pick et al. 1990). Similar obsersation has also been reported by Greenfield et al. (1987) with yeast.

The phosphate metabolism of *Platymonas subcordifirmis* was investigated by ^{31}P-NMR over a wide range of external pH (Kugel et al. 1987). Under unaerobic condition, Kugel et al. found that the polyP chain starts to disintegrate under high pH conditions. Since polyP responds to cytoplasmic pH change, the authors concluded that polyP may be fairly accessible to cytoplasm.

4.5
Plant and Animal Cells

PolyP can be detected in cultured cells by radiolabeling and polyacrylamide gel electrophoresis (Cowling and Birnboin 1994). The estimated level of polyP is about 0.02 fmol/cell compared to 0.5–3 fmol ATP/cell, much lower than that in microorganisms. Kumble and Kornberg (1995) then unambiguously demonstrated the presence of polyP in mammalian tissues and cells and in various subcellular fractions. They used pure polyP kinase to convert ATP to polyP and hydrolysis of polyP to orthophosphate by a pure exopolyphosphatase as a means to estimate the amount and the size of polyP. PolyP in various rat tissues has a chain length of about 50–800 residues and a concentration range of 20–100 µM, which is roughly 0.2–1 % that of DNA when expressed as nucleotides. It has been shown that aging for rats or apoptosis in HL-60 cells may perturb polyP pools (Lorenz et al. 1997). Despite these elegant studies, no serious attempt has been made to examine whether phosphorus NMR can be adopted to study polyP in animal or plant tissues and cells.

5
Conclusions

High-resolution ^{31}P-NMR provides the only means to monitor polyP at real time in intact tissues and cells that are maintained in various physiological states. Phosphorus NMR spectra cover a wide chemical shifts range and reveal most biological phosphorus compounds in a snapshot. Thus, one can get a dynamic picture of the interconversion of almost all the phosphorus metabolites, including the turnover of polyP and the equilibrium between NMR-visible and -invisible polyP. From the area of the resonance peak, one can estimate the concentration of intracellular polyP on the basis of phosphate unit. By comparing the area of inner P resonance peak and terminal P resonance peak, one can estimate the average length of polyP. The major limitation of using NMR in polyP research is its inability to detect insoluble or immobile polyP due to line broadening. In addition, it remains a technical challenge to bring the detection sensitivity down to micromolar range. Despite these limitations, the potential of phosphorus NMR as a research tool for probing the metabolism and functions of polyP has yet to be fully explored.

Most of the studies discussed above relied on ^{31}P-NMR as a convenient tool to quantitative the soluble polyP in intact cells under different physiological conditions. Such comparison is based on the assumption that the chemical environments of polyP remain the same so that the signal intensities can be used directly to compare polyP levels. Line broadening can occur when polyP is complexed with paramagnetic ions or when the viscosity of the environment increases and polyP becomes less mobile. In this context, it is worth noting that Sianoudis et al. (1986) found that addition of EDTA or NaOH to *Chlorella fusca* cell culture causes a sharpening and an increase of the polyP signals, indicating that polyP is probably membrane bound. Further, they showed that intracellular alkalization induces the conversion of NMR-invisible polyP to NMR-visible polyP. Using non-penetrating cations, like UO_2^{2+} and Eu^{3+}, Tijssen and Steveninck (1984) showed that 20% of polyP is membrane bound.

As can be seen from this review, the techniques and strategies of using ^{31}P-NMR to study polyP have not changed very much over the past 15 years. Sophisticated multiple pulse approach including two-dimensional and multiple quantum spectroscopy has not been employed by investigators yet in polyP research. In the past, there is also a lack of systematic comparison of polyP metabolism and function in different organisms using phosphorus NMR. In the future, phosphorus NMR study of intact cells should perhaps emphasize not only the in-depth study of a particular organism, in conjunction with biochemical and molecular biological tools, but also the broader comparative studies of different organisms, particularly other well-established experimental multicellular organisms such als *C. elegans* and *Drosophila*.

Acknowledgements. The author wishes to thank Dr. Ching-Nien Chen, National Institutes of Health, and Dr. Lou-Sing Kan, Academia Sinica, for their critical comments and discussions. The author also thanks Li-Yan Choi for the graphic work. Due to the limits of space, the author wishes to apologize to those whose work is not cited here.

References

Akiyama M, Crooke E, Kornberg A (1992) The polyphosphate kinase gene of *Escherichia coli*: isolation and sequence of the *ppk* gene and membrane location of the protein. J Biol Chem 267: 22556–22561

Akiyama M, Crooke E, Kornberg A (1993) An exopolyphosphatase of *Escherichia coli*: the enzyme and its *ppx* gene in a polyphosphate operon. J Biol Chem 268: 633–639

Beauvoit B, Rigoulet M, Raffard G, Canioni P, Guerin B (1991) Differential sensitivity of the cellular compartments of *Saccharomyces cerevisiae* to protonophoric uncoupler *under fermentative and respiratory energy supply. Biochemistry 30: 11212–11220*

Bental M, Pick U, Avron M, Degani H (1990) Metabolic studies with NMR spectroscopy of the alga *Dunaliella salina* trapped within agarose beads. Eur J Biochem 188: 111–116

Bental M, Pick U, Avron M, Degani H (1991) Polyphosphate metabolism in the alga *Dunaliella salina* studied by ^{31}P-NMR. Biochim Biophys Acta 1092: 21–28

Bonting CF, Kortstee GJ, Zehnder AJ (1991) Properties of polyphosphate: AMP phosphotransferase of *Acinetobacter* strain 210A. J Bacteriol 173: 6484–6488

Bourne RM (1990) A ^{31}P-NMR study of phosphate transport and compartmentation in *Candida utilis*. Biochim Biophys Acta 1055: 1–9

Cassone A, Carpinelli G, Angiolella L, Maddaluno G, Podo F (1983) ^{31}P nuclear magnetic resonance study of growth and dimorphic transition in *Candida albicans*. J Gen Microbiol 129: 1569–1575

Castro CD, Meehan AJ, Koretsky AP, Domach MM (1995) In situ ^{31}P nuclear magnetic resonance for observation of polyphosphate and catabolite responses of chemostat-cultivated *Saccharomyces cerevisiae* after alkalinization. Appl Environ Microbiol 61: 4448–4453

Cowling RT, Birnboim HM (1994) Incorporation of [^{32}P]orthophosphate into inorganic polyphosphates by human granulocytes and other human cell types. J Biol Chem 269: 9480–9485

Cramer CL, Davis RH (1984) Polyphosphate–cation interaction in the amino acid-containing vacuole of *Neurospora crassa*. J Biol Chem 259: 5152–5157

Gillies RJ, Ugurbil K, den Hollander JA, Shulman RG (1981) ^{31}P-NMR studies of intracellular pH and phosphate metabolism during cell division cycle of *Saccharomyces cerevisiae*. Proc Natl Acad Sci USA 78: 2125–2129

Greenfield NJ, Hussain M, Lenard J (1987) Effects of growth state and amines on cytoplasmic and vacuolar pH, phosphate and polyphosphate levels in *Saccharomyces cerevisiae*: a 31P-nuclear magnetic resonance study. Biochim Biophys Acta 926: 205–214

Greenfield NJ, McKenzie MA, Adebodun F, Jordan F, Lenard J (1988) Metabolism of D-glucose in a wall-less mutant of *Neurospora crassa* examined by ^{13}C and ^{31}P nuclear magnetic resonances: effects of insulin. Biochemistry 27: 8526–8533

Grimmecke HD, Meyer H, Scheller D, Reuter G (1981) Structure of the cell wall polysaccharide in the food protein yeast *Candida spec* H. III. Characterization of different phosphate bonds in the mannan–protein–phosphate complex. Z Allg Mikrobiol 21: 201–210

Harold FM (1966) Inorganic polyphosphates in biology: structure, metabolism, and function. Bacteriol Rev 30: 772–794

Holahan PK, Knizner SA, Gabriel CM, Swenberg CE (1988) Alterations in phosphate metabolism during cellular recovery of radiation damage in yeast. Int J Radiat Biol 54: 545–562

Huang R, Reusch RN (1996) Poly(3-hydroxybutyrate) is associated with specific proteins in the cytoplasm and membranes of *Escherichia coli*. J Biol Chem 271: 22196–22202

Kjieldstad B, Johnsson A (1987) A ^{31}P-NMR study of *Propionibacterium acnes*, including effects caused by near-ultraviolet irradiation. Biochim Biophys Acta 927: 184–189

Kjieldstad B, Johnsson A, Furuheim KM, Bergan AS, Krane JZ (1989) Hyperthermia induced polyphosphate changes in *Propionibacterium acnes* as studied by ^{31}P-NMR. Naturforschung C44: 45–48

Kugel H, Mayer A, Kirst GO, Leibfritz (1987) In vivo P-31 NMR measurements of phosphate metabolism in *Platymonas subcordiformis* as related to external pH. Eur Biophys J 14: 461–470

Kulaev IS (1979) The biochemistry of inorganic polyphosphates. Wiley, New York

Kulaev IS, Vagabov VM (1983) Polyphosphate metabolism in micro-organisms. Adv Microb Physiol 24: 83–171

Kumble KD, Kornberg A (1995) Inorganic polyphosphate in mammalian cells and tissues. J Biol Chem 270: 5818–5822

Lee PC, Bochner BR, Ames BN (1983) AppppA, heat shock stress, and cell oxidation. J Biol Chem 258: 6827–6834

Lorenz B, Munkner J, Oliveira MP, Kuusksalu A, Leitao JM, Muller WE, Schroder HC (1997) Changes in metabolism of inorganic polyphosphate in rat tissues and human cells during development and apoptosis. Biochim Biophys Acta 1335: 51–60

Loureiro-Dias MC, Santos H (1989) Effects of 2-deoxyglucose on *Saccharomyces cerevisiae* as observed by in vivo ^{31}P-NMR. FEMS Microbiol Lett 48: 25–28

Lyngstad M, Grasdalen H (1993) A new NMR airlift bioreactor used in ^{31}P-NMR studies of itaconic acid producing *Aspergillus terreus*. J Biochem Biophys Methods 27: 105–116

Magness JE, McFarland EW (1997) A radiobiological probe for simultaneous NMR spectroscopy and 192Ir gamma irradiation of *Saccharomyces cerevisiae*. Biochem Biophys Res Commun 233: 238–243

Mannazzu I, Guerra E, Strabbioli R, Masia A, Maestrale GB, Zoroddu MA, Fatichenti F (1997) Vanadium affects vacuolation and phosphate metabolism in *Hansenula polymorpha*. FEMS Microbiol Lett 147: 23–28

Meehan AJ, Eskey CJ, Koretsky AP, Domach MM (1992) Cultivator for NMR studies of suspended cell cultures. Biotechnol Bioeng 40: 1359–1366

Phillips NF, Horn PJ, Wood HG (1993) The polyphosphate- and ATP-dependent glucokinase from *Propionibacterium shermanii*: both activities are catalyzed by the same protein. Arch Biochem Biophys 300: 309–319

Pilatus U, Mayer A, Hildebrandt A (1989) Nuclear polyphosphate as a possible source of energy during the sporulation of *Physarum polycephalum*. Arch Biochem Biophys 275: 215–223

Pick U, Bental M, Chitlaru E, Weiss M (1990) Polyphosphate-hydrolysis – a protective mechanism against alkaline stress? FEBS Lett 274: 15–18

Rao NN, Torriani A (1988) Utilization by *Escherichia coli* of a high-molecular-weight, linear polyphosphate: roles of phosphatases and pore proteins. J Bacteriol 170: 5216–5223

Reidl HH, Grover TA, Takemoto JY (1989) 31P-NMR evidence for cytoplasmic acidification and phosphate extrusion in syringomycin-treated cells of *Rhodotorula pilimanae*. Biochim Biophys Acta 1010: 325–329

Reusch RN, Sadoff HL (1988) Putative structure and functions of a poly-beta-hydroxybutyrate/calcium polyphosphate channel in bacterial plasma membranes. Proc Natl Acad Sci USA 85: 4176–4180

Roberts MF (1987) Polyphosphates. In: Bert CT (ed) Phosphorus NMR in biology. CRC Press, Boca Raton, FL, pp 85–94

Salhany JM, Yamane T, Shulman RG, Ogawa S (1975) High resolution ^{31}P nuclear magnetic resonance studies of intact yeast cells. Proc Natl Acad Sci USA 72: 4966–4970

Sharfstein ST, Keasling JD (1994) Polyphosphate metabolism in *Escherichia coli*. Ann N Y Acad Sci 745: 77–91

Shirahama K, Yazaki Y, Sakano K, Wada Y, Ohsumi Y (1996) Vacuolar function in the phosphate homeostasis of the yeast *Saccharomyces cerevisiae*. Plant Cell Physiol 37: 1090–1093

Sianoudis J, Kusel AC, Mayer A, Grimme LH, Leibfritz D (1986) Distribution of polyphosphate in cell compartments of *Chlorella fusca* by 31-P-NMR spectroscopy. Arch Microbiol 144: 48–54

Tijssen JP, Van Steveninck J (1984) Detection of a yeast polyphosphate fraction localized outside the plasma membrane by the method of phosphorus-31 nuclear magnetic resonance. Biochem Biophys Res Commun 119: 447–451

Van Dien SJ, Keyhani S, Yang C, Keasling JD (1997) Manipulation of independent synthesis and degradation of polyphosphate in *Escherichia coli* for investigation of phosphate secretion from the cell. Appl Environ Microbiol 63: 1689–1695

Van Veen HW, Abee T, Kortstee GJ, Pereira H, Konings WN, Zehnder AJ (1994) Generation of a proton motive force by the excretion of metal-phosphate in the polyphosphate-accumulating *Acinetobacter johnsonii* strain 210A. J Biol Chem 269: 29509–29514

Van Wazer JR, Ditchfield R (1987) Phosphorus compounds and their ^{31}P chemical shifts. In: Bert CT (ed) Phosphorus NMR in biology. CRC Press, Boca Raton, pp 1–24

Wood HG, Clark JE (1988) Biological aspects of inorganic polyphosphates. Annu Rev Biochem 57: 235–260

Wurst H, Shiba T, Kornberg A (1995) The gene for a major exopolyphosphatase of *Saccharomyces cerevisiae*. J Bacteriol 177: 898–906

Yang YC, Bastos M, Chen KY (1993) Effects of osmotic stress and growth stage on cellular pH and polyphosphate metabolism in *Neurospora crassa* as studied by 31P nuclear magnetic resonance spectroscopy. Biochim Biophys Acta 1179: 141–147

Polyphosphate-Accumulating Bacteria and Enhanced Biological Phosphorus Removal

G. J. J. Kortstee[1] and H. W. van Veen[2]

1
Introduction

More than 50 years ago, Jeener and Brachet (1944) observed that addition of inorganic phosphate (Pi) to a suspension of yeast cells previously subjected to phosphate starvation induced massive accumulation of a basophilic substance within the cells. Soon afterwards this substance was isolated and identified as inorganic polyphosphate (polyP) (Schmidt et al. 1946; Wiame 1948). This was by no means the first isolation of polyP from a microorganism. As early as 1888 Liebermann had obtained polyP from yeast, probably metaphosphate. Over the next 50 years hardly any paper on this polyP from yeast appeared. The work of Wiame and others marks the beginning of biological research on inorganic polyP.

During the following decade numerous papers describe the occurrence of polyP in a wide variety of micro-organisms and its accumulation under nutritional conditions unfavourable to growth. The polyP encountered in these studies was found to be a polymer of orthophosphate, with phosphoanhydride linkages thermodynamically equivalent to the energy-rich phosphate bonds of ATP. Upon hydrolysis of the P-O-P bond in polyP, approximately $10\,kCal\,mol^-$ is liberated, that is about the same amount as is liberated in the hydrolysis of the terminal phosphoric anhydride bonds in ATP (Harold 1966; Dawes and Senior 1973). During recent years a number of reviews on the biochemistry (Wood and Clark 1988), biology (Kortstee et al. 1994) and molecular biology (Kornberg 1995) of polyP metabolism have been published. Therefore, the present chapter aims at a comprehensive coverage of almost only the most recent literature. It contains sections on: (1) three presumed functions of polyP in bacteria; (2) the biosynthesis and degradation of polyP in bacteria, in particular in the polyP-accumulating *Acinetobacter johnsonii* 210A; (3) the uptake and efflux of Pi in Gram-negative bacteria, including *A. johnsonii* 210A; and (4) the biochemical and ecologi-

[1] Department of Microbiology, Agricultural University, Hesselink van Suchtelenweg 4, 6703 CT, Wageningen, The Netherlands.
[2] Department of Microbiology, University of Groningen, Kerklaan 30, 9751 NN, Haren, The Netherlands.

Progress in Molecular and Subcellular Biology, Vol. 23
H. C. Schröder, W. E. G. Müller (Eds.)
© Springer-Verlag Berlin Heidelberg 1999

cal aspects of enhanced biological phosphorus removal, preceded by a few comments on the chemistry of polyP.

2
Chemistry of Inorganic Condensed Phosphates

The term inorganic condensed phosphates comprises pentavalent phosphorus compounds in which various numbers of tetrahedral PO_4 groups are linked together by oxygen bridges (Thilo 1962). These compounds fall into the following three classes.

1. Cyclic Condensed Phosphates. These have the following general formula, $M_nP_nO_{3n}$ (M is a monovalent cation), and are correctly called metaphosphates. The most well-known species are tri- and tetrametaphosphate which occur in certain melts. Strong alkaline conditions convert metaphosphates to the corresponding linear polyphosphate and when heated in acid they are hydrolyzed to Pi.

2. Linear Condensed Polyphosphates. These phosphates have an elementary composition according to the formula $M_{n+2}P_nO_{3n+1}$ and are called polyphosphates (polyP). The length of the unbranched chain may vary from 2 (pyrophosphate) to around 10^4 phosphate residues. All alkali metal salts of polyP are soluble in water although the polyPs of Ba^{2+}, Pb^{2+} and Mg^{2+} only dissolve to a limited extent in aqueous solutions. In the absence of divalent cations, linear polyP is stable to alkali. However, polyP is completely hydrolyzed by 1 N acid, within 15 min at 100 °C.

3. Cross-Linked Condensed Phosphates. In these polyphosphates phosphate groups occur in which three oxygen atoms are shared with neighbouring phosphate groups. However, these branching points are readily hydrolyzed in water and therefore these polyphosphates unlikely occur in biological tissues.

The methods developed over the years to detect polyP in microorganisms have been reviewed by Bonting (1993). From his overview and also from his own work it appears that the non-invasive ^{31}P-NMR technique which is able to monitor the middle, ultimate and penultimate phosphate residues of polyphosphate chains is an elegant tool to detect polyP when mobile, soluble pools of polyP are present. However, polyP is often complexed in granular structures and in this form it cannot be detected in a ^{31}P-NMR spectrum (Roberts 1987). These granules can easily be detected by electron microscopy because they are very electron dense. The elemental composition of these granules can be determined by the use of energy dispersive X-ray analysis (EDAX) in conjunction with scanning transmission electron microscopy (STEM). It is emphasized that staining of cells with basic dyes such as methylene blue and neutral red as well as Neisser staining give no conclusive evidence of the presence of polyP since these dyes may also stain other polymers.

Many of the methods applied in the chemical estimation of polyP in bacteria depend in their final analysis on the acid hydrolysis of polyP to Pi. Since this hydrolysis will also release Pi from other cellular compounds, polyP must be separated from these compounds in order to obtain a real estimation of the content of polyP. The genetics of the enzymes polyphosphate kinase (PPK) and polyphosphatase (PPX) of *Escherichia coli* (Kornberg 1995) has provided the tools to measure the amount of polyP accurately once the polyP is extracted, free in solution and accessible to both enzymes.

All the methods developed over the years for extraction of polyP from bacteria have been critically reviewed by Bonting (1993). The relatively new procedure for the isolation of polyP was claimed to be complete and non-hydrolytic (Clark et al. 1986). However, in our hands the extraction of polyP from a number of bacteria according to this method is not entirely complete. This has been concluded from ^{31}P-NMR spectra. The other new method of Clark and Wood (1987) appears to be a useful method for sizing microgram quantities of polyphosphates ranging from 3 to nearly 1000 P residues.

3
Functions of PolyP in Bacteria

In recent reviews of polyP metabolism the various functions of polyP have been summarized briefly (Kornberg 1995; Kortstee et al. 1998). Three of the possible general functions of polyP in bacteria remained somewhat underexposed; its function as an energy source, its involvement in the acquisition of competence, and its involvement in the uptake of Ca^{2+}. Therefore these three functions are covered in this chapter in some detail.

3.1
Energy Source

In vitro PPK catalyses the formation of long-chain polyP in a reversible reaction: $ATP + polyP_n \leftrightarrow ADP + polyP_{n+1}$. It has therefore been suggested that PPK may be involved in the formation of ATP when polyP is degraded in vivo (Kornberg 1995). However, the evidence presently available does not entirely support this suggestion. Mutants of *Klebsiella aerogenes* (formerly named *Aerobacter aerogenes*) devoid of PPX, catalysing the reaction $polyP_n + H_2O \rightarrow polyP_{n-1} + Pi$, were blocked in polyP degradation in spite of the fact that PPK was present (Harold and Harold 1965; Harold 1966). These results suggest that at least in *K. aerogenes*, and possibly in other bacteria as well, PKK is not involved in the degradation of polyP in vivo and polyP does not function as an anergy source in the way suggested above.

That metabolic energy liberated from the cleavage of polyP can be conserved via a direct enzymatic synthesis of ATP was demonstrated in the polyP-accumulating, strictly aerobic *Acinetobacter johnsonii* 210A (Van Groenestijn et al. 1987). Cell-free extracts possess polyphosphate:AMP-phosphotransferase activity, catalysing the reaction $polyP_n + AMP \rightarrow polyP_{n-1} + ADP$. The thus formed ADP is converted to ATP by the potent and ubiquitous adenylate kinase that catalyses the reaction $2\ ADP \leftrightarrow ATP + AMP$ and simultaneously regenerates the acceptor molecule AMP. This regeneration of AMP can keep polyP degradation going for a long time where considerable amounts of polyP are available, such as in *A. johnsonii* 210A (Bonting et al. 1991, 1992a).

Convincing evidence that polyP can act as an energy source in vivo in bacteria was produced only recently. High-Pi-grown cells of *A. johnsonii* 210A contain polyP as small granules and one or two large granules in the cytoplasm (Bonting et al. 1993a) while low-Pi-grown cells contain no polyP granules. During aerobic incubation for 2 h without an exogenous carbon and energy source, high-Pi-grown cells hardly excreted Pi, indicating that polyP was not degraded aerobically (Fig. 1). Their ATP content and membrane potential were similar to those in low-Pi-grown cells. During the subsequent anaerobic incubation, the high-Pi-grown cells degraded polyP, resulting in Pi excretion at a rate of about 3 nmol per min per mg protein. These cells maintained a significant ATP level and membrane potential for at least 8 h, whereas in low-Pi-grown cells both dropped to a low level within 1 h (Van Veen et al. 1994a).

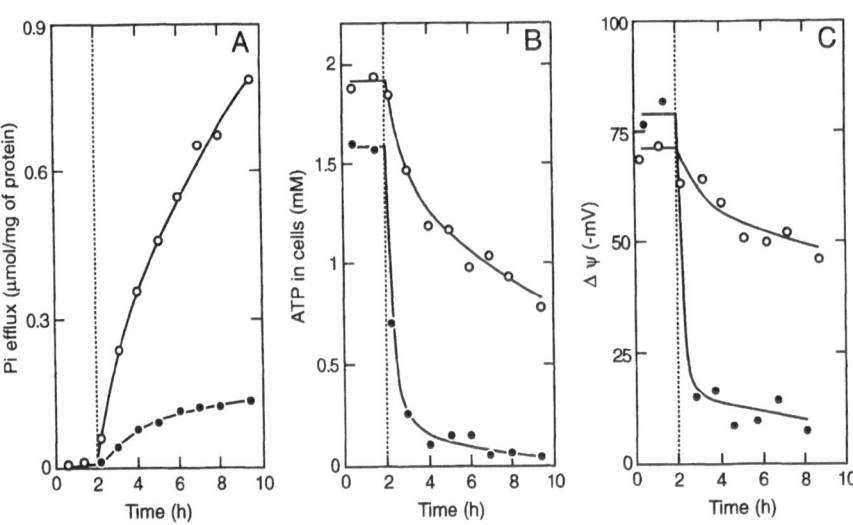

Fig. 1. Conservation of metabolic energy from polyphosphate degradation in *Acinetobacter johnsonii* 210A. Pi efflux (**A**), cellular ATP concentration (**B**) and membrane potential (**C**) were measured in cells with (*open circles*) and without (*closed circles*) polyphosphate during an aerobic period of 2 h and, subsequently, during an anaerobic period of 8 h. (Van Veen et al. 1994a)

The polyphosphate:AMP phosphotransferase/adenylate kinase pathway is not the only way in which *A. johnsonii* 210A maintains a certain proton motive force and ATP level during polyP degradation. The organism also contains a PPX (Bonting et al. 1993b) and the mere excretion of Pi as a metal phosphate (MeHPO$_4$) complex (see Sect. 5.2) occurs in symport with an H$^+$ (Van Veen et al. 1993a) and results in a proton motive force, and consequently in ATP formation. *Acinetobacter johnsonii* 210A contains several cation-amino acid transport systems which couple amino acid translocation to the proton motive force. When in Pi-loaded vesicles of the organism an outwardly directed Pi-gradient was imposed, the amino acids L-lysine and L-proline (via a sodium/proton antiporter) were accumulated, but unloaded vesicles failed to do so (Van Veen et al. 1994a, b). It should be realized that during polyP degradation an appreciable outwardly directed Pi gradient is built up: the internal Pi concentration can reach values between 150 and 200 mM.

In organisms like *K. aerogenes* polyP degradation is most likely catalysed by PPX, and not by PPK or polyphosphate:AMP phosphotransferase (Harold and Harold 1965). In these organisms only an MeHPO$_4$ efflux-induced proton motive force may occur as a result of polyP degradation.

3.2
Acquisition of Competence

Competence is the word that is used for the state of bacterial cells in which they are able to take up exogenous DNA. The result of uptake of foreign DNA is a genetic transformation. In spite of the universal use of bacterial transformation, for a long time little was known about the molecular basis of competence. In 1988 it was proposed (Reusch and Sadoff 1988) that a complex of poly-β-hydroxybutyrate (PHB), Ca^{2+} and polyP was involved in competence. This complex was detected not only in *E. coli*, but also in other genetically transformable bacteria such as *Azotobacter* and *Bacillus* species (Reusch and Sadoff 1983) and in eukaryotic organisms (Reusch 1989). The complex is located in the cytoplasmic membrane and the molar PHB/polyP/Ca^{2+} ratio was estimated to be 2:2:1. The presence of PHB in the complex was unequivocally established by chemical and immunological assays and by ^{1}H-NMR, the presence of polyP with specific enzyme assays and the presence of Ca^{2+} by atomic absorption chromatography (Castuma et al. 1995). During log-phase growth, the complexes are present in low numbers, but increase considerably in number in the stationary phase of growth and, in particular, when the cells are incubated in ice-cold calcium buffers to make them genetically competent (Reusch et al. 1986). There was a strong correlation between the degree of competence and the quantities of the complex. The complexes give rise to a sharp thermotropic transition at about 56 °C, attributed to an increase in membrane viscosity due to dissociation of the complexes (Reusch et al. 1986). This was measured with the fluorescent probe *N*-phenyl-1-naphtylamine (NPN). With 1,6-diphenyl-1,3,5-hexatriene as a fluorescent probe, a steep loss in fluorescence anisotropy was seen when the cell membranes undergo a transition from the rigid to the fluid phase at about 25 °C. The rise in relative fluorescence activity with NPN was not seen in non-competent cells; it was, however, present in *ppk⁻* cells but to a somewhat lesser extent (Castuma et al. 1995). Since *ppk⁻* cells lack the high molecular polyP of about 1000 Pi residues, one is tempted to conclude that polyP ist not involved in competence. However, the polyP of the PHB/polyP/Ca^{2+} complex in the cytoplasmic membrane appeared to be low molecular polyP of 60–70 Pi residues. Thus, this type of polyP is present in the PHB/polyP/Ca^{2+} complex of wild type cells and in that of *ppk⁻* cells whereas the high molecular weight polyP is only present in the cytoplasma of wild type cells. The observations of Castuma et al. (1995) clearly show that membranes of non-competent cells, either derived from wild type cells or from *ppk⁻* cells, hardly contained polyP, indicating that competence depends on polyP. However, how the PHB/polyP/Ca^{2+} complex in the cytoplasmic membrane is involved in competence is not yet clear. It has been suggested that this complex perturbs the conformation of the lipid matrix, resulting in a greater permeability of charged molecules and thus allowing the entry of DNA. No studies are available with mutants of *E. coli* which lack the genes encoding the enzymes catalysing the synthesis of the low molecular weight PHB and low molecular weight polyP to confirm the involvement of these compounds in bacterial competence.

3.3
Uptake of Ca^{2+}

The composition, location and putative structure of the PHB/polyP/Ca^{2+} complex found in competent cells of *E. coli* and other transformable bacteria led Reusch and Sadoff (1988) to suggest that the complex also functions as a Ca^{2+} channel, in addition to being involved in the acquisition of competence. Voltage-activated Ca^{2+} channels composed of PHB, polyP and Ca^{2+} were shown to be present in the membranes of *E. coli* (Reusch et al. 1995). The channels extracted from cells and incorporated into bilayers of synthetic phosphatidylcholine displayed many of the characteristics of protein Ca^{2+} channels, such as voltage-activated, selective for divalent over monovalent cations and blocked by Co^{2+}, Cd^{2+} and Mg^{2+}. Channels reconstituted from synthetic Ca-polyP and PHB isolated from *E. coli* showed comparable single channel currents. These observations support the conclusion that in *E. coli* biological non-proteinaceous Ca^{2+} channels are available for Ca^{2+} entry. The existence of a Ca^{2+} channel in *E. coli* was already earlier suggested by the inhibition of the calcium-related chemotaxis by the calcium channel blocker ω-conotoxin (Tisa et al. 1993). In addition to the above Ca^{2+} channel for the uptake of Ca^{2+}, *E. coli* probably contains three distinct secondary systems involved in Ca^{2+} extrusion (for a review, see Van Veen et al. 1994c). For other functions of polyP the reader is referred to chapters 1, 3, 4, 8 and 9 in this volume.

4
Biosynthesis and Biodegradation of PolyP in *Acinetobacter* spp.

4.1
Biosynthesis

Until recently, the mechanism of polyP formation in bacteria belonging to the genus *Acinetobacter* was unclear. Very low activities, ranging from 0 to 4.7 nmol min^{-1} mg $protein^{-1}$, have been recorded for PPK in a number of polyP-accumulating strains (T'Seyen et al. 1985; Van Groenestijn et al. 1989). The enzyme could also not be detected spectroscopically in the most studied polyP-accumulating acinetobacter, strain *A. johnsonii* 210A, when the polyP-dependent conversion of ATP into ADP was assayed. Only Bayly et al. (1991) reported readily detectable PPK activities in polyP-accumulating *Acinetobacter* spp. isolated from activated sludge. These values, up to 43 nmol min^{-1} mg $protein^{-1}$, were obtained by assaying the PPK activity with ADP and polyP as substrates. Since ADP is also a substrate for adenylate kinase, a complete inhibition of adenylate kinase is essential for obtaining the true activity of PPK. In view of the possible incomplete inhibition of adenylate kinase by P^1,P^5-di(adenosine-5^1)-pentaphosphate, the values reported by Bayly and colleagues may be overestimations. Bark et al. (1992) attained a similar conclusion during the study of their polyP-accumulating *Acinetobacter* isolates. When low Pi-grown cells of *A. johnsonii* 210A, which do not contain polyP and only a very limited content of PHB (Bonting et al. 1992a), were energized with glucose and pyrrolo-quinoline quinone (the coenzyme of the glucose dehydrogenase present in acinetobacters), a polyP signal emerged in the ^{31}P-NMR spectrum (Fig. 2) and the extracellular Pi signal disappeared. No other Pi-containing compounds accumulated and small peaks of adenine nucleotides were detectable, indicating that Pi was only converted into polyP in these non-growing cell suspensions. When these cell suspensions were pre-incubated with the H^+-ATPase inhibitor N,N-dicyclohexylcarbodiimide, 7.5 µg mg $protein^{-1}$, ATP formation was completely impaired and no polyP was formed, indicating that ATP was involved in polyP formation. In spite of its low activity in vitro, Bonting (1993) therefore suggested that PPK was the most likely candidate for polyP formation in *A. johnsonii* 210A. Recently, the gene *ppk* has been found in this organism (H.-Y. Kim, pers. comm.). Based on the ^{31}P-NMR experiments, a specific activity of 60–120 nmol min^{-1} mg $protein^{-1}$ for PPK was calculated in Pi-deficient cells, the higher values approaching the Pi uptake rate of

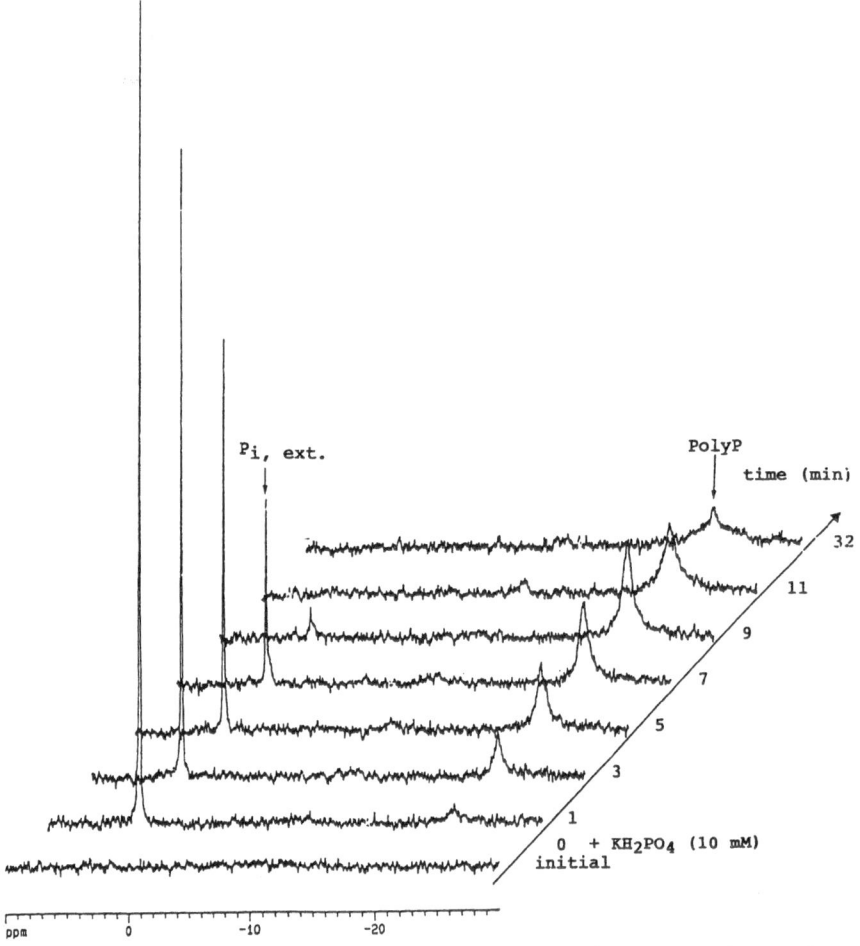

Fig. 2. Formation of polyphosphate (PolyP) by phosphate (Pi)-deficient cells of *Acinetobacter johnsonii* 210A as monitored by in vivo ^{31}P-NMR. Note immobilization of polyP in the period 11–32 min

150–200 nmol min^{-1} mg protein^{-1} (Bonting et al. 1992a). It would be a big step forward if methods would become available that maintain the PPK activity during extract preparation from all types of microbial cells.

Mutant studies with *K. aerogenes* (Harold 1966) and *E. coli* (Crooke et al. 1994) have shown that PPK is vital for the synthesis of polyP with about 1000 Pi residues. However, *ppk*⁻ mutants of *E. coli* synthesize low molecular weight polyP of 60–70 Pi residues, indicating a second pathway for polyP formation distinct from the PPK pathway (Castuma et al. 1995). Two alternative systems have been put forward for polyP formation (Bonting 1993): (1) a phosphorylated compound other than ATP acts as the phosphate donor in polyP formation; (2) a membrane-bound, proton-translocating polyP synthase which acts in the same way as the proton-translocating ATPase. In fact, in *Rhodospirillum rubrum* a proton-translocating, pyrophosphatase has been observed (Nyren et al. 1991). It would be interesting to see whether a proton-translocating polyP synthase is involved in the formation of low molecular weight polyP, not only in *E. coli* but in particular in activated sludges that contain appreciable amounts of low molecular weight polyP (Mino et al. 1985; Appeldoorn et al. 1992b). The properties of PPK of *E. coli* and the gene *ppk* are dealt with in other chapters in this volume (see Chaps. 1 and 9).

4.2
Biodegradation

Linear condensed polyP can be biochemically attacked in three ways at the terminal Pi residues: with AMP, glucose (fructose) and water. These reactions are catalysed by polyphosphate:AMP phosphotransferase, polyphosphate glucokinase (fructokinase) and PPX. A number of these enzymes isolated from bacteria have been reviewed by Wood and Clark (1988). When grown in carbon-limited batch or chemostat cultures, *A. johnsonii* 210A contains two enzymes catalysing the degradation of polyP: polyphosphate:AMP phosphotransferase and PPX (Bonting et al. 1992a).

The molecular weight of this polyphosphate:AMP phosphotransferase could not be determined in a strightforward manner because of the delicate stability of the enzyme. However, the native enzyme is most likely a monomer of about 60 kDa (Bonting et al. 1991). The highest activities were found with polyP molecules of 18–44 Pi residues and the enzyme showed a broad pH optimum, ranging from 6.5 to 8.5. No activity was seen at pH values below 5.5 and above 10. The polyphosphate chain was completely converted to ADP and the mechanism of degradation is processive as in a number of other polyP-transforming enzymes (Wood and Clark 1988). Thus, once bound, the polyphosphate chain does not dissociate from the enzyme, resulting in the formation of free, shorter polyP intermediates. However, although pyro-, tri- and tetraphosphate could not serve as substrate for the enzyme, they inhibited the enzyme. This is probably due to binding of the smaller molecules in the neighbourhood of the catalytic site, resulting in a change of the catalytic activities of the enzyme. Kinetic studies revealed apparent Km values of 0.26 mM for AMP and 0.8 μM for polyP with an average chain length of 35 Pi residues, indicating that the enzyme possesses a high affinity for both substrates and as a result may function well in the energy metabolism of the organism.

Purified (exo)PPX of *A. johnsonii* 210A is a monomer of 55 kDa (Bonting et al. 1993b). The enzyme catalyses the degradation of long polyphosphate chains to short chains of 5, 4 and 3 Pi residues in a processive manner since no long-chain intermediates were formed. However, since tri- and tetraphosphate are intermediates during the degradation of polyP with 5 Pi residues, short-chain polyphosphate chains are degraded to Pi in a non-processive manner. The apparent Km value for high molecular weight polyP with an average chain length of 64 residues was 5.9 µM and for tetraphosphate 1.2 mM. The activity towards tri- and tetraphosphate was dependent on certain ions. In the presence of 300 mM NH_4^+ and 10 mM Mg^{2+} the enzyme was active with tri- and tetraphosphate; in the presence of 0.1 M K^+ and 2 mM Mg^{2+} no activity with triphosphate was seen whereas the activity with tetraphosphate was drastically reduced. Enzyme activity with high molecular weight polyP as substrate was partially inhibited by pyrophosphate and triphosphate added to a final concentration of 2 mM. This inhibition and the two mechanisms of polyP degradation could be explained by assuming two binding sites for the polyP chain, as described by Bonting et al. (1993b). The PPX of *A. johnsonii* 210A is clearly different from the enzymes of *Corynebacterium xerosis*, *K. aerogenes* and *E. coli*: the enzymes from the latter three organisms failed to produce tetra-, tri- and pyrophosphate during degradation of long-chain polyphosphates.

Located 7 bp downstream of the *ppk* gene in *E. coli*, a gene *ppx* was found that encodes the PPX of this organism. Transcription of the *ppx* gene was found to be almost completely dependent of the *ppk* promoter, indicating a polyP operon of *ppk* and *ppx* (Akiyama et al. 1993). A second polyphosphatase has been found in mutants of *E. coli* which lack *ppx*. This enzyme appeared to be guanosine pentaphosphate phosphohydrolase (GPP) (Hara and Sy 1983; Keasling et al. 1993). The significance of GPP as a polyphosphatase and of polyP in the regulation of stress and survival in bacteria, in particular in *E. coli*, has been discussed by Kornberg (1995).

Little is known about the regulation of the enzymes involved in polyP synthesis and polyP degradation, also in relation to phosphate deprivation in polyP-accumulating bacteria. For *E. coli* it has been suggested that alkaline phosphatase, triphosphatase and PPX are regulated by a single metabolite, possess a common regulatory gene and constitute a single regulon (Kulaev 1979). Harold (1966) suggested that in *K. aerogenes*, PPK, PPX and alkaline phosphatase share a common regulatory gene but do not fall into single operon. In *A. johnsonii* 210A the phosphate uptake system and alkaline phosphatase are derepressed during phosphate starvation whereas polyphosphate:AMP phosphotransferase and PPX are repressed (Bonting et al. 1992a). This organism contains a considerable PPK activity, as measured by ^{31}P-NMR, under Pi-deficient growth conditions, but no straight-forward values for this enzyme activity could be calculated from experiments with cells grown under high Pi conditions, possibly due to a much lower activity of the Pi uptake system or a repressed PPK, or to both. Polyphosphate:AMP phosphotransferase and PPX seem to be regulated by intracellular polyP in *A. johnsonii* 210A, rather than by the extracellular Pi concentration. As soon as polyP formation is possible, when more Pi is present in the growth medium than the amount required for growth, the activity of both enzymes increases. There is no evidence to support the idea that both enzymes or one of these two enzymes are involved in polyP formation in this organism.

5
Phosphate Transport in *Acinetobacter johnsonii* 210A

5.1
Uptake and Efflux

Since polyP-accumulating acinetobacters are often encountered in treatment plants showing enhanced biological phosphorus removal (Toerien et al. 1990; Bonting et al. 1992b) and the polyP metabolism of one of these strains, *A. johnsonii* 210A, has been relatively well studied and takes up Pi aerobically and releases it anaerobically (Van Groenestijn 1988), the transport of Pi in this polyP-accumulating organism was examined (Van Veen 1994).

Pi is taken up against a concentration gradient by energy-dependent, carrier-mediated processes. The organism contains two uptake systems, an inducible high-affinity unidirectional transport system belonging to the group of ATP-driven, binding-protein-dependent transport systems and a low-affinity, bidirectional system which is constitutive and driven by the proton motive force. Their Km values are 0.7 ± 0.2 and 9 ± 1 µM, respectively. The inducible system is 6- to 10-fold stimulated upon transfer of cells grown in the presence of excess Pi to Pi-free medium. The constitutive low-affinity system is inhibited by uncouplers and can mediate counterflow, indicating its reversible secondary nature. The presence of an inducible high-affinity uptake system and the ability to decrease the free internal Pi pool by forming polyP enable this organism to reduce the Pi concentration in an aerobic environment to very low levels. Anaerobically, polyP is degraded again and Pi is released via the low-affinity secondary transport system (Van Veen et al. 1993b). This secondary transport system of *A. johnsonii* 210A was further studied in membrane vesicles and proteoliposomes in which the transport system was functionally reconstituted. Surprisingly, Pi uptake appeared to be dependent on divalent cations, like Mg^{2+}, Ca^{2+}, Mn^{2+} or Co^{2+} (Fig. 3; Van Veen et al. 1993a). These ions form a metal phosphate ($MeHPO_4$) complex with up to 87 % of the Pi present in the external medium, suggesting that divalent cations and Pi are co-transported via a metal phosphate chelate. The uptake is driven by the proton motive force and the metal phosphate/proton stoichiometry was close to unity. The transport system mediates efflux and homologous exchange of metal phosphate but not heterologous exchange of metal phosphate and glycerol-3-P or glucose-6-P. The exchange was essentially pH-independent, but the uptake and efflux increased with increasing pH. The efflux was inhibited by the proton motive force and the exchange by the membrane potential only. The results with *A. johnsonii* 210A indicate that the efflux of Pi is limited by deprotonation of

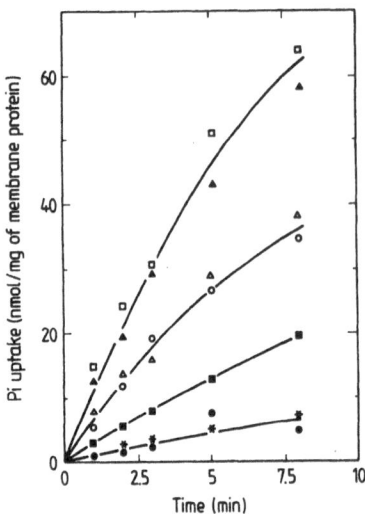

Fig. 3. Effect of divalent cations (2 mM) on proton motive force-driven transport of Pi in proteoliposomes in which the Pit carrier of *Acinetobacter johnsonii* 210A was functionally reconstituted. *Open circles* Mg^{2+}; *open triangles* Ca^{2+}; *solid triangles* Co^{2+}; *open squares* Mn^{2+}; *solid circles* 0.5 mM EDTA; *solid squares* without added divalent cations; *asterisks* in the absence of a proton motive force. (Van Veen et al. 1993a)

the carrier protein at the outer surface of the membrane and that the carrier recycles in a protonated form during exchange (Van Veen et al. 1993a).

The observations made with *A. johnsonii* 210A lead to a re-evaluation of the phosphate inorganic transport (Pit) system of *E. coli*. Based on (1) extensive complexation of Pi and divalent cations to a neutral metal phosphate complex, (2) co-transport of a divalent cation and Pi in a molar ratio of 1:1, (3) electrogenic proton/metal phosphate symport and (4) a metal phosphate/proton stoichiometry of unity, it was concluded that the Pit system of *E. coli* mediates the uptake of metal phosphate rather than Pi, similar to the situation in *A. johnsonii* 210A. Besides, the Pit system of *E. coli* also mediates efflux and homologous exchange of metal phosphate (Van Veen et al. 1994d).

The finding that in *E. coli* as well as in *A. johnsonii* 210A divalent cations and Pi are cotransported by the Pit system suggests that this is a general mechanism for the transport of divalent cations and Pi in Gram-negative bacteria and provides these organisms whith a new mechanism for the entry of Ca^{2+}. However, in Gram-positive bacteria such as *Micrococcus lysodeikticus* and *Bacillus cereus* Pi transport is also stimulated by Mg^{2+}, and in other Gram-positive bacteria the uptake of divalent cations is stimulated by Pi (for a review, see Van Veen et al. 1944c), suggesting that the simultaneous uptake of Pi and divalent cations is not restricted to Gram-negative bacteria only but may be a more general feature amongst bacteria.

In contrast to the constitutive, reversible, secondary Pi transport system, the primary Pst (phosphate specific transport) system in *A. johnsonii* 210A and other Gram-negative bacteria is unidirectional. It is only involved in uptake and not in efflux. This binding protein-dependent system most likely mediates the translocation of HPO_4^{2-} and $H_2PO_4^-$, but not of $MeHPO_4$ (Van Veen et al. 1994e). At pH values between 5.5 and 8.0 under conditions where divalent metal ions are in excess of Pi, the prevailing phosphate species are $H_2PO_4^-$, KPO_4^{2-} and $MeHPO_4$. When operating in concert, both transport systems in *A. johnsonii* 210A enable the organism to efficiently acquire Pi from its environment through uptake of the predominant Pi species.

The literature on transport of Pi in bacteria other than *A. johnsonii* 210A and *E. coli* and a few other Gram-negative bacteria has been summarized by Rosenberg (1987) and, more recently, by Van Veen et al. (1994c). Generally speaking, Gram-positive bacteria use either of the two Pi transport systems found in Gram-negative bacteria. For instance, *Streptococcus feacalis* possesses a phosphate bond energy-dependent, unidirectional transport system, but *Paracoccus denitrificans*, *Micrococcus lysodeikticus* and an alkaliphilic *Bacillus* species seem to contain a bidirectional system, allowing uptake and efflux of Pi.

5.2
Pi Efflux as an Energy-Recycling Mechanism

Work with vesicles of *A. johnsonii* 210A has clearly demonstrated that both components of the proton motive force, the membrane potential and the pH gradient, are generated by the electrogenic excretion of $MeHPO_4$ and H^+ (Van Veen et al. 1993a, 1994a). Presently, it is unknown which of the two systems, the polyphosphate:AMP phosphotransferase/adenylate kinase system or the electrogenic excretion of $MeHPO_4$ in symport with a proton, is more important in maintaining a proton motive force during anaerobiosis in this strictly aerobic organism. However, the in vitro studies demonstrate the potential of $MeHPO_4/H^+$ efflux as an energy-recycling mechanism (Michels et al. 1979) in *A. johnsonii* 210A and that the so-called Pit system provides Gram-negative bacteria with an additional transport system for the uptake of divalent cations, including Ca^{2+}. In Fig. 4 the $MeHPO_4$ efflux-induced pro-

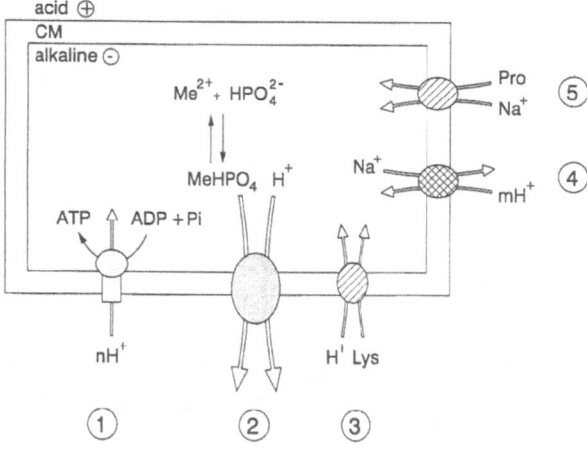

Fig. 4. Coupling of $MeHPO_4/H^+$ efflux to several energy-requiring processes in *Acinetobacter johnsonii* 210A; *1* H^+-ATPase; *2* secondary $MeHPO_4/H^+$ carrier; *3* lysine/H^+ symporter; *4* Na^+/H^+ antiporter; *5* proline/Na^+ symporter. (Van Veen et al. 1994a)

ton motive force is coupled to several metabolic energy-requiring processes in *A. johnsonii* 210A: synthesis of ATP, uptake of lysine, efflux of sodium and uptake of proline.

6
Enhanced Biological Phosphorus Removal (EBPR)

6.1
The Process

The removal of phosphorus from wastewater is an essential feature of sewage treatment facilities because of the threat of eutrophication of lakes and inland seas. Activated sludge processes with alternating anaerobic and aerobic phases have been widely adopted to eliminate phosphorus from waste-

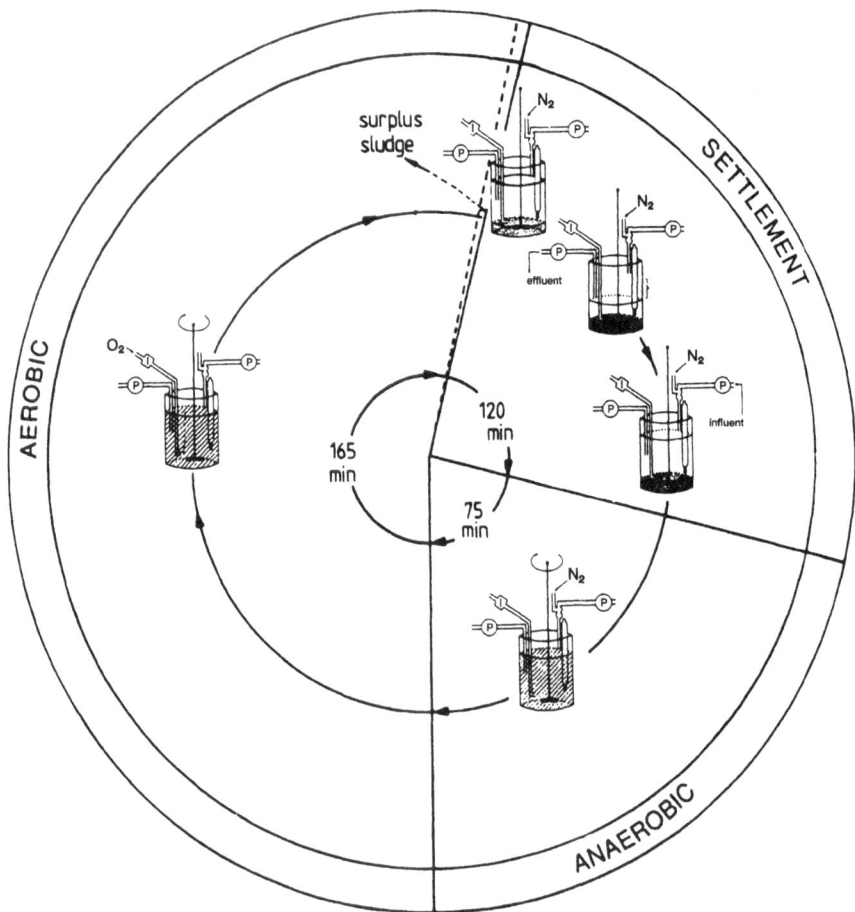

Fig. 5. Operational cycle of the fill-and-draw system. (Appeldoorn et al. 1992a)

water flows (Toerien et al. 1990). The biological removal of phosphorus in these processes is achieved by storage of excess Pi as polyP by bacteria present in activated sludge (Kulaev 1995). The very essence of EBPR is that Pi is removed from huge volumes of wastewater under aerobic conditions and is released anaerobically from these bacteria during stripping into a small volume of water, resulting in a high concentration of phosphate.

In order to learn more about the process of EBPR a simple, one-vessel lab-scale, fill-and-draw (F&D) system was designed (Fig. 5; Appeldoorn et al. 1992a). The activated sludge was exposed to cycles with three distinct consecutive periods: first an anaerobic period where the sludge is mixed with fresh mineral salts medium that contained glucose and acetic acid as carbon and energy sources, then an aerobic period and finally a settlement period in which the surplus sludge was removed and the medium was partly refreshed.

After running this F&D system for 6–8 weeks, activated sludge with a phosphorus content of up to 110 mg g dry biomass^{-1} was obtained and almost all bacteria contained polyP. With other synthetic wastewaters phosphorus contents as high as 180 mg g dry weight^{-1} have been obtained (Appeldoorn et al. 1992a). These values represent appreciable accumulations of phosphate compared to the normal P content of bacteria of 15 mg P g dry biomass^{-1}. During the anaerobic period a part of the stored polyP was hydrolyzed and released as Pi, and taken up again together with the Pi in the fresh medium in the subsequent aerobic period (Fig. 6). Glucose and acetic acid were completely taken up during the anaerobic period and the sludge concentration remained between 3 and 4 g dry weight l^{-1} and the residence time of the sludge was 8.3 days.

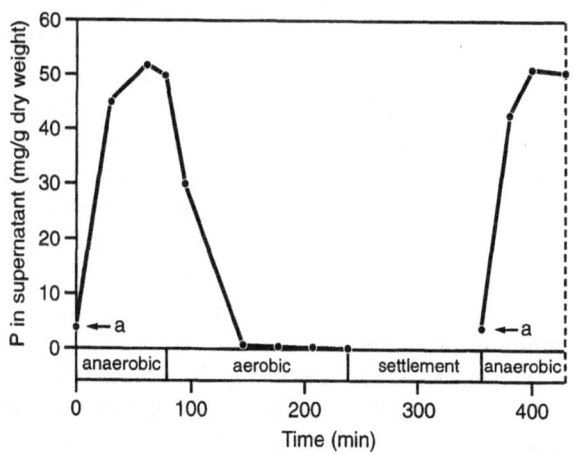

Fig. 6. Phosphate release and uptake in the fill-and-draw system. *a* Amount of phosphorus in refreshed medium. (Appeldoorn et al. 1992a)

6.2
The Biochemical Model

Around 1985 Comeau and coworkers and Wentzel and coworkers designed two biochemical models (Comeau et al. 1986; Wentzel et al. 1986) for EBPR which resemble each other. In these models fatty acids, in particular acetate, are taken up and stored with the aid of the energy released by the degradation of polyP, resulting in the excretion of Pi. The lower fatty acids are produced anaerobically by fermenting bacteria from complex substrates and are converted by the polyP-accumulating bacteria into osmotically inert poly-β-hydroxyalkanoates (PHA), poly-β-hydroxybutyrate (PHB) and poly-β-hydroxy-valerate (PHV) being the predominant ones. The conversion of substrates into PHA requires reducing equivalents in the form of $NADH_2$. Originally it was thought that the citric acid cycle provided the reducing equivalents for the conversion of acetyl-CoA into PHB. However, the almost complete lack of $^{14}CO_2$ formation from ^{14}C-labelled acetate (Bordacs and Chiesa 1989) and the significant decrease in cellular glycogen during the anaerobic period (Mino et al. 1987) point to a production of reducing equivalents by glycolysis. The acetyl-CoA formed during glycolysis is thought to be converted into PHB. This model of carbon metabolism has recently been confirmed by Smolders et al. (1994). By using phosphorus-removing sludge from a well-controlled sequencing batch reactor it was shown that under really anaerobic conditions (1) extracellular acetate was completely converted to PHA, (2) the molar 3-hydroxybutyrate/acetate ratio was 1.2–1.3, (3) cellular glycogen was consumed during the conversion of acetate into PHB and (4) all cells contained glycogen at the start of the anaerobic period (see also Smolders 1995). These data show also that some glycogen is converted into PHB during anaerobiosis. Thus, glycogen serves not only as a source of reducing equivalents but also as an energy source and the resultant acetyl-CoA is converted into PHA.

The lab-scale system developed by Smolders et al. (1994) for EBPR under well-controlled conditions is, like our own system, a relatively simple EBPR system, in particular by using a mineral salts medium with acetate as the sole organic carbon compound, which is not often the case in full-scale treatment systems. The results obtained with this simple system allow, however, the important conclusion that in this system, and possibly in many other treatment system showing EBPR as well, the polyP-accumulating bacteria can also accumulate polyP-β-hydroxyalkanoates, in particular PHB, and glycogen. Thus, all three types of polymers, polyP, PHA and glycogen, are present in one type of organism, and not distributed over two or three types of organisms.

When the sludge is entering the aerobic phase, in all properly functioning lab-scale phosphorus removing treatment systems no extracellular, easily degradable carbon compounds are left (see, for instance, De Vries and

Rensink 1985; Appeldoorn 1993; Smolders 1995). This means that only intracellular compounds serve as substrates for growth of the heterotrophic, strictly aerobic polyP-accumulating bacteria. Growth under these conditions is a rather slow process; the maximum growth rate of the biological phosphorus removing organisms is in the range of $0.04 h^{-1}$ (Smolders 1995). During growth of these organisms Pi is taken up in considerable amounts and largely stored as polyP and glycogen is resynthesized (see, for instance, Smolders 1995). Why these organisms spend a significant part of the intracellular PHB on accumulation of polyP and glycogen, and not merely on cell growth is presently unknown and is, as far as we know, a unique and fascinating property of these polyP-accumulating bacteria. Apparently the biosynthesis of certain cellular components is slow compared to the conservation of metabolic energy and the excess of energy is spent on transport of Pi, conversion of Pi into polyP and accumulation of glycogen from PHB. Alternatively, the genes encoding these reactions are for some unknown reasons overexpressed, and as a result, less energy is left for growth. Very recently, the phosphate uptake kinetics in relation to PHB were studied (Petersen et al. 1998). The uptake of Pi increased the rate of PHB degradation significantly and at low cellular PHB concentrations hardly any Pi was taken up. As much as 30 % of the intracellular PHB was estimated to be used for uptake and polymerization of Pi. Recently, it was firmly confirmed that some of the bacteria involved in EBPR in wastewater treatment systems can use nitrate as terminal electron acceptor instead of molecular oxygen (Kuba et al. 1995).

Polyphosphate are strongly negatively charged molecules and are complexed in bacteria almost exclusively with K^+ and Mg^{2+} (Bonting et al. 1993a; Van Veen et al. 1994e) unless unusually high concentrations of other metals in the growth medium are provided. PopyP metabolism in activated sludge is also accompanied by the accumulation/release of both ions. Sometimes polyP is complexed also slightly with Ca^{2+} in activated sludge. From observations with a variety of phosphorus removing sludges it appears that K^+ and Mg^{2+}, and sometimes Ca^{2+} to a minor extent, are co-transported with Pi molecules in a total molar ionic charge ratio of about one, irrespective of the origin of the sludge and the direction of transport (summarized by Comeau et al. 1986; see also Rickard and McClintock 1992). In EBPR sludge obtained by enrichment using the mineral salts medium with acetate as the sole carbon compound (Smolders 1995), a metabolically rather homogenous group of bacteria might be expected. In this sludge too, a total charge ratio between cations and Pi of about unity was observed and the molar Mg^{2+}/Pi and K^+/Pi ratios were 0.42 and 0.34, respectively (Kuba et al. 1995). In the polyP-accumulating A. johnsonii 210A the molar Mg^{2+}/Pi ratio was, however, close to unity during Pi efflux (Clijssen and Kortstee, unpubl. observ.) as could be expected from the work of Van Veen et al. (1993a) which shows that Pi is excreted as a neutral metal phosphate ($MgHPO_4$) by the the reversible phosphate inorganic transport (Pit) system. Thus, in spite of the fact that K^+

and Mg^{2+} are the counterions of polyP in this organism, during anaerobiosis Mg^{2+} and Pi are excreted in a molar ratio of about 1:1.

6.3
The Bacteria Involved

In spite of the impressive number of treatment systems showing EBPR from which *Acinetobacter* spp. have been isolated, the importance of *Acinetobacter* spp. in EBPR remains unclear. On the basis of ubiquinone and polyamine profiles, the oxidation of glucose to gluconic acid and from the results obtained with 16S rRNA-targeted oligonucleotide probes specific for several bacterial groups, it was concluded that the isolated *Acinetobacter* spp. do not play an important part in EBPR in some full-scale wastewater treatment systems and in a number of well-defined lab-scale systems (Appeldoorn 1993; Smolders et al. 1994; Kortstee et al. 1998). An additional argument against the involvement of these strictly aerobic bacteria in EBPR is their failure to accumulate PHB from acetate during anaerobiosis in spite of the fact that intracellular polyP is degraded and Pi is excreted under these conditions.

In properly functioning EBPR systems no extracellular carbon and energy sources are available anymore when the sludge is entering the aerobic phase. Growth, Pi uptake (Fig. 6), polyP synthesis and glycogen synthesis in the polyP-accumulating bacteria only occur with intracellular carbon and energy sources. Attempts to isolate the bacteria from activated sludge which are responsible for EBPR should therefore comprise the following two criteria: (1) ability to accumulate PHB (from extracellular acetate) or another polymer intracellularly during anaerobiosis coupled to Pi excretion and (2) ability to accumulate polyP intracellularly under aerobic (or denitrifying) conditions in the absence of an extracellular carbon and energy source. Until recently the isolates were only examined for their ability to accumulate polyP aerobically in the presence of relatively high concentrations of an extracellular carbon and energy source.

By applying the two above criteria, two polyP-accumulating bacteria were isolated, *Microlunatus phosphovorus* (Nakamura et al. 1995a) and an unidentified, coccoid organism (Ubukata and Takii 1994). Their phosphorus contents were 16.6 and 28 %, respectively. The latter value is surprisingly high. The survival of *Microlunatus phosphovorus* in a wastewater treatment system with alternating anaerobic and aerobic periods appears easy to understand: during the anaerobic period polyP is hydrolyzed and glucose can be taken up and stored as a polymer, and during the subsequent aerobic period with no extracellular carbon available anymore the glucose polymer is used as a carbon and energy source and Pi is accumulated as polyP. Up to now no data are available on the type of polyP synthesized, the enzymes involved in polyP biosynthesis and biodegradation, the regulation of both types of enzymes and the genes encoding these enzymes in this organism.

Surprisingly, *Microlunatus phosphovorus* is not able to accumulate acetate, propionate and lactate anaerobically in the form of PHA (Nakamura et al. 1995b). Of the other substrates tested, glucose, L-glutamate, peptone and yeast extract were taken up. This probably indicates that *Microlunatus phosphovorus* ist not identical to the organism responsible for EBPR in Pi-removing sludges which release Pi anaerobically upon the addition of acetate and propionate but not or hardly by adding glucose. The latter organism seems much more universal in wastewater treatment systems showing EBPR than *Microlunatus phosphovorus*. However, both organisms belong to the Gram-positive bacteria with a high G+C DNA content which are most likely responsible for EBPR in (at least some) full-scale wastewater treatment systems (Wagner et al. 1994). The isolation of the polyP-accumulating bacterium that can anaerobically accumulate PHB from acetate and aerobically polyP from Pi in the absence of an extracellular, oxidizable carbon compound may be executed by testing the isolates for these abilities.

7
Conclusions and Outlook

Bacteriological research on polyP during the last decade has been mainly concentrated on three types of bacteria. The investigations into *E. coli*, an organism which does not accumulate polyP in visible amounts, have resulted in the purification of PPK and PPX, the discovery of the corresponding genes *ppk* and *ppx*, enzymatic tools for analytical and preparative work, and, above all, in evidence supporting the involvement of polyP in the regulation of stress and survival of at least some bacteria. Particularly in the latter field exciting new functions of polyP may be disclosed in the years to come. Also the PPK-independent synthesis of low molecular weight polyP has so far escaped solution.

Work done on the polyP-accumulating *A. johnsonii* 210A has shown that polyP can act as an energy source by the polyphosphate:AMP phosphotransferase/adenylate kinase pathway and by the efflux of Pi in symport with a proton. The organism possesses two transport systems for Pi: a constitutive, bidirectional, proton-motive-driven, low affinity transport system involved in uptake and efflux of phosphate and an inducible, unidirectional, ATP-driven, high affinity system that is only involved in phosphate uptake during phosphate deficient conditions. The substrate for the low affinity system is a metal phosphate complex ($MgHPO_4$, $CaHPO_4$, $MnHPO_4$ or $CoHPO_4$) rather than Pi and the substrates for the high affinity system are $H_2PO_4^-$ and HPO_4^{2-}. The low affinity phosphate transport system (the so-called Pit system) of *E. coli* reacted similarly to that of *A. johnsonii* 210A. It provides Gram-negative bacteria with a new entry system for Ca^{2+}. When acting in concert, both Pi transport systems enable *A. johnsonii* 210A to efficiently acquire Pi from its habitats through uptake of the predominant phosphate species.

Research on the third of organism, the Gram-positive, high percent G+C DNA-containing, polyP-accumulating organism(s) in activated sludge responsible for enhanced biological phosphorus removal, has so far almost only been carried out at the activated sludge level (mixed culture level). So far no pure culture is available which shows the same phosphate characteristics as the majority of the sludges: acetate-mediated polyP degradation, PHB formation and Pi efflux under anaerobic conditions and intracellular PHB-dependent growth and polyP accumulation under aerobic conditions. The isolation of this type of organism will be a milestone in bacterial polyP research and of great biotechnological importance. Once isolated, the organism may be examined according to the lines followed in *E. coli* and *A. johnsonii* 210A. This chapter contains suggestions as to the isolation of these fairly unique bacteria.

The recently identified *Microlunatus phosphovorus* shows, essentially, the right phosphate characteristics, but unfortunately its Pi uptake and release rates are comparatively low. In addition, it fails to sequester acetate and/or propionate anaerobically and, for the time being, this glucose-sequestering organism seems much less universal than the organism which shows the acetate-mediated polyP degradation, phosphate release, PHB formation and glycogen reduction.

References

Akiyama M, Crooke E, Kornberg A (1993) An exopolyphosphatase of *Escherichia coli*. The enzyme and its *ppx* gene in a polyphosphate operon. J Biol Chem 268: 633–639

Appeldoorn KJ (1993) Ecological aspects of the biological phosphate removal from wastewaters. PhD Thesis, Agricultural University of Wageningen

Appeldoorn KJ, Kortstee GJJ, Zehnder AJB (1992a) Biological phosphorus removal by activated sludge under defined conditions. Water Res 26: 453–460

Appeldoorn KJ, Boom AJ, Korststee GJJ, Zehnder AJB (1992b) Contribution of precipitated phosphates and acid-soluble polyphosphate to enhanced biological phosphate removal. Water Res 26: 937–943

Bark K, Sponner A, Kämpfer P, Grund S, Dott W (1992) Differences in polyphosphate accumulation and phosphate absorption by *Acinetobacter* isolates from wastewater producing polyphosphate:AMP phosphotransferase. Water Res 26: 1379–1388

Bayly RC, Duncan A, May JW, Schembri M, Semertjis A, Vasiliadis G, Raper WGC (1991) Microbiological and genetic aspects of the synthesis of polyphosphate by species of *Acinetobacter*. Water Sci Technol 23: 747–754

Bonting CFC (1993) Polyphosphate metabolism in *Acinetobacter johnsonii* 210A. PhD Thesis, Agricultural University of Wageningen

Bonting CFC, Kortstee GJJ, Zehnder AJB (1991) Properties of polyphosphate:AMP phosphotransferase of *Acinetobacter* strain 210A. J Bacteriol 173: 6484–6488

Bonting CFC, Van Veen HW, Taverne A, Kortstee GJJ, Zehnder AJB (1992a) Regulation of polyphosphate metabolism in *Acinetobacter* strain 210A grown in carbon- and phosphate-limited continuous cultures. Arch Microbiol 158: 139–144

Bonting CFC, Willemsen BMF, Akkermans-Van Vliet W, Bouvet PJM, Kortstee GJJ, Zehnder AJB (1992b) Additional characteristics of the polyphosphate-accumulating *Acinetobacter* strain 210A and its identification as *Acinetobacter johnsonii*. FEMS Microbiol Ecol 102: 57–64

Bonting CFC, Kortstee GJJ, Boekestein A, Zehnder AJB (1993a) The elemental composition dynamics of large polyphosphate granules in *Acinetobacter* strain 210A. Arch Microbiol 159: 428–434

Bonting CFC, Kortstee GJJ, Zehnder AJB (1993b) Properties of polyphosphatase of *Acinetobacter johnsonii* 210A. Antonie van Leeuwenhoek 64: 75–81

Bordacs K, Chiesa SC (1989) Carbon flow patterns in enhanced biological phosphorus accumulating activated sludge cultures. Water Sci Technol 21: 387–396

Castuma CE, Huang R, Kornberg A, Reusch RN (1995) Inorganic polyphosphates in the acquisition of competence in *Escherichia coli*. J Biol Chem 270: 12980–12983

Clark JE, Wood HG (1987) Preparation of standards and determinations of sizes of long-chain polyphosphates by gel electrophoresis. Anal Biochem 161: 280–290

Clark JE, Beegen H, Wood HG (1986) Isolation of intact chains of polyphosphate from *Propionibacterium shermanii* grown on glucose or lactate. J Bacteriol 168: 1212–1219

Comeau Y, Hall KJ, Hancock REW, Oldham WK (1986) Biochemical model for enhanced biological phosphorus removal. Water Res 20: 1511–1521

Crooke E, Akiyama M, Rao NN, Kornberg A (1994) Genetically altered levels of inorganic polyphosphate in *Escherichia coli*. J Biol Chem 269: 6290–6295

Dawes EA, Senior PJ (1973) The role and regulation of energy reserve polymers in microorganisms. Adv Microb Physiol 10: 135–266

De Vries HP, Rensink JH (1985) Biological phosphorus removal at low sludge loadings by partial stripping. In: Lester JN, Kirk PWW (eds) Proceedings of the International Conference on Management Strategies for Phosphorus in the Environment. Seeper, London, pp 54–65

Hara A, Sy J (1983) Guanosine 5'-triphosphate, 3'-diphosphate, 5'-phosphohydrolase. J Biol Chem 258: 1678–1683

Harold FM (1966) Inorganic polyphosphates in biology: structure, metabolism and function. Bacteriol Rev 30: 772: 788

Harold FM, Harold RL (1965) Degradation of inorganic polyphosphate in mutants of *Aerobacter aerogenes*. J Bacteriol 89: 1262–1269

Jeener R, Brachet J (1944) Ribonucleic acid of yeast, microdetermination and relation of growth conditions to synthesis. Enzymologia 11: 222–234

Keasling JD, Bertsch L, Kornberg A (1993) Guanosine pentaphosphate phosphohydrolase of *Escherichia coli* is a long-chain exopolyphosphatase. Proc Natl Acid Sci USA 90: 7029–7033

Kornberg A (1995) Inorganic polyphosphate: toward making a forgotten polymer unforgettable. J Bacteriol 177: 491–496

Kortstee GJJ, Appeldoorn KJ, Bonting CFC, Van Niel EWJ, Van Veen HW (1994) Biology of polyphosphate-accumulating bacteria involved in enhanced biological phosphorus removal. FEMS Microbiol Rev 15: 137–153

Kortstee GJJ, Appeldoorn KJ, Bonting CFC, Van Niel EWJ, Van Veen HW (1998) Ecological aspects of biological phosphorus removal in activated sludge systems. Adv Microb Ecol (in press)

Kuba T, Van Loosdrecht MCM, Heijnen JJ (1995) Biological phosphate removal under denitrifying conditions. In: Toekomstige generatie rioolwaterzuiveringsinrichtingen RWZI 2000 (94–12), RIZA/STOWA, Lelystad/Utrecht, pp 1–83

Kulaev IS (1979) The biochemistry of inorganic polyphosphates. Wiley, Chichester

Kulaev IS (1995) Evolutionary biochemistry of inorganic polyphosphates. Mol Biol 29: 716–720

Michels PAM, Michels JPJ, Boonstra K, Konings WN (1979) Generation of an electrochemical proton gradient in bacteria by the excretion of metabolic end products. FEMS Microbiol Lett 5: 357–364

Mino T, Kawakami T, Matsuo T (1985) Location of phosphorus in activated sludge and function of intracellular polyphosphates in biological phosphorus removal process. Water Sci Technol 17: 93–106

Mino T, Arun V, Tsuzuki Y, Matsuo T (1987) Effect of phosphorus accumulation on acetate metabolism in the biological phosphorus removal process. In: Ramadori R (ed) Biological phosphate removal from wastewaters. Advances in water pollution control vol 4. Pergamon Press, Oxford, pp 27–38

Nakamura K, Hiraishi A, Yoshimi Y, Kawaharasaki M, Masuda K, Kamagata Y (1995a) *Microlunatus phosphovorus* gen. nov., sp. nov., a new Gram-positive polyphosphate-accumulating bacterium isolated from activated sludge. Int J Syst Bacteriol 45: 17–24

Nakamura K, Ishikawa S, Kawaharasaki M (1995b) Phosphate uptake and release activity in immobilized polyphosphate-accumulating *Microlunatus phosphovorus* strain NM-1. J Ferment Bioeng 80: 377–382

Nyren P, Nore BF, Strid A (1991) Proton-pumping N,N^l-dicyclohexylcarbodiimide-sensitive inorganic pyrophosphate synthase from *Rhodospirillum rubrum*: purification, characterization, and reconstitution. Biochemistry 30: 2883: 2887

Petersen B, Temmink H, Henze M, Isaacs S (1998) Phosphate kinetics in relation to PHB under aerobic conditions. Water Res 32: 91–100

Reusch RN (1989) Poly-β-hydroxybutyrate/calcium polyphosphate complexes in eukaryotic membranes. Proc Soc Exp Biol Med 191: 377–381

Reusch RN, Sadoff HL (1983) D-(-)-poly-β-hydroxybutyrate in membranes of genetically competent bacteria. J Bacteriol 156: 778–788

Reusch RN, Sadoff HL (1988) Putative structure and function of poly-β-hydroxybutyrate/calcium polyphosphate channel in bacterial plasma membranes. Proc Natl Acad Sci USA 85: 4176–4180

Reusch RN, Hiske TW, Sadoff HL (1986) Poly-β-hydroxybutyrate membrane structure and its relationship to genetic transformability in *Escherichia coli*. J Bacteriol 168: 553–562

Reusch RN, Huang R, Bramble LL (1995) Poly-3-hydroxybutyrate/polyphosphate complexes form voltage-activated Ca^{2+} channels in the plasma membranes of *Escherichia coli*. Biophys J 69: 754–766

Rickard LF, McClintock SA (1992) Potassium and magnesium requirements for enhanced biological phosphorus removal from wastewater. Water Res 26: 2203–2206

Roberts M (1987) Polyphosphates. In: Tyler Burt C (ed) Phosphorus NMR in biology. CRC Press, Boca Raton, F, pp 85–94

Rosenberg H (1987) Phosphate transport in prokaryotes. In: Rosen BP, Silver S (eds) Ion transport in prokaryotes. Academic Press, New York, pp 205–248

Schmidt G, Hecht L, Thannhauser J (1946) The enzymatic formation and the accumulation of large amounts of a metaphosphate in baker's yeast under certain conditions. J Biol Chem 166: 775: 776

Smolders GJF (1995) A metabolic model of the biological phosphorus removal stoichiometry, kinetics and dynamic behaviour. PhD Thesis, Delft University of Technology, Delft

Smolders GJF, Van der Mey J, Van Loosdrecht MCM, Heijnen JJ (1994) Model of the anaerobic metabolism of the biological phosphorus removal process: stoichiometry and pH influence. Biotechnol Bioeng 43: 461–470

T'Seyen J, Malnou D, Block JC, Faup G (1985) Polyphosphate kinase activity during phosphate uptake by bacteria. Water Sci Technol 17: 43–56

Thilo E (1962) Condensed phosphates and arsenates. Adv Inorg Chem Radiochem 4: 1–77

Tisa LS, Olivera BM, Adler J (1993) Inhibition of *Escherichia coli* chemotaxis by ω-conotoxin, a calcium channel blocker. J Bacteriol 175: 1235–1238

Toerien DF, Gerber A, Lötter LH, Cloete TE (1990) Enhanced phosphorus removal systems in activated sludge systems. Adv Microb Ecol 11: 173–230

Ubukata Y, Takii S (1994) Induction ability of excess phosphate accumulation for phosphate removing bacteria. Water Res 28: 247–249

Van Groenestijn JW (1988) Accumulation and degradation of polyphosphate in *Acinetobacter* sp. PhD Thesis, Agricultural University of Wageningen

Van Groenestijn JW, Deinema MH, Zehnder AJB (1987) ATP production from polyphosphate in *Acinetobacter* strain 210A, Arch Microbiol 148: 14-19

Van Groenestijn JW, Bentvelzen MMA, Deinema MH, Zehnder AJB (1989) Polyphosphate degrading enzymes in *Acinetobacter* spp. and activated sludge. Appl Environ Microbiol 55: 219-223

Van Veen HW (1994) Energetics and mechanisms of phosphate transport of *Acinetobacter johnsonii*. PhD Thesis, Agricultural University of Wageningen

Van Veen HW, Abee T, Kortstee GJJ, Konings WN, Zehnder AJB (1993a) Mechanism and energetics of the secondary phosphate transport system of *Acinetobacter johnsonii* 210A. J Biol Chem 268: 19377-19383

Van Veen HW, Abee T, Kortstee GJJ, Konings WN, Zehnder AJB (1993b) Characterization of two phosphate transport systems in *Acinetobacter johnsonii* 210A. J Bacteriol 175: 200-206

Van Veen HW, Abee T, Kortstee GJJ, Konings WN, Zehnder AJB (1994a) Generation of a proton motive force by the excretion of metal phosphate in the polyphosphate-accumulating *Acinetobacter johnsonii* strain 210A. J Biol Chem 269: 29509-29514

Van Veen HW, Abee T, Kleefsman AWF, Melgers B, Kortstee GJJ, Konings WN, Zehnder AJB (1994b) Energetics of alanine, lysine, and proline transport in cytoplasmic membranes of the polyphosphate-accumulating *Acinetobacter johnsonii* 210A. J Bacteriol 176: 2670-2676

Van Veen HW, Abee T, Kortstee GJJ, Konings WN, Zehnder AJB (1994c) Phosphate inorganic transport (Pit) system in *Escherichia coli* and *Acinetobacter johnsonii*. In: Torriani-Gorini A, Yagil Y, Silver S (eds) Phosphate in micro-organisms: cellular and molecular biology. American Society for Microbiology, Washington DC, pp 43-49

Van Veen HW, Abee T, Kortstee GJJ, Konings WN, Zehnder AJB (1994d) Translocation of metal phosphate via the phosphate inorganic transport (Pit) system of *Escherichia coli*. Biochemistry 33: 1766-1770

Van Veen HW, Abee T, Kortstee GJJ, Konings WN, Zehnder AJB (1994e) Substrate specificity of the two transport systems of *Acinetobacter johnsonii* 210A in relation to phosphate speciation in its aquatic environment. J Biol Chem 269: 16212-16216.

Wagner M, Erhart R, Manz W, Amann R, Lemmer H, Wedi D, Schleiffer K-L (1994) Development of an rRNA-targeted oligonucleotide probe specific for the genus *Acinetobacter* and its application for in situ monitoring in activated sludge. Appl Environ Microbiol 56: 1919-1925

Wentzel MC, Lötter LH, Loewenthal RE, Marais GvR (1986) Metabolic behaviour of *Acinetobacter* spp. in enhanced biological phosphorus removal – a biochemical model. Water SA (Protoria) 12: 209-224

Wiame J (1948) The occurrence and physiological behavior of two metaphosphate fractions in yeast. J Biol Chem 178: 919-929

Wood HG, Clark JE (1988) Biological aspects of inorganic polyphosphates. Annu Rev Biochem 57: 235-260

Genetic Improvement of Bacteria for Enhanced Biological Removal of Phosphate from Wastewater

H. Ohtake, A. Kuroda, J. Kato and T. Ikeda[1]

1
Introduction

Phosphorus (P) is an essential constituent in all types of living organisms. It is present in nucleic acids, phospholipids and various cytoplasmic solutes. Microorganisms use inorganic phosphate (P_i) as the preferred P source (Wanner 1996). P_i removal from wastewaters has received considerable attention, since P_i is believed to be responsible for nuisance growth of algae in lakes and waterways (Codd and Bell 1985). Algal blooms degrade water quality by producing an offensive odor and taste. The nuisance growth of algae renders boating and fishing difficult and discourages swimming. Excessive growth of algae consumes dissolved oxygen, when the algae are decomposed by aerobic bacteria, causing mass mortality of fish and other aquatic organisms. Algal toxin production is also a serious problem in drinking water supplies (Wicks and Thiel 1990).

Activated sludge processes, which are commonly used for treating wastewaters, are very effective in removing organic pollutants, but they remove P_i relatively poorly. Under ordinary operating conditions, activated sludges are capable of removing an average of only 20 to 40 % of the P_i concentrations normally found in municipal wastewaters (Carberry and Tenney 1973). Municipal wastewaters are relatively low in sources of carbon, and this limits the removal of P_i by activated sludge whose P content is typically 1 to 2 % on a dry weight basis. Accordingly, to make activated sludge more effective in removing P_i, it appears essential to enable sludge microorganisms to take up and store P_i in excess of their requirements for growth (Ohtake et al. 1985).

Many microorganisms accumulate excess P_i in the form of polyphosphate (polyP) (Kulaev 1975). Biologically synthesized polyP is a linear polymer of P_i with a chain length of up to 1000 or more. The biological roles of polyP are still uncertain. However, since polyP can serve as a P source for the biosynthesis of nucleic acids and phospholipids during P_i starvation conditions (Harold 1966), it is possible that polyP functions as a P_i reservoir with

[1] Department of Fermentation Technology, Hiroshima University, Higashi-Hiroshima, Hiroshima 739-8527, Japan.

Progress in Molecular and Subcellular Biology, Vol. 23
H. C. Schröder, W. E. G. Müller (Eds.)
© Springer-Verlag Berlin Heidelberg 1999

osmotic advantages. In the present review, we describe the current efforts to genetically improve the ability of bacteria to accumulate polyP. Genetic improvement of bacterial polyP accumulation will serve as the first step to make sludge microorganisms more effective in removing P_i from wastewaters (Kato et al. 1993a).

2
Strategy for Genetic Improvement of *Escherichia coli*

The ability of *E. coli* MV1184 (Vieira and Messing 1987) to accumulate polyP was improved by manipulating the genes involved in the P_i transport and polyP metabolism (Fig. 1). In *E. coli*, P_i is transported through the cytoplasmic membrane via either the high-affinity P_i-specific transport (Pst) system or the low-affinity P_i inorganic transport (Pit) system (Wanner 1996). The Pst system of *E. coli* is a periplasmic protein-dependent transporter similar to those of histidine, maltose and ribose. These transporters belong to the superfamily of ABC (ATP-binding cassette) transporters (Higgins 1992). The Pst system of *E. coli* comprises four distinct subunits encoded by the *pstS*, *pstA*, *pstB* and *pstC* genes (Amemura et al. 1985). These genes, together with the *phoU* gene, form the *pst* operon. Two recombinant plasmids, pEP02.2 and pEP02.5, both of which carry the *E. coli pst* operon from MV1184, were constructed (Fig. 2; Kato et al. 1993a). The *pst* genes were expressed at high levels from the *tac* promoter in pEP02.2 and from the *tet* promoter in pEP02.5 under conditions of P_i excess.

The enzyme responsible for polyP synthesis is the homotetrameric polyP kinase (PPK). Plasmid pBC29 (Akiyama et al. 1992), which contains the *E. coli ppk* gene together with its own promoter, was used to increase the dosage of the *ppk* gene. The *ppk* gene was expressed from its own promoter. PPK polymerizes the terminal P_i of ATP into polyP in a freely reversible reaction ($nATP \rightleftharpoons nADP + polyP_n$). Acetate kinase (ACK) was employed as an ATP regeneration system for polyP synthesis. ACK catalyzes the formation of ATP and acetate in the presence of acetyl phosphate and ADP (Thauer et al. 1977). Plasmid pEP01 contains the *E. coli ackA* encoding ACK and its own promoter (Lee et al. 1990). Although the transcriptional direc-

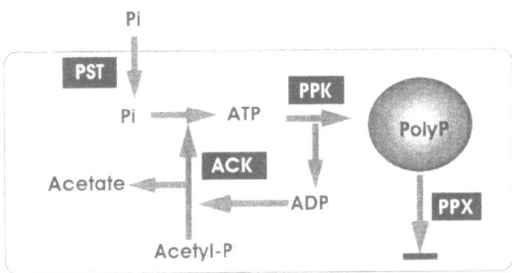

Fig. 1. Strategy for genetic improvement of *E. coli* for enhanced polyP accumulation. Acetate kinase was employed as an ATP regeneration system for polyP synthesis. *ACK* Acetate kinase; *PPK* polyP kinase; *PPX* exopolyPase; *PST* P_i-specific transport system

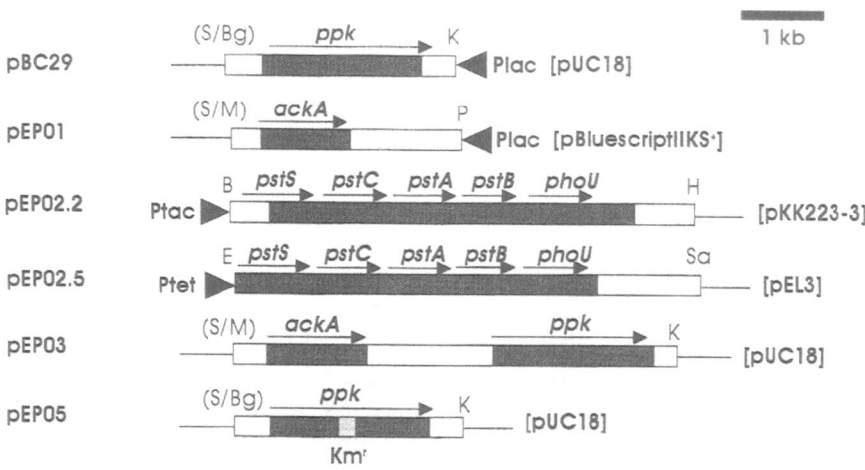

Fig. 2. Recombinant plasmids used for genetic improvement. *Open boxes* (with *shaded regions*) indicate DNAs cloned from *E. coli*, and *thin lines* are vector DNAs. Transcriptional directions of cloned genes are shown by *arrows*. Sites and directions of *lac*, *tac* and *tet* promoters are also indicated by *arrowheads*. Restriction site abbreviations: *B*, *Bam*HI; *Bg*, *Bgl*I; *E*, *Eco*RI; *H*, *Hin*-dIII; *K*, *Kpn*I; *M*, *Mlu*I; *P*, *Pst*I; *S*, *Sma*I; *Sa*, *Sal*I. Sites in *parentheses* were destroyed during construction. (Kato et al. 1993a)

tion of the *ackA* gene was the reverse of that of the *lac* promoter in pEP01, the levels of ACK activity in MV1184 (pEP01) were more than 100-fold higher than the parental level. Plasmid pEP03 contains both the *ppk* and *ack* genes.

The utilization and degradation of polyP is catalyzed by polyPases and by specific kinases including polyPglucokinase and polyPfructokinase (Wood and Clark 1988). The *ppx* gene, which encodes an exopolyPase (PPX), is located immediately downstream of the *ppk* gene in *E. coli* (Akiyama et al. 1993). Plasmid pEP05 was constructed to disrupt the *ppk* gene on the *E. coli* chromosome. Insertion of a kanamycin resistance (Km^r) gene cassette into the *ppk* gene on the chromosome of *E. coli* NM522 (*recA*$^+$) (Gough and Murray 1983) disrupted not only the PPK but also PPX activities.

3
Rate-Limiting Step for PolyP Accumulation

Escherichia coli does not accumulate appreciable amounts of polyP, especially those of high molecular weight types, despite its possession of PPK activity (Kulaev 1975). Since the kinetics of bacterial polyP accumulation are not fully understood, it is unclear what limits polyP accumulation in *E. coli*. P$_i$ uptake experiments were performed with *E. coli* strain MV1184 (pUC18), a control strain, and its recombinant derivatives containing pBC29 (carrying *ppk*), pEP01 (*ackA*), pEP2.2 (*pst* operon), pEP03 (*ppk* and *ack*) or both pEP2.2 and pBC29 (Fig. 3). Cells were grown in Luria broth (L broth) and used for P_i uptake experiments in T medium (Harold 1963) without being subjected to P_i starvation (Kato et al. 1993a). Growth of recombinant strains, except for MV1184 (pBC29 and pEP02.2), was almost equivalent to that of the control strain. The control strain removed about 20 % of the P_i from the medium during the first 3 h. However, no P_i uptake occurred after growth stopped. MV1184 containing pBC29 (*ppk*) removed twice as much P_i from the medium as did the control strain. As a result, increasing the dosage of the *ppk* gene alone doubled the P content of *E. coli* MV1184 (Fig. 4).

Strain MV1184 bearing pEP01 (*ackA*) alone took up less P_i, even compared with the control strain. However, the rate of P_i removal greatly increased when pEP03 (*ppk* and *ackA*) was introduced into MV1184, indicating that ACK functioned as an effective system for ATP regeneration. This recombinant strain removed approximately 90 % of the P_i from the medium within 4 h. Until 3 h, growth of this recombinant was almost equivalent to that of the control strain. However, unlike the control strain, the cell density of MV1184 (pEP03) decreased after reaching a maximum at 3 h. The decrease in cell density was concomitant with an increase in P_i in the medium, indicating that P_i was released by cell lysis (Fig. 3). Interestingly, strain MV1184 bearing pEP02.2 (*pst* operon) removed more P_i from the medium than did MV1184 (pBC29). The P content of MV1184 (pEP02.2) was approximately twice that of MV1184 (pBC29) (Fig. 4). This finding may indicate that P_i transport across the cell membrane is a rate-limiting step for P_i accumulation in *E. coli* MV1184 under conditions of P_i excess.

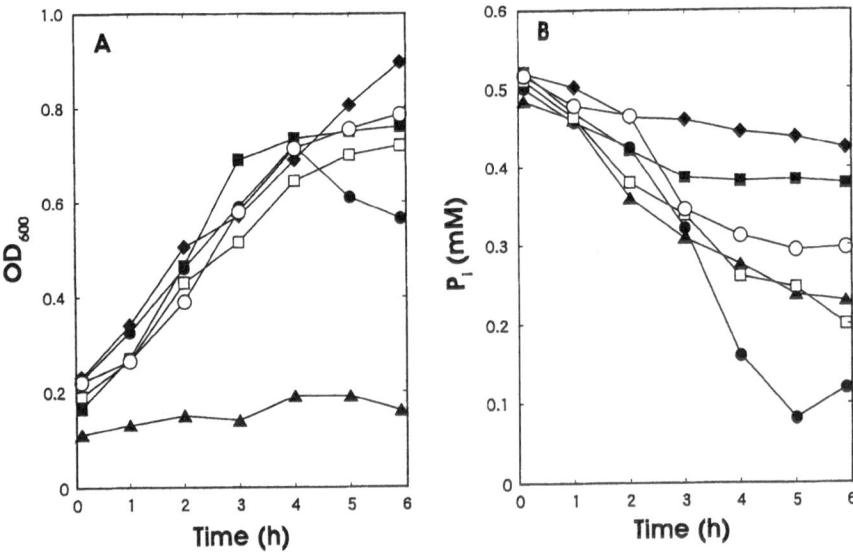

Fig. 3. A Growth of and **B** P_i removal by *E. coli* MV1184 containing pUC18 (*solid squares*), pBC29 (*open circles*), pEP01 (*diamonds*), pEP02.2 (*open squares*), pEP03 (*solid circles*) and pEP02.2 plus pBC29 (*triangles*) in T medium containing 0.5 mMP_i. (Kato et al. 1993a)

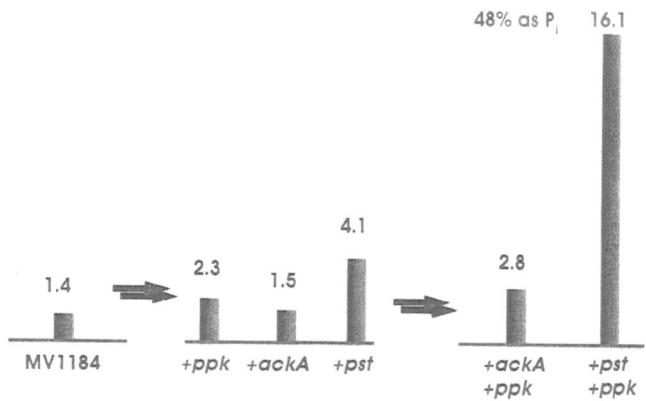

Fig. 4. Phosphorus content of MV1184 (pUC18) and the five recombinant derivatives. The *number* above each bar represents the content of cellular P expressed as a percentage of dry cell weight

Drastic changes in growth and P_i uptake were observed when pBC29 (*ppk*) and pEP02.2 (*pst* operon) were simultaneously introduced into MV1184. Growth of this recombinant was limited in T medium, even though it grew well in L broth. Nevertheless, this recombinant removed approximately threefold more P_i than did the control strain. Consequently, the P content of this recombinant reached 16 % (48 % as P_i) on a dry weight basis, or approximately ten fold more than that of the control strain. Such a high level of P_i accumulation in bacteria has never been reported before. The fractionation of cellular P revealed that acid-soluble and acid-insoluble polyPs accounted for approximately 65 % of the total cellular P of MV1184 (pBC29 and pEP02.2) (Kato et al. 1993a). No detectable release of P_i was observed with this recombinant. In pEP02.2, the *pst* genes were placed under the control of the *tac* promoter, and isopropyl-β-D-thio-galactopyranoside (IPTG) was required for inducing their expression. Plasmid pEP02.5 was constructed to solve this problem by placing the *pst* genes under the (constitutive) control of the *tet* promoter. MV1184 (pEP02.5) removed P_i at a rate similar to that seen with MV1184 (pEP02.2) without addition of IPTG (Hardoyo et al. 1994).

Strain MV1184 containing both pEP02.2 (*pst* operon) and pEP03 (*ppk* and *ack*) was also constructed. However, introduction of both plasmids appeared to be detrimental to MV1184. The cell viability of this recombinant decreased soon after the cells entered the stationary phase of growth in L broth. As a result, subsequent P_i uptake experiments in T medium could not be performed with this recombinant. Strain NM522 (pBC29) removed P_i at a rate similar to that of MV1184 (pBC29), suggesting that disruption of PPX activity is not important for improving the ability of *E. coli* to accumulate polyP. This is probably because the *ppk* gene was carried on a high-copy-number plasmid, pUC18, and several enzymes other than PPX were also responsible for polyP degradation (Wood and Clark 1988).

4
PolyP Accumulation Capacity

When *E. coli* cells accumulated excessive levels of polyP, they released polyP into the medium. This seems to be the mechanism by which a further increase in cellular polyP is limited. The release of polyP was first observed during P_i uptake experiments with MV1184 (pBC29 and pEP02.2) (Kato et al. 1993a). In these experiments, it was observed that the total P content in the culture supernatant increased at 4 h after the start of incubation, even though the P_i concentration continued to decrease. As a consequence, the P content of this recombinant, which reached a maximum of 16 % (48 % as P_i) at 4 h, decreased to approximately 12 % by 6 h. Most of the P compounds, released to the medium, were acid labile (1 N HCl for 7 min at 100 °C) and were hydrolyzed by bacterial alkaline phosphatase (Bap), liberating P_i. When the culture supernatant was applied to an anion-exchange column (DEAE-Toyopearl M650), approximately 93 % of the acid-labile P was retained in the column. The acid-labile P could be eluted with 1 M NaCl, being concentrated approximately 100-fold compared with that in the culture supernatant. To identify polyP, high-resolution ^{31}P-NMR spectra at 109.4 MHz were recorded (Hardoyo et al. 1994). The ^{31}P-NMR spectra identified the concentrated acid-labile P as polyP (–26 to –28 ppm). This assignment was based on the results obtained with sodium polyP (type 75; average chain length longer than 75 P_i molecules; Sigma Chemical Co., St. Louis).

The polyP release did not accompany the decrease of cell density. The rate of polyP release, estimated from the increase in the concentration of acid-labile P in the culture supernatant, was dependent on that of P_i uptake. The amount of polyP release was nearly proportional to that of P_i removed from the medium unless cell lysis occurred. No polyP release was observed after the cells completely removed P_i from the medium, but it resumed soon after P_i was added to the medium (Fig. 5). Once *E. coli* cells accumulated excessive levels of polyP, the rate of polyP release became essentially equivalent to that of P_i uptake (Hardoyo et al. 1994). Nevertheless, these cells did not release a detectable amount of polyP unless P_i was added to the culture. When P_i uptake was inhibited by 0.1 mM carbonyl cyanide m-chlorophenylhydrazone (CCCP), no polyP release occurred. In addition, neither P_i uptake nor polyP release was observed at 4 °C. However, there is no evidence that the *E. coli* recombinants actively efflux polyP from the cells. Since the chemicals which inhibited polyP release also inhibited P_i uptake, it was not possible to determine whether polyP is actively released into the medium.

PolyP was detected in intact cells of MV1184 (pBC29 and pEP02.2) by means of high-resolution ^{31}P-NMR spectroscopy (Hardoyo et al. 1994). The

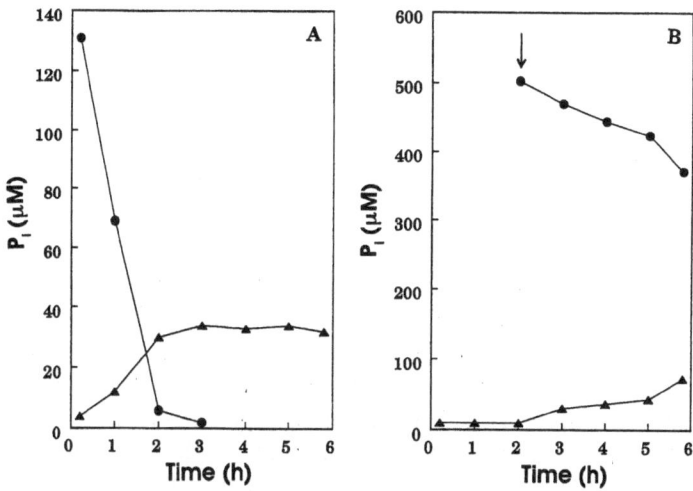

Fig. 5. Time course of P_i (*circles*) and acid-labile phosphorus (*triangles*) concentrations in culture supernatants during growth of *E. coli* MV1184 (pBC29 and pEP02.2) in T medium with 0.13 mM P_i (**A**) and P_i-free medium (**B**). For P_i-free T medium, 0.5 mM P_i was added at 2 h (*arrow*) (Hardoyo et al. 1994)

intensity of the polyP signal increased when cells were treated with EDTA, a membrane-impermeable chelator. In contrast, the addition of a shift reagent, praseodymium (0.1 mM), greatly reduced the polyP resonance. It has been shown that in vivo high-resolution ^{31}P-NMR spectroscopy monitors mobile, soluble pools of polyP (Rao et al. 1985). Cytoplasmic polyP, linked to either proteins or nucleic acids, cannot be detected under normal spectra conditions. Therefore, evidence from the ^{31}P-NMR studies suggests that a mobile, soluble pool of polyP exists on the surface of *E. coli* recombinants. The presence of surface polyP was also suggested by polyP metachromacy with *E. coli* intact cells. When MV1184 (pBC29 and pEP02.2) cells were incubated with toluidine blue, the absorption maximum at 630 nm was observed to shift to 580 nm. At low concentrations, toluidine blue does not penetrate the cell membrane in microorganisms, and the absorption maximum of the dye shifts from 630 to 580 nm, when it is incubated with a negatively charged polyelectrolyte such as purified polyP (Halvorson et al. 1987). No significant shift was detected with the control strain MV1184 (pUC18).

It is highly likely that the mobile, soluble pool of surface polyP is the source of polyP released into the medium. However, the mechanism for the formation of surface polyP is not clear. In *E. coli*, PPK preferentially attaches to the outer membrane, even though ATP is the substrate for the enzyme (Akiyama et al. 1992). This location of PPK and the lack of a leader sequence to translocate it to the outer membrane suggest that the enzyme may be present in Bayer patches (Bayer 1968), described as fusions of inner and outer membrane communicating directly between the cell exterior and interior compartments. If this is the case, the orientation of PPK may lead to the formation of surface polyP. Accumulation of surface polyP has been also reported with *Acinetobacter lwoffi* (Halvorson et al. 1987).

5
PolyP Accumulation in *Klebsiella aerogenes*

Although many bacteria are known to accumulate polyP (Wood and Clark 1988), genetic analysis of polyP metabolism has been done exclusively with *E. coli*. However, *E. coli* does not accumulate appreciable amounts of polyP, especially those of high molecular types (Kulaev 1975). *Escherichia coli* may not be a good system for investigating biological P_i removal. *Klebsiella aerogenes* (formerly named *Aerobacter aerogenes*), which is closely related to *E. coli*, exhibits extensive polyP accumulation (Harold 1966). It has been shown that polyP accumulation in *K. aerogenes* falls into two distinct patterns (Harold 1966). When growth and nucleic acid synthesis are blocked by depriving the organism of sulfate, assimilation of P_i from the medium continues resulting in slow accumulation of polyP. This phenomenon, called luxury uptake (Fuhs and Chen 1975), is now known to occur in many bacteria, including sludge bacteria, when growth is arrested by lack of a nutrient other than P_i. Addition of P_i to *K. aerogenes* cells previously subjected to P_i starvation also induces rapid and extensive accumulation of polyP (Fig. 6). Potassium, magnesium and a source of energy are required for polyP accumulation. Upon resumption of growth and nucleic acid synthesis, the polyP

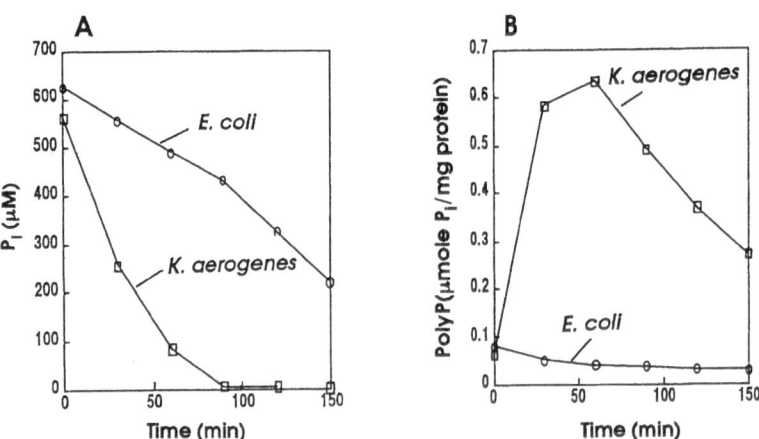

Fig. 6. A P_i uptake and **B** polyP accumulation in *E. coli* MV1184 and *K. aerogenes* ATCC9621. Cell were grown overnight in P_i-sufficient (1.3 mM P_i) MOPS medium (Neidhardt et al. 1974), inoculated (a 1 % inoculum) into P_i-limiting (0.1 mM P_i) MOPS medium and incubated for 6 h at 37 °C with shaking. P_i-uptake experiments were started by adding P_i to the cultures at approximately 0.6 mM

is gradually degraded by conversion to nucleic acids. This pattern of polyP accumulation is called polyP overplus (Harold 1966).

The pathway of polyP metabolism in *K. aerogenes* has been investigated by means of mutants blocked in the biosynthesis and degradation of polyP (Harold and Harold 1965). However, until recently, no investigation has been undertaken to identify and characterize the genes involved in the polyP metabolism in this organism. The PPK-encoding gene (*ppk*) has been cloned from *K. aerogenes* ATCC9621 (Kato et al. 1993b). Nucleotide sequence analysis revealed that it codes for a polypeptide of 685 amino acids, and this polypeptide shared 93 % of identical amino acid residues with the *E. coli* PPK protein.

Like the *E. coli ppk-ppx* operon, the *ppx* gene, which encodes an exopolyPase PPX, existed immediately downstream of the *ppk* gene. However, the nucleotide sequence of the promoter region of the *K. aerogenes ppk* gene differed from that of the *E. coli ppk* gene. A putative *pho* box sequence was found in the promoter region of the *K. aerogenes ppk* gene (Kato et al. 1993b). There was a 12/18 bp match with the consensus *pho* box sequence (Makino et al. 1986). Unlike the *E. coli ppk-ppx* operon, the *K. aerogenes ppk-ppx* operon is likely to be under control of the PhoB and PhoR proteins which constitute the two-component regulatory system for the P_i regulon genes (Wanner 1996). The expression of multicopy *lacZ* fusion of the *K. aerogenes ppk* promoter in *E. coli* MV1184 has been shown to increase under conditions of P_i limitation. However, in *E. coli* ANCS3 (PhoB⁻) (Makino et al. 1986), the expression levels were not elevated by subjecting the cells to P_i starvation.

If the *ppx* gene is cotranscribed with the *ppk* gene, the question then arises of the mechanism by which *K. aerogenes* can show polyP overplus. To answer this question, the PPK and PPX activities were determined before and after polyP overplus took place (Fig. 7). As expected, the PPK activity increased responding to P_i starvation and decreased upon addition of P_i. However, unlike PPK, the PPX activity did not increase under conditions of

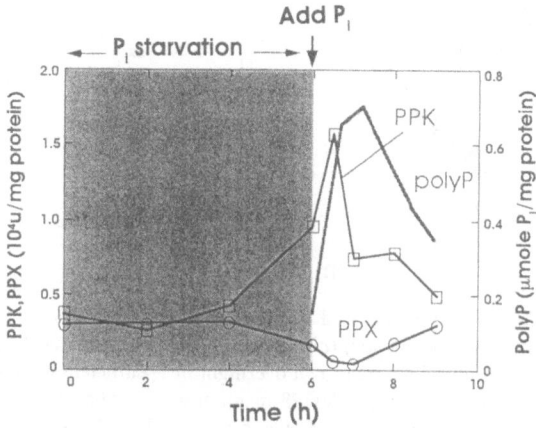

Fig. 7. Time course of PPK and PPX activities, as well as cellular levels of polyP, before and after addition of P_i to *K. aerogenes* cells. *Klebsiella aerogenes* ATCC9621 was grown overnight in P_i-sufficient MOPS medium, inoculated (a 1 % inoculum) into P_i-limiting MOPS medium and incubated for 6 h at 37 °C with shaking. P_i was then added to the culture at approximately 0.6 mM. PolyP was determined by the method described by Crooke et al. (1994)

P_i starvation. Most surprisingly, the PPX activity significantly decreased upon the addition of P_i. Although the mechanism for controlling the PPX activity is unknown, it is clear that both increased polyP synthesis and decreased polyP degradation are responsible for polyP overplus in *K. aerogenes*. Unlike *K. aerogenes*, no significant changes in the PPK and PPX activities were detected with *E. coli* before and after P_i addition.

P_i uptake experiments were also performed with *K. aerogenes* ATCC9621 and its recombinant derivatives bearing either pKP06 (*ppk*) or pKP07 (*ppk* and *ppx*) (Fig. 8). Plasmid pKP06 was constructed to subclone the *K. aerogenes ppk* gene and its own promoter into a vector plasmid pSTV29 (Takara Shuzo Co., Kyoto), while pKP07 contained the entire *K. aerogenes ppk-ppx* operon. In both the recombinant plasmids, the cloned genes were expressed by their own promoters. Cells were subjected to P_i starvation for 6 h, and P_i uptake experiments were started by adding P_i at approximately 0.5 mM. The growth of the recombinant strains was almost equivalent to that of the control strain (Fig. 8). Strain ATCC9621 (pKP06) removed approximately 80 % of the P_i from the medium within 2 h, while 50 % of the P_i was removed by the control strain ATCC9621 (pSTV29). Strain ATCC9621 (pKP06) accumulated 0.9 μmole P_i per mg protein as polyP, or approximately threefold more polyP than did the control strain. Thus, increasing the dosage of the *ppk* gene improved the ability of *K. aerogenes* to remove P_i. The copy number of pKP06 was between 10 and 15 per genome of *K. aerogenes*. Further improvement would be possible if a high copy-number plasmid is available. Strain

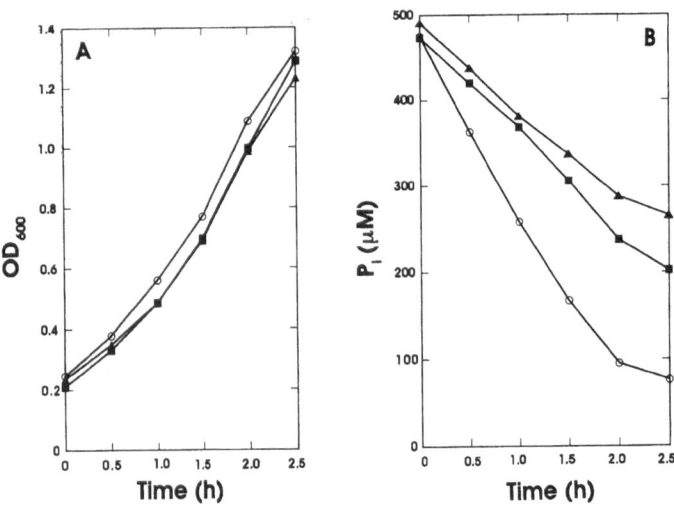

Fig. 8. **A** Growth of and **B** P_i removal by *K. aerogenes* ATCC9621 containing pSTV29 (*squares*), pKT06 (carrying *ppk*) (*circles*) or pKT07 (*ppk* and *ppx*) (*triangles*) in MOPS medium. *Klebsiella aerogenes* cells were grown overnight in P_i-sufficient MOPS medium, inoculated (a 1 % inoculum) into P_i-limiting MOPS medium and incubated for 6 h at 37 °C with shaking. P_i-uptake experiments were started by adding P_i to the cultures at approximately 0.5 mM

ATCC9621 (pKP07), which carries both the *K. aerogenes ppk* and *ppx* genes, removed less P_i than did ATCC9621 (pSTV29). It is likely that the increased dosage of the *ppx* gene elevated the PPX activity in ATCC9621 (pKP07), even though its specific activity decreased upon the addition of P_i (Fig. 7).

6
Concluding Remarks

From the data presented, it is evident that the ability of *E. coli* to take up and store excess P_i can be improved by modifying the genetic regulation and increasing the dosage of the genes involved in the key steps of P_i transport and polyP metabolism. *Escherichia coli* MV1184 (pBC29 and pEP02.2) accumulated a maximum of 16 % of its dry weight as P (48 % as P_i). This level of P content is surprisingly high and even exceeds those of natural phosphorite deposits (typically less than 43 % as P_i). Therefore, the genetic approach presented in this review should ultimately provide a new technology for removing P_i from wastewaters and converting it into "biophosphorites". The production of biophosphorite should contribute not only to pollution control but also to conservation of valuable P resources.

PPK is present in numerous microorganisms (Kulaev 1975). The *ppk* gene has been cloned from a variety of microorganisms (Geissdorfer et al. 1995; Tinsley and Gotschlich 1995; Kaneko et al. 1996). The consensus sequences derived from the reported *ppk* genes can be used for making PCR primers to pick up the *ppk* genes from a wide range of bacterial species. This evidence supports to potential for genetic improvement of more useful microorganisms, including sludge bacteria, for enhanced P_i removal.

Escherichia coli recombinants are able to overproduce PPK and accumulate polyP. Market opportunities of polyP can be also exploited in several ways. PolyP can be potentially employed to generate ATP using PPK in industrial processes. PolyP is also used commercially as an antibacterial agent in processed meat. Furthermore, it is possible to use polyP as a component of chemical fertilizers that release P_i under conditions of P_i limitation. Creating such market opportunities may be of great help for promoting the use of engineered microbes for biological P_i removal from wastewaters.

References

Akiyama M, Crooke E, Kornberg A (1992) The polyphosphate kinase gene of *Escherichia coli*. J Biol Chem 267: 22556–22561

Akiyama M, Crooke E, Kornberg A (1993) An exopolyphosphatase of *Escherichia coli*: the enzyme and its *ppx* gene in a polyphosphate operon. J Biol Chem 268: 633–639

Amemura M, Shinagawa H, Makino K, Otsuji N, Nakata A (1985) Cloning of and complementation tests with alkaline phosphatase regulatory genes (*phoS* and *phoT*) of *Escherichia coli*. J Bacteriol 152: 692–701

Bayer ME (1968) Areas of adhesion between wall and membrane of *Escherichia coli*. J Gen Microbiol 53: 395–404

Carberry JB, Tenney MW (1973) Luxury uptake of phosphate by activated sludge. J Water Pollut Control Fed 45: 2444–2462

Codd GA, Bell SG (1985) Eutrophication in freshwaters. J Water Pollut Control Fed 84: 225–232

Crooke E, Akiyama M, Rao NN, Kornberg A (1994) Genetically altered levels of inorganic polyphosphate in *Escherichia coli*. J Biol Chem 269: 6290–6295

Fuhs GW, Chen M (1975) Microbiological basis of phosphorus removal in the activated sludge process for the treatment of wastewater. Microb Ecol 2: 119–138

Geissdorfer W, Frosch SC, Haspel G, Ehrt S, Hillen W (1995) Two genes encoding proteins with similarities to rubredoxin and rubredoxin reductase are required for conversion of dodecane to lauric acid in *Acinetobacter calcoaceticus*. Microbiology 141: 1425–1432

Gough JA, Murray NE (1983) Sequence diversity among related genes for recognition of specific targets in DNA molecules. J Mol Biol 166: 1–19

Griffin JB, Davidian NM, Penniall R (1965) Studies of phosphorus metabolism by isolated nuclei. VII. Identification of polyphosphate as a product. J Biol Chem 240: 4427–4434

Haeusler PA, Dieter L, Rittle KJ, Shepler LS, Paszkowski AL, Moe OA (1992) Catalytic properties of *Escherichia coli* polyphosphate kinase: an enzyme for ATP regeneration. Biotechnol Appl Biochem 15: 125–133

Halvorson HO, Suresh N, Roberts MF, Coccia M, Chikarmane HM (1987) Metabolically active surface polyphosphate pool in *Acinetobacter lwoffi*. In: Torriani-Gorini A, Rothman FG, Silver S, Wright A, Yagil E (eds) Phosphate metabolism and cellular regulation in microorganisms. American Society for Microbiology, Washington, DC, pp 220–223

Hardoyo, Yamada K, Shinjo H, Kato J, Ohtake H (1994) Production and release of polyphosphate by a genetically engineered strain of *Escherichia coli*. Appl Environ Microbiol 60: 3485–3490

Harold FM (1963) Accumulation of inorganic polyphosphate in *Aerobacter aerogenes*. I. Relationship to growth and nucleic acid synthesis. J Bacteriol 86: 216–221

Harold FM (1966) Inorganic polyphosphates in biology: structure, metabolism, and function. Bacteriol Rev 30: 772–794

Harold FM, Harold RL (1965) Degradation of inorganic polyphosphate in mutants of *Aerobacter aerogenes*. J Bacteriol 89: 1262–1270

Higgins CF (1992) ABC transporters: from microorganisms to man. Annu Rev Cell Biol 8: 67–113

Kaneko T, Sato S, Kotani H, Tanaka A, Asamizu E, Nakamura Y, Miyajima N, Hirosawa M, Sugiura M, Sasamoto S, Kimura T, Hosouchi T, Matsuno A, Muraki A, Nakazaki N, Naruo K, Okumura S, Shimpo S, Takeuchi C, Wada T, Watanabe A, Yamada M, Yasuda M, Tabata S (1996) Sequence analysis of the genome of the unicellular cyanobacterium *Synechocystis* sp. strain PCC6803. II. Sequence determination of the entire genome and assignment of potential protein-coding regions (supplement). DNA Res 30: 185–209

Kato J, Yamada K, Muramatsu A, Hardoyo, Ohtake H (1993a) Genetic improvement of *Escherichia coli* for the enhanced biological removal of phosphate. Appl Environ Microbiol 59: 3744–3749

Kato J, Yamamoto T, Yamada K, Ohtake H (1993b) Cloning, sequence and characterization of the polyphosphate kinase-encoding gene (*ppk*) of *Klebsiella aerogenes*. Gene 137: 237–242

Kulaev IS (1975) Biochemistry of inorganic polyphosphates. Rev Physiol Biochem Pharmacol 73: 131–158

Lee T-Y, Makino K, Shinagawa H, Nakata A (1990) Overproduction of acetate kinase activates phosphate regulon in the absence of the *phoR* and *phoM* functions in *Escherichia coli*. J Bacteriol 172: 2245–2249

Makino K, Shinagawa H, Amemura M, Nakata A (1986) Nucleotide sequence of the *phoB* gene, the positive regulatory gene for the phosphate regulon of *Escherichia coli* K12. J Mol Biol 190: 37–44

Neidhardt FC, Bloch PL, Smith DF (1974) Culture medium for Enterobacteria. J Bacteriol 119: 736–747

Ohtake H, Takahashi K, Tsuzuki Y, Toda K (1985) Uptake and release of phosphate by a pure culture of *Acinetobacter calcoaceticus*. Water Res 19: 1587–1594

Rao NN, Roberts MF, Torriani A (1985) Amount and chain length of polyphosphates in *Escherichia coli* dependent on cell growth conditions. J Bacteriol 162: 205–211

Thauer RK, Jungermann K, Decker K (1977) Energy conservation in chemotrophic anaerobic bacteria. Bacteriol Rev 41: 100–180

Tinsley CR, Gotschlich EC (1995) Cloning and characterization of the meingococcal polyphosphate kinase gene: production of polyphosphate synthesis mutants. Infect Immun 63: 1624–1630

Vieira J, Messing J (1987) Production of single-stranded plasmid DNA. Methods Enzymol 153: 3–11

Wanner BL (1996) Phosphorus assimilation and control of the phosphate regulon. In: Neidhardt FC, Curtiss R III, Ingraham JL, Lin ECC, Low KB, Magasanik B, Reznikoff WS, Riley M, Schaechter M, Umbarger HE (eds) *Escherichia coli* and *Salmonella typhimurium*: cellular and molecular biology, 2nd edn. American Society for Microbiology, Washington, DC, pp 1357–1381

Wicks JR, Thiel PG (1990) Environmental factors affecting the production of peptide toxins in floating scums of the cyanobacterium *Microcystis aeruginosa* in a hypertrophic African reservoir. Environ Sci Technol 24: 1413–1418

Wood HG, Clark JE (1988) Biological aspects of inorganic polyphosphates. Annu Rev Biochem 57: 235–260

Subject Index